# International Conference on Computer Simulation in Physics and Beyond (CSP2017)



Journal of Physics: Conference Series Volume 955

Moscow, Russia
9 – 12 October 2017

ISBN: 978-1-5108-5755-1
ISSN: 1742-6588

**Printed from e-media with permission by:**

Curran Associates, Inc.
57 Morehouse Lane
Red Hook, NY 12571

**Some format issues inherent in the e-media version may also appear in this print version.**



Printed by Curran Associates, Inc. (2018)

For permission requests, please contact the Institute of Physics at the address below.

Institute of Physics
Dirac House, Temple Back
Bristol BS1 6BE UK

Phone: 44 1 17 929 7481
Fax: 44 1 17 920 0979

techtracking@iop.org

**Additional copies of this publication are available from:**

Curran Associates, Inc.
57 Morehouse Lane
Red Hook, NY 12571 USA
Phone: 845-758-0400
Fax: 845-758-2633
Email: curran@proceedings.com
Web: www.proceedings.com

# Table of contents

## Volume 955

**Computer Simulations in Physics and beyond (CSP2017)**

**9–12 October 2017, Moscow, Russian Federation**

**Accepted papers received: 9 January 2018**
**Published online: 5 February 2018**

### Preface



### Papers

# Preface

Computer simulations are nowadays a firmly established third pillar of modern natural sciences, complementing experimentation and paper-and-pencil theoretical studies. Simulations, experiments *in silico*, prove indispensable in diverse areas of research in physics and other natural sciences.

This volume collects papers based on presentations delivered at the Second International Conference on Computer Simulations in Physics and beyond (CSP2017), which took place October 9-12, 2017 in Moscow. The Conference, which continues a biannual tradition started by an innaugural conference in 2015, took place on campus of A.N. Tikhonov Moscow Institute of Electronics and Mathematics, was jointly organized by the National Research University Higher School of Economics, the Landau Insitute for Theoretical Physics and Science Center in Chernogolovka.

As the name implies, the Conference is a multidisciplinary meeting, with a focus on computational physics and related subjects. Indeed, methods of computational physics prove useful in a broad spectrum of research in multiple branches of natural sciences, and this volume provides a sample.

We hope that this volume will interest a wide range of readers, and we are already looking forward for the next conference in this biannual series.

Moscow, Russia
November, 2017

E.A. Burovski
S.A. Krashakov
L.N. Shchur

IOP Conf. Series: Journal of Physics: Conf. Series **955** (2017) 011002 doi:10.1088/1742-6596/955/1/011002

# Peer review statement

All papers published in this volume of *Journal of Physics: Conference Series* have been peer reviewed through processes administered by the proceedings Editors. Reviews were conducted by expert referees to the professional and scientific standards expected of a proceedings journal published by IOP Publishing.

# From inverse problems to learning: a Statistical Mechanics approach

**Carlo Baldassi[1], Federica Gerace[2], Luca Saglietti[2] and Riccardo Zecchina[1]**

[1] Bocconi Institute for Data Science and Analytics, Bocconi University, Milan, Italy
[2] Dipartimento di Scienza Applicata e Tecnologia, Politecnico di Torino, Italy

E-mail: riccardo.zecchina@unibocconi.it

**Abstract.** We present a brief introduction to the statistical mechanics approaches for the study of inverse problems in data science. We then provide concrete new results on inferring couplings from sampled configurations in systems characterized by an extensive number of stable attractors in the low temperature regime. We also show how these result are connected to the problem of learning with realistic weak signals in computational neuroscience. Our techniques and algorithms rely on advanced mean-field methods developed in the context of disordered systems.

## 1. Introduction

The enormous amount of data which is rapidly accumulating across scientific disciplines has given to inverse problems a renewed fundamental role. For instance, the problem of inferring effective or direct interactions in systems composed of many heterogeneous degrees of freedom is at the root of many fields, ranging from natural sciences to social sciences. Quite often, the systematic experimental measurements of the interactions are difficult to perform, and to estimate them one needs to resort to indirect statistical inference methods. These need to be computationally efficient in order to process experimental datasets which are typically huge. During the last decade, statistical inference challenges have stimulated lots of research efforts in the statistical physics community [1] in the attempt to adapt rigorous theories and algorithms to the problem of going from observables to parameters, somehow the inverse of what statistical physics has mostly been developed for.

Spectacular examples of inverse problems can be found in the biological sciences where experimental data on neuronal populations [2, 3], proteins [4], biomolecules and genetic interactions [5], have largely surpassed the understanding of the underlying complex chemical mechanisms. In physics too there exist many examples of inverse statistical problems, e.g. the design of many-body systems with particular desired properties (e.g. [6], see refs. [1]) or to learn from data some effective parametrizations of wave-functions (e.g. [7]). The technological progress is rapidly opening up many more applications.

Inverse problems and statistical inference are closely linked to machine learning. Here we will focus on unsupervised learning, i.e. extracting information from unlabeled data [8]. The vast majority of the data are indeed unlabeled, and unsupervised learning is possibly the most challenging branch of machine learning, which has a direct mapping onto maximum entropy ideas [9] and statistical mechanics methods.

Adopting the physics notation, the unsupervised learning problem can be stated as follows. We assume that a set of $M$ i.i.d. sampled configurations $D = \left\{\sigma^{(\mu)}\right\}_{\mu=1,\dots,M}$ drawn from an unknown distribution $P(\sigma) = \frac{1}{Z} e^{-H(\sigma)}$ are given. The configurations components $\sigma_i^{(\mu)}$ can be either continuous or discrete. We want to identify a probability distribution over the *energy* functions $H(\sigma)$ which is compatible with the data. The likelihood of the model given the data is defined as the posterior distribution $P(H|D) \propto \exp\left(-M\left(\langle H\rangle + \log Z\right)\right) P(H)$ , where $\langle H\rangle = \frac{1}{M}\sum_{\mu=1}^{M} H\left(\sigma^{(\mu)}\right)$ is the average over the data and $P(H)$ is the prior about $H$. The number of samples $M$ plays a crucial role, quite similar to inverse temperature: when $M$ is large, $P(H|S)$ concentrates on the maximums with respect to $H$ of the *log-likelihood* $\mathcal{L} = \langle H\rangle + \log Z$. Inferring the most probable $H$ from the data amounts at identifying the maximum of $\mathcal{L}$. This is a formidable optimization problem, due to the difficulty of estimating $Z$ on a candidate solution for $H$.

In what follows, for the sake of simplicity, we assume that the dynamical variables $\{\sigma\}$ are binary, i.e. Ising spins. The simplest non-trivial scenario assumes that we can have access to random configurations which allow us to measure non vanishing pairwise correlations between spins. In absence of correlations, the problem trivializes. In a general setting, the couplings characterizing the $H(\sigma)$ we want to infer will not all be positive. Moreover they will be embedded in some high-dimensional space. A concrete working example is provided by a system of $N$ spins $\{\sigma_i \equiv s_i = \pm 1\}$ which interact according to a generic (unknown) spin glass like Hamiltonian, i.e. spins are coupled by pairwise couplings $J_{ij}$ and are subject to external magnetic fields $h_i$. The unknown distribution from which the random configurations (the data) are extracted is a Boltzmann distribution

$$P(\{s_i\}) = \frac{1}{Z}\, e^{\sum_{i<j} J_{ij} s_i s_j + \sum_i h_i s_i} \tag{1}$$

where we have included the inverse temperature factor into the couplings and fields. The problem we want to solve is the determination of the most probable values for the couplings $J_{ij}$ and local fields $h_i$, given a set of $M$ observed spin configurations. Depending on the particular nature of the system at hand, the functional form of the Hamiltonian may be need to include higher order terms. Or even the choice of the form of the probability distribution may not be appropriate (e.g. if the systems are out of equilibrium). Still, the basic ideas are already captured by the above Ising like choice. It is in fact well known that the distribution (1) is not only the equilibrium distribution of the Ising model but, more importantly in our context, it is also the form of the distribution which maximizes the (Gibbs) entropy

$$S[P] = -\sum_{\{s_i\}} P(\{s_i\})\, lnP(\{s_i\}) \tag{2}$$

with $P(\{s_i\})$ satisfying the constraints of normalization and of having first and second moments (magnetizations and correlations) that match some given experimentally measured values. The maximum entropy $P(\{s_i\})$ is thus the probability distribution which matches the constraints given by some observables, without making any further assumption on the data, i.e. being the least biased distribution [10]. In settings where higher moments are not captured by (1) one must resort to maximum entropy distributions which match such moments and have multi-body interactions. Finding the parameters of (1) is a delicate computational problem: any small change in the couplings affects correlations between many spin variables and vice versa. Determining correlations for a given value of the parameters is a difficult computational problems which needs to be repeated multiple times in the search for the optimal couplings. Several techniques can be used, ranging from Monte Carlo chains, to variational methods to mean-field techniques. Inverse statistical problems are obviously not new in statistical mechanics as well as in statistics and applied probability. Language barriers quite often have hampered the communication between disciplines.

In what follows we will provide some preliminary results in which we discuss how the above ideas can be used to infer couplings in systems characterized by multiple attractors. As a by-product we will show how the overall technique can lead to some interesting learning mechanisms which might be of interest in modeling neural systems. We will perform sampling by a mean-field technique developed in the context of disordered systems[11]. Still other approaches, e.g. Monte Carlo Markov chains, could be used.

## 2. Attractor networks

The celebrated Hopfield model [12] suggests a way of modeling associative memory as a process through which a set of $M$ memories $\{\xi^\mu\}_{\mu=1}^M$ are stored within a recurrent neural network as the minima of an Ising-like Hamiltonian. The network is modeled as consisting of $N$ binary stochastic units representing the neurons. At each time $t$, a unit state $s_i^t \in \{-1, 1\}$ (representing whether the neuron is active or quiescent) is determined from the other units according to the probability distribution

$$P\left(s_i^t\right) \propto \exp\left(\beta s_i^t \left(\sum_{j\neq i} J_{ij} s_j^{t-1} + h_i^{\text{ext}}\right)\right) \tag{3}$$

where $\beta$ controls the level of stochasticity, the quantities $J_{ij}$ model the synaptic connection strengths and $h_i^{\text{ext}}$ some external input. As $\beta \to \infty$, the neurons become deterministic and their state is simply given by the sign of the total local field $h_i^{\text{tot}} = \sum_{j\neq i} J_{ji} s_j^{t-1} + h_i^{\text{ext}}$ acting upon them. The above equation implements a parallel dynamics, in which time is discretized and all units are updated simultaneously; it is straightforward to implement other update protocols (e.g. random sequential); in such case, and if the synaptic connections are assumed to be symmetric, $J_{ij} = J_{ji}$, the probability distribution of the whole network matches Eq. (1) with $h = h^{\text{ext}}$. This insight allows one to use the tools from statistical mechanics to study the behavior or the network for given $J$ and $h^{\text{ext}}$.

In its original formulation, the Hopfield model assumed that the synaptic couplings are shaped according to the Hebb's learning rule, reinforcing the connections inducing positive correlations among neurons and depressing the couplings leading to negative ones. Then, if the network is presented, during a training phase, with a number $M$ of binary input patterns $\xi^\mu$ (by setting in succession $h_i^{\text{ext}} = \lambda \xi_i^\mu$ for all $\mu$, with $\lambda \gg 1$ such that the external fields completely clamp the state of the network in Eq. (3) to $s^t = \xi^\mu$) the resulting couplings will have the symmetric form $J_{ij} \propto \sum_\mu \xi_i^\mu \xi_j^\mu$. It can then be shown that, for sufficiently high $\beta$ and sufficiently low $M$, there exist a regime in which this model behaves as an attractor network, meaning that initializing the state of the network in the proximity of any of the patterns, i.e. $s^0 = \xi^\mu + \nu$ with $\nu$ some binary noise that inverts some fraction of the units, the dynamics of the network, Eq. (3), spontaneously converges towards the basin of attraction of the memory to be stored, $s^t \sim \xi^\mu$ for $t \gg 1$. The amount of noise $\nu$ for which this is possible determines the size of the basin of attraction of the patterns. In this way, the input patterns $\{\xi^\mu\}$ are said to be stored as (associative) memories in the network, since they are encoded in the $J$ matrix and can be recovered from partial or corrupted inputs.

The Hopfield model with the Hebb's learning rule has notoriously a number of problematic features [13], among which the most prominent are: a rather low capacity (the maximum $M$ scales as $0.14N$, while it is known that a network could reach $2N$ if $\beta \to \infty$ and if the symmetry constraint on $J$ is dropped); the existence of attractors of the dynamics that were not in $\{\xi^\mu\}$

(called "spurious" attractors); the phenomenon of *catastrophic forgetting* (when $M$ exceeds $0.14N$ the network enters a spin-glass phase in which all the $\xi^\mu$ attractors are suddenly forgotten and an exponential number of spurious attractors are present instead); the inability to deal in general with correlated input patterns (although Hebb's rule can be amended in some simpler cases, e.g. for biased patterns); the requirement of very strong external fields during the training phase (expressed as $\lambda \gg 1$ above), which is somewhat biologically implausible, while the learning rule requires that each unit should somehow be aware of both its current state (clamped to $\xi_i^\mu$) and the state that it would have in absence of its own external field, which is even more biologically problematic.

The choice of using Hebb's rule was motivated by its simplicity and by biological considerations; moving away from this choice allows to amend many of the above-mentioned issues (see e.g. Refs. [14, 15]). Here, we extend the learning protocol proposed in Ref. [16], which specifically addressed the problem of learning with small external fields, to fully-connected networks.

As shown in more detail in the next section, the approach frames the learning problem as an inverse problem, in which the memories $\{\xi^\mu\}$ represent the given data $D$, and we seek to find a coupling matrix $J$ such that the memories would be typical samples from the distribution of Eq. (1), i.e. representatives of the basins of attraction of the final model.

## 3. Description of the learning protocol

Generalizing the work of Ref. [16], we study the problem of learning stable attractors by employing finite signals, in a fully-connected recurrent neural network. The inverse problem, i.e. the unsupervised learning task, is defined as follows: each pattern $\xi^\mu$, $\mu = 1, ..., \alpha N$, is presented to the network in the form of an external field of intensity $\lambda$, oriented along the direction of the pattern itself, namely $h^{\text{ext}} = \lambda \xi^\mu$. The relative balance between the intensities of the recurrent and the external contributions to the local field $h_i = \sum_{j \neq i} J_{ij} s_j + h^{\text{ext}}$, can produce a dynamics which goes from a clamped regime, where the state of the network is set by the external field in a deterministic fashion, to a noise dominated regime, where the dynamics of the network becomes fully stochastic. The goal of the learning rule is that of adapting the coupling matrix $J$ (i.e., the synaptic couplings), so that the driving effect of the external field is effectively absorbed in the recurrent interaction of the network. In this way, the freely evolving dynamics of the network is required to maximally resemble the driven dynamics.

This prescription naturally translates into a simple learning rule, inspired by the close connection between learning and inference in inverse problems:

$$J_{ij}^{t+1} = J_{ij}^t + \eta \left( \langle s_i s_j \rangle_{\lambda_1} - \langle s_i s_j \rangle_{\lambda_2 = 0} \right) \tag{4}$$

where the parameter $\eta$ plays the role of a learning rate.

The analogy with the Inverse Ising problem is immediately evident: the difference between the empirical correlations and the ones estimated from the partition function of the inferred model, is here replaced by the correlation difference between two models with the same synaptic couplings $J$, but subject to different external field intensities, $\lambda_1$ and $\lambda_2$. However, we should note that, contrary to what happens in inference problems, the two correlations will never perfectly match.

As in Ref. [16], we note that in the limit of high temperatures and very strong biasing signals, i.e. $\lambda_1 \to \infty$, we can recover the classic Hebb rule, since $\langle s_i s_j \rangle_0 \simeq 0$ and $\Delta J_{ij} \propto \langle s_i s_j \rangle_\infty = \xi_i^\mu \xi_j^\mu$. At lower temperatures, in the same clamping limit, the learning protocol translates into the so-called Contrastive Divergence algorithm Ref. [17].

The computational problem associated with this learning strategy is that of the correlations estimation: in the following sections we propose a mean-field method developed in the Spin Glass theory framework.

**Input**: parameters: $N$, $p$, $\eta$, $T_{\mathrm{L}}$, $t_{\mathrm{L}}$, $\lambda_2^{\mathrm{rate}}$
Randomly initialize $\xi^\mu$, $\mu = 1, ..., p$ with $\xi_i^\mu = \{-1, 1\}$;
Initialize $J_s$ to zero;
Choose a field intensity $\lambda_1$;
Set a field intensity $\lambda_2^{\mathrm{init}} = \lambda_1/2$;
**for** $T = 1$ **to** $T_L$ **do**
  **while** $\lambda_2 > 0$ **do**
    $\lambda_2 \leftarrow \lambda_2 - \lambda_2^{\mathrm{init}}/\lambda_2^{\mathrm{rate}}$;
    **for** $\mu$ *in random permutation of p patterns* **do**
      Initialize magnetization close to the choosen pattern $\mu$;
      Compute $< s_i s_j >_{\lambda_1}$;
      **for** $t = 1$ **to** $t_L$ **do**
        Compute $< s_i s_j >_{\lambda_2}$;
        $J_s \leftarrow J_s + \eta(< s_i s_j >_{\lambda_1} - < s_i s_j >_{\lambda_2})$;
      **end**
    **end**
  **end**
**end**

**Algorithm 1:** Learning

## 4. Correlations recording

In the previous section we have seen how the analogy between inference and learning leads to a new learning protocol, according to which the parameters of the network, namely the synaptic couplings $J_{ij}$, have to be modulated in order to allow the storage of a set of $p$ memories. The resulting update rule relies on the matching between the pair-wise correlations, estimated in presence of an external field of intensity $\lambda_1$, and the ones recorded in presence of a weaker external field $\lambda_2$, progressively vanishing.

The central issue is then how to compute the pair-wise correlations. In some contexts, more biologically inspired, the evolution in time of the network is simulated through the Glauber dynamics. The correlations among neuronal state variables are then estimated by means of the Monte Carlo method. However, it is known that this procedure becomes unfeasible for too large network sizes: the time needed for reaching a stationary state consistently increases when a bigger number of network components comes into play. We thus need to resort to more efficient methods.

In this work, we take advantage of the Belief Propagation algorithm (BP). This mean-field cavity method provides an estimation of the marginals associated to the variables that characterize a given physical system. Despite the fact that this estimation is exact only in tree-like graphical models, BP can be still successfully employed on loopy-graphs, when they can be locally approximated as trees.

By computing the (cavity) marginals, other interesting physical observables, such as the pair-wise correlations, can then be estimated. The BP algorithm accomplishes this task by defining a set of messages (also called *cavity beliefs*) running across the edges of the graph and corresponding to the marginals of a given variable in absence of a specific node of the graph (i.e. in presence of a "cavity"). The marginals can then be obtained as the product of the messages over all the cavity sites. BP provides a set of self-consistent equations that need to be solved iteratively, in order to find a closed expression for the cavity beliefs.

In the model of a fully-connected neural network, the BP equations of the messages are known, and given by:

$$\pi_{i\to j}(s_i) \;=\; \sum_{\{s_k\}\backslash j} \exp\left(\beta\left(\sum_{k\neq j}\frac{J_{ki}}{\sqrt{N}}s_i s_k + \lambda\xi_i^{\mu}\right)\right)\prod_{k\neq j}\pi_{k\to i}(s_k) \tag{5}$$

The cavity message $\pi_{i\to j}(s_i)$ then determines the marginal of the $s_i$ variable in absence of the node $j$ in the network. In the thermodynamic limit, we can get the asymptotic expression:

$$\pi_{i\to j}(s_i) \;\simeq\; \prod_{k\neq j}\exp\left(\beta\frac{J_{ki}}{\sqrt{N}}s_i m_{k\to i}\right) \tag{6}$$

where we have defined $m_{i\to j}$ as the cavity magnetization:

$$m_{i\to j} = \sum_{s_i} s_i\pi_{i\to j}(s_i) = \tanh\left(\beta\left(\lambda\xi_i^{\mu} + \sum_{k\neq j}\frac{J_{ki}}{\sqrt{N}}m_{k\to i}\right)\right) \tag{7}$$

This self-consistent system of equations can be solved by iteration. Once a fixed point is reached, the pair-wise connected correlations can be recovered, on the basis of the cavity magnetization, as:

$$\langle s_i s_j\rangle = \tanh(\beta J_{ij})\frac{\left(1-m_{j\to i}^2\right)\left(1-m_{i\to j}^2\right)}{\left(1+\tanh(\beta J_{ij})\, m_{i\to j}m_{j\to i}\right)^2} + m_i m_j \tag{8}$$

### *4.1. The TAP Approach*

The BP algorithm is a very powerful tool. As we have seen, once the cavity messages are known, we are able to derive the main interesting observables characterizing a physical system, such as the pair-wise correlations.

However, one of the drawbacks of this method is that BP requires the storage of a number $\mathcal{O}\left(N^2\right)$ of cavity messages, thus considerably affecting the computational performances. In order to overcome this difficulty, we can resort to the Thouless-Anderson-Palmer (TAP) approach. It can be derived through the high temperature expansion, or, equivalently, the limit of small synaptic couplings, of the mean field equations, on which the BP algorithm is based.

By performing the expansion, we get the TAP self-consistent equations, to be solved iteratively:

$$m_i^{t+1} \simeq \tanh\left[\beta\left(\lambda\xi_i^{\mu} + \sum_{k\epsilon i} m_k^t\, J_{ki} - \left(1-\left(m_k^t\right)^2\right) m_i^{t-1}\, J_{ki}^2\right)\right] \tag{9}$$

where the time indices have been assigned so as to ensure the convergence of the iterative scheme [18]. We can notice that, instead of involving the cavity magnetizations, the TAP equations refer directly to the magnetizations, thus reducing the number of equations to be solved to $\mathcal{O}(N)$.

In this framework, the pair-wise correlations can be computed as:

$$\langle s_i s_j\rangle = \frac{\beta J_{ij} + m_i m_j - \left(1-m_i^2\right)\beta J_{ij} m_j^2 - \left(1-m_j^2\right)\beta J_{ij} m_i^2}{1+\beta J_{ij} m_i m_j} \tag{10}$$

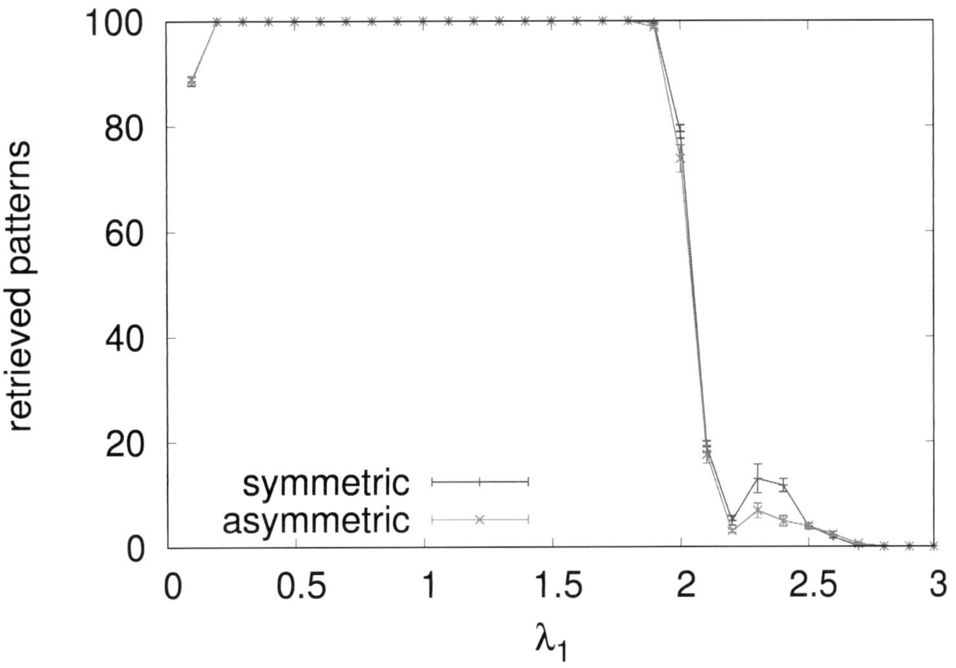


**Figure 1. Memory retrieval as a function of the external field intensity** $\lambda_1$, in a network with $N = 201$ neurons. The purple and green curves show, respectively, the case of symmetric synapses, where the network learns by matching the equal time correlations, and the case of asymmetric synapses, where the time delayed correlations are instead employed. In both cases the correlations are obtained in the TAP approximation at a fixed inverse temperature $\beta = 2$. When running the simulations at a fixed number of learning cycles $T_L$, we can observe that learning with weaker fields is not only possible, but allows the network to reach higher storage loads. The points are obtained by averaging over 20 samples.

## 5. Asymmetric Synaptic Couplings

The Hebb's learning rule is characterized by the unrealistic feature of symmetric synaptic couplings, namely $J_{ij} = J_{ji}$: there is no reason to assume that the signal sent by neuron $i$ towards neuron $j$ has to be of the same strength of the one sent by neuron $j$ towards neuron $i$. In fact, kinetically persistence states can be still observed in asymmetric networks.

In this section we thus consider a recurrent and fully connected neural network characterized by asymmetric synaptic couplings. The main difficulty in dealing with such a kinetic model is represented by the impossibility of defining a Hamiltonian. Therefore, in principle, we are not allowed to naively resort to the mean field approach. Here we follow the solution proposed in Ref. [19], where the authors show how the TAP equations can still be derived from a Kullback-Leibler divergence minimization even in the context of asymmetric synaptic couplings. We extend their results to the case of a parallel neuronal dynamics. The fixed point equations for the marginal probabilities of the neuronal states read:

$$m_i = \tanh\left(h_i + \sum_j J_{ij} m_j - m_i \sum_j \left(1 - m_j^2\right) J_{ij}^2\right) \tag{11}$$

The TAP approach can be thus exploited in the asymmetric context for the computation of the pair-wise correlations, as in [19]. However, in this specific case, we need to slightly modify the update rule of Eq. (4), in favor of the time-delayed correlations $\left\langle s_i^{t+1} s_j^t \right\rangle$:

$$\langle s_i^{t+1} s_j^t \rangle = \left(1 - m_j^2\right)\left(1 - m_i^2\right)\left(J_{ij} + \sum_l J_{il} J_{jl}\left(1 - m_l^2\right) + 2 m_i m_j J_{ij}^2\right) + m_i m_j \quad (12)$$

Indeed, by definition, the equal time correlations considered so far are not the proper quantity for learning in the asymmetric scenario.

Here, we present the results we obtained by applying the new learning protocol to a recurrent neural network, both in the symmetric and in the asymmetric cases. In the retrieval phase, we say that a pattern $\xi^\mu$ has been stored if the overlap of the stationary state, reached at convergence of the TAP equations, is consistently higher than all the other overlaps with all the other patterns $\xi^\nu$, with $\nu \neq \mu$.

In order to guarantee convergence of the TAP equations, instead of directly comparing the $\lambda_1$ intensity with $\lambda_2 = 0$, we slowly decrease the magnitude of the second external field, starting from an initial value of $\lambda_1/2$, as detailed in Algorithm 1.

Fig. 1 shows the number of patterns that have been retrieved by the network, as a function of the intensity of the reference external field $\lambda_1$. Notice that we work at storage loads that are higher then the critical capacity estimated for the Hebb's rule. We can notice that, the network is able to learn even in the presence of weak external field.

## 6. Learning correlated patterns

As pointed out above, one of the major drawbacks of the classic Hebb rule is in the inability of embedding correlated memories. We test the learning protocol of Eq. (4) in this setting, introducing an inter-pattern correlation inspired to the hierarchical scheme as discussed e.g. in Ref [20]. In this scheme, the patterns are organized in $L$ categories, each defined by an independently generated random prototype. Each category contains $\alpha N/L$ patterns, obtained as independent perturbations of the corresponding prototype, with a fraction $1 - x$ of randomly flipped components. By increasing $x$, one can thus enhance the correlations in the dataset. The purple curve in Fig. 2 shows the maximum storage load $\alpha_c$ at which perfect memorization can be achieved, in the case of $N = 201$ and $L = 5$, as a function of the correlation $x$. The reference field intensity $\lambda_1$ was set to 0.3. As the learning problem becomes harder, the performance of the networks remains stable before suffering from a sudden drop, for values of $x > 0.5$.

In order to overcome this problem, we slightly modified the learning prescription, by updating the synapses as:

$$J_{ij}^{t+1} = J_{ij}^t + \eta\left(\langle s_i s_j \rangle_{\lambda_1} - \langle s_i s_j \rangle_{\lambda_2 = -\lambda_1}\right). \quad (13)$$

The main intuition behind this adjustment is the following: we are not only requiring the free dynamics of the network to resemble the biased dynamics, but we are also inducing some robustness into the memorization process. In fact, in this case, the network tries to match the initial correlations with the one measured in the presence of an antagonist external field, whose effect needs to be overcompensated by the recurrent couplings of the network. The green curve in Fig. 2 shows the results with the new update rule, in the same settings as above: by implementing this protocol, the network is indeed able to learn even in presence of stronger correlations.

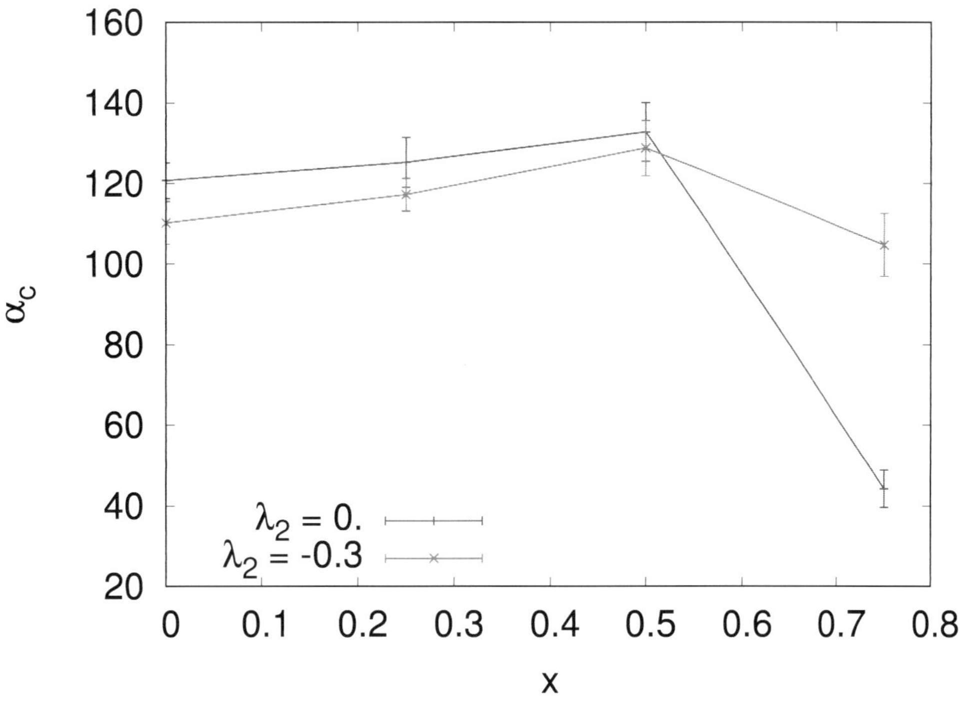


**Figure 2. Storage of correlated patterns**, with a comparison between rules Eq. (4) and Eq. (13), in a symmetric network with $N = 201$ neurons. The purple curve shows the learning behavior when the weaker external field $\lambda_2$ tends to zero: after the pattern correlation exceeds a given threshold ($x = 0.5$), the ability of the network to store stable attractors sharply deteriorates. The green curve, instead, shows the case where $\lambda_2$ tends to a weak negative value ($-0.3$): the performance in the highly correlated regime is clearly improved. All the points are obtained by averaging over 20 samples.

## Conclusions

We have shown how statistical physics methods can be used to infer couplings in systems characterized by multiple attractors, given relatively few samples. These results are of direct interest for computational neuroscience in that they lead to explicit learning protocols for weak realistic signals.

## References

[1] Nguyen H C, Zecchina R and Berg J 2017 *Advances in Physics* **66** 197–261
[2] Schneidman E, Berry M J, Segev R and Bialek W 2006 *Nature* **440** 1007–1012
[3] Cocco S, Leibler S and Monasson R 2009 *Proceedings of the National Academy of Sciences* **106** 14058–14062
[4] Morcos F, Pagnani A, Lunt B, Bertolino A, Marks D S, Sander C, Zecchina R, Onuchic J N, Hwa T and Weigt M 2011 *Proceedings of the National Academy of Sciences* **108** E1293–E1301
[5] Lezon T R, Banavar J R, Cieplak M, Maritan A and Fedoroff N V 2006 *Proceedings of the National Academy of Sciences* **103** 19033–19038
[6] Miskin M Z, Khaira G, de Pablo J J and Jaeger H M 2016 *Proceedings of the National Academy of Sciences* **113** 34–39
[7] Carleo G and Troyer M 2017 *Science* **355** 602–606
[8] MacKay D J 2003 *Information theory, inference and learning algorithms* (Cambridge university press)
[9] Jaynes E T 1957 *Physical review* **106** 620
[10] Jaynes E T 1986 *Maximum-Entropy and Bayesian Methods in Applied Statistics* (Cambridge university press)

[11] Mezard M and Montanari A 2009 *Information, physics, and computation* (Oxford University Press)
[12] Hopfield J J 1982 *Proc Natl Acad Sci U S A* **79** 2554–2558 ISSN 0027-8424 6953413[pmid] URL `http://www.ncbi.nlm.nih.gov/pmc/articles/PMC346238/`
[13] Amit D J 1989 *Modeling Brain Function: the World of Attractor Neural Networks* (New York, NY, USA: Cambridge University Press) ISBN 0-521-36100-1
[14] Amit D J, Gutfreund H and Sompolinsky H 1987 *Physical Review A* **35** 2293
[15] Alemi A, Baldassi C, Brunel N and Zecchina R 2015 *PLOS Computational Biology* **11** 1–23 URL `http://dx.doi.org/10.1371/journal.pcbi.1004439`
[16] Braunstein A, Ramezanpour A, Zecchina R and Zhang P 2011 *Physical Review E* **83** 056114 URL `https://doi.org/10.1103/PhysRevE.83.056114`
[17] Ackley D H, Hinton G E and Sejnowski T J 1985 *Cognitive science* **9** 147–169
[18] Bolthausen E 2014 *Communications in Mathematical Physics* **325** 333–366
[19] Kappen H J and Spanjers J J 2000 *Phys. Rev. E* **61**(5) 5658–5663 URL `http://link.aps.org/doi/10.1103/PhysRevE.61.5658`
[20] Mezard M 2017 *Phys. Rev. E* **95**(2) 022117 URL `http://link.aps.org/doi/10.1103/PhysRevE.95.022117`

# The study of the effects of sea-spray drops on the marine atmospheric boundary layer by direct numerical simulation

**O Druzhinin[1] Yu Troitskaya[1] and S Zilitinkevich[2]**

[1]Institute of Applied Physics RAS, Nizhny Novgorod, Russia

[2]Finnish Meteorological Institute, Helsinki, Finland

E-mail: droujmail@mail.ru

**Abstract.** The detailed knowledge of turbulent exchange processes occurring in the atmospheric marine boundary layer are of primary importance for their correct parameterization in large-scale prognostic models. These processes are complicated, especially at sufficiently strong wind forcing conditions, by the presence of sea-spray drops which are torn off the crests of sufficiently steep surface waves by the wind gusts. Natural observations indicate that mass fraction of sea-spray drops increases with wind speed and their impact on the dynamics of the air in the vicinity of the sea surface can become quite significant. Field experiments, however, are limited by insufficient accuracy of the acquired data and are in general costly and difficult. Laboratory modeling presents another route to investigate the spray-mediated exchange processes in much more detail as compared to the natural experiments. However, laboratory measurements, contact as well as Particle Image Velocimetry (PIV) methods, also suffer from inability to resolve the dynamics of the near-surface air-flow, especially in the surface wave troughs. In this report, we present a first attempt to use Direct Numerical Simulation (DNS) as tool for investigation of the drops-mediated momentum, heat and moisture transfer in a turbulent, droplet-laden air flow over a wavy water surface. DNS is capable of resolving the details of the transfer processes and do not involve any closure assumptions typical of Large-Eddy and Reynolds Averaged Navier-Stokes (LES and RANS) simulations. Thus DNS provides a basis for improving parameterizations in LES and RANS closure models and further development of large-scale prognostic models. In particular, we discuss numerical results showing the details of the modification of the air flow velocity, temperature and relative humidity fields by multidisperse, evaporating drops. We use Eulerian-Lagrangian approach where the equations for the air-flow fields are solved in a Eulerian frame whereas the drops dymanics equations are solved in a Largangain frame. The effects of air flow and drops on the water surface wave are neglected. A point-force approximation is employed to model the feed-back contributions by the drops to the air momentum, heat and moisture transfer.

## 1. Introduction

Detailed knowledge of the interaction of turbulent wind with surface water waves is necessary for correct parameterization of turbulent exchange at the air-sea interface in prognostic models. At sufficiently strong winds, sea-spray droplets interfere with the wind-wave interaction. The results of field experiments and laboratory measurements show that mass fraction of air-borne spume water droplets increases with the wind speed and their impact on the carrier air-flow may become significant [1].

Until recently, two approaches were mainly employed in numerical modeling of spray-mediated turbulent transfer in marine atmospheric boundary layer. The first approach relies on phenomenological modeling in the framework of RANS equations. This approach considers Reynolds-averaged equations for the air flow where the effects of spray drops are modeled by introducing source terms accounting for mass, momentum, and heat exchange between the air and drops [2-4]. To a large extent, the predictions of this approach depend on the assumptions made about the details of the interaction between drops and the carrier air flow.

The dynamics of droplets in marine atmospheric boundary layer was also extensively studied with the use of Largangian stochastic approach [5-7]. In these models, the droplet equation of motion is solved numerically in a Lagrangian framework. The surrounding, mean air-flow component is prescribed as a logarithmic boundary-layer approximation whereas the stochastic component is obtained as a numerical solution to a Langevin equation with random forcing that mimics turbulent fluctuations in the air-flow "seen" by the drop. Largrangian stochastic models provide very important information on droplets dynamics and estimates of droplet-mediated exchange in the marine atmospheric boundary layer. However, it is recognized that these models are unable to reproduce the two-way interaction between the droplets and turbulent boundary-layer eddy structures [6].

A first attempt to perform DNS of a droplet-laden air-flow over *waved* water surface was performed only recently [8]. In that study, a turbulent Couette air-flow over waved water boundary was considered as a model of marine atmospheric boundary layer. However, only the momentum exchange between the drops and the carrier air-flow was considered whereas the effects of drops on the humidity and temperature of the carrier air were not taken into account.

In the present paper, we present results of DNS of droplet-laden, two-way-coupled, turbulent Couette air-flow over waved water surface and, in particular, take into account effects of air-water latent heat exchange related to droplets evaporation, and droplet-mediated sensible heat exchange which generally occurs in case of a finite temperature difference between air and water. To our knowledge, this is the first DNS study of such flow. The carrier, turbulent Couette-flow configuration in DNS is similar to that used in previous numerical studies (cf. [10] and references therein). Discrete spherical droplets are tracked in a Lagrangian framework, and their impact on the carrier flow is modeled with the use of a point-force approximation. The droplets parameters in DNS are matched to the typical known spume-droplets parameters in laboratory and field experiments. Section 2 below presents the governing equations and numerical method. Numerical results are discussed in Section 3, and final conclusions and discussion are provided in Section 4.

## 2. Numerical method

We perform direct numerical simulation of turbulent, droplet-laden Couette flow above waved water surface. The schematic of the numerical experiment is similar to that used in our previous DNS study of turbulent, droplet-laden air-flow over waved surface [8] (Fig. 1). A Cartesian framework is considered where $x$-axis is oriented along the mean wind, $z$-axis is directed vertically upwards and $y$-axis is orthogonal to the mean flow and parallel to the wave front. We prescribe two-dimensional, $x$-periodic water wave with amplitude $a$, wavelength $\lambda$

and phase velocity *c*. The maximum wave slope considered in our DNS is $ka = 2\pi a/\lambda = 0.2$. The rectangular computational domain has sizes $L_x = 6\lambda$, $L_y = 4\lambda$ and $L_z = \lambda$ in the *x*-, *y*-, and *z*-directions, respectively, and the air flow is assumed to be periodical in the *x*- and *y*-directions. DNS is performed in a reference frame moving with the wave phase velocity, *c*, so

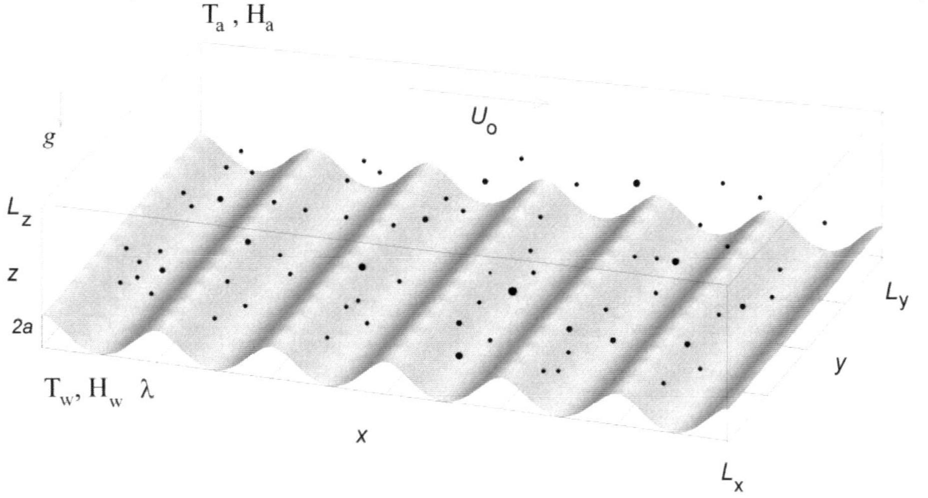


Fig. 1. Schematic of numerical experiment: $L_x, L_y, L_z$ are the domain sizes in the horizontal (*x*), spanwise (*y*), and vertical (*z*) directions; *a* and *λ* are the surface water wave amplitude and length; $T_w, H_w$ and $T_a, H_a$ are the temperature and fractional relative humidity at the water surface and at the top (air) boundary, respectively; $U_0$ the bulk velocity of the air-flow, and *g* is the acceleration due to gravity. Drops are denoted by black dots.

that the horizontal coordinate in the moving framework is $x = x' - ct$, where $x'$ is the coordinate in the laboratory reference frame. Then the lower boundary, representing the wave surface, is stationary in the moving reference frame. The no-slip boundary condition is prescribed at the lower boundary, so that the wind velocity at the boundary coincides with the velocity in the water wave. The no-slip boundary condition is also prescribed at the upper horizontal plane moving in *x*-direction with bulk velocity $U_0$. This condition provides an external source of momentum due to the viscous shear stress, which compensates viscous dissipation in the boundary layer and makes the flow statistically stationary. Air and water surface temperatures (*T*, measured in Kelvins, K) and fractional relative humidity (*H*) are prescribed at the top and bottom boundary planes, respectively, as $T_w$ and $H_w$ at $z = 0$ and $T_a$ and $H_a$ at $z = L_z$; typically $T_w > T_a$ and $H_w > H_a$.

The numerical algorithm is based on the integration of full, 3D Navier-Stokes equations for the carrier air-flow velocity and equations for the air temperature and relative humidity coupled with the Lagrangian equations for each drop coordinate, velocity, temperature and mass. The equations for the air flow velocity, temperature and relative humidity are written under the Boussinesq approximation in the well-known form [9]:

*air-flow velocity:*

$$\frac{\partial U_i}{\partial t} + \frac{\partial (U_i U_j)}{\partial x_j} = -\frac{1}{\rho_a}\frac{\partial P}{\partial x_j} + \nu \frac{\partial^2 U_i}{\partial x_j \partial x_j} + \delta_{iz} g \frac{T}{T_a} + \sum_{n=1}^{N_d} f_{Ui}^n, \qquad (1)$$

*continuity:*

$$\frac{\partial U_j}{\partial x_j} = 0, \tag{2}$$

*air temperature:*

$$\frac{\partial T}{\partial t} + \frac{\partial (TU_j)}{\partial x_j} + \frac{\partial (T_{ref} U_3)}{\partial x_3} = \kappa \frac{\partial^2 T}{\partial x_j \partial x_j} + \sum_{n=1}^{N_d} f_T^n, \tag{3}$$

*relative humidity:*

$$\frac{\partial H}{\partial t} + \frac{\partial (HU_j)}{\partial x_j} + \frac{\partial (H_{ref} U_3)}{\partial x_3} = D \frac{\partial^2 H}{\partial x_j \partial x_j} + \sum_{n=1}^{N_d} f_H^n, \tag{4}$$

where $x_i = x,\ y,\ z$, $U_i (i = x,\ y,\ z)$ are the velocity components, $P$ is pressure, $\rho_a$ the air density at temperature $T_a$ and pressure $P_a$; $T$ and $H$ are instantaneous deviations of the air temperature and fractional relative humidity from their respective initial, *reference* profiles, $T_{ref}(z)$ and $H_{ref}(z)$; $\nu$ and $\kappa$ are the air kinematic viscosity and thermal diffusivity; and $D$ is the diffusivity of water vapor. We consider linear reference profiles of air temperature and relative humidity in the form:

$$T_{ref}(z) = T_w + (T_a - T_w)\frac{z}{\lambda}, \tag{5}$$

$$H_{ref}(z) = H_w + (H_a - H_w)\frac{z}{\lambda}, \tag{6}$$

The last terms on the right hand side of Eqs. (1), (3) and (4), $f_{Ui}^n$, $f_T^n$, and $f_H^n$, are feedback contributions (to be defined below) of $n$-th droplet ($n = 1, \ldots, N_d$) to the rate of change of air momentum, temperature, and humidity; $N_d$ is the total, constant number of droplets considered in DNS.

Discrete spherical drops of diameter $d_n$ and temperature $T_n$ are tracked in a Lagrangian framework. Thus, for each $n$-th droplet we solve the equation for the coordinate and velocity, $r^n$ and $V^n$, and also take into account droplet evaporation/condensation and heat exchange with the surrounding air. Thus the following equations are solved simultaneously with Eqs. (1)-(4) for each ($n$-th) drop [2]:

*drop coordinate:*

$$\frac{dr_i^n}{dt} = V_i^n, \tag{7}$$

*drop velocity:*

$$\frac{dV_i^n}{dt} = \frac{1}{\tau_n}\left(U_i(r^n) - V_i^n\right)\left(1 + 0.15\,\mathrm{Re}_d^{0.687}\right) - \delta_{iz} g, \tag{8}$$

*drop temperature:*

$$m_n c_w \frac{dT_n}{dt} = 2\pi\kappa' d_n \left(T_a(r^n) - T_n\right)\left(1 + 0.25\,\mathrm{Re}_n^{0.25}\right) + L_v \frac{dm_n}{dt}, \tag{9}$$

*drop mass* ($m_n = \rho_n \pi d_n^3 / 6$)*:*

$$\frac{dm_n}{dt} = 2\pi D' d_n \rho_{sat}^v \left(H(r^n) - H_n^s\right)\left(1 + 0.25\,\mathrm{Re}_n^{0.25}\right). \tag{10}$$

In Eqs. (7) - (10), $r_i^n = x_n, y_n, z_n$ are Cartesian drop coordinates, and derivative over time ($d/dt$) is taken along the droplet trajectory.

In Eq. (8), $U_i(r^n)$ is the instantaneous air-flow velocity at the location of $n$-th drop, $g$ is the gravitational acceleration, and $\tau_n$ is the drop response time,

$$\tau_n = \frac{d_n^2}{18\nu}\frac{\rho_n}{\rho_a}, \tag{11}$$

where $\rho_n$ is the salt solution density of the $n$-th drop. The correction of the viscous drag force on the drop by the surrounding air in Eq. (8) is introduced taking into account that particle Reynlods number, defined as [10]

$$\mathrm{Re}_n = \frac{|U(r^n) - V^n|\, d_n}{\nu}, \tag{12}$$

is finite.

Equation (9) describes the rate of change of the drop temperature, $T_n$ (assumed to be uniform throughout the drop volume), due to the heat flux from/to the surrounding air and latent heat flux due to drop evaporation/condensation. The former flux is proportional to the instantaneous difference of the air temperature at the drop location, $[T_a(r^n) = T(r^n) + T_{ref}(z_n)]$, and $T_n$, and thermal conductivity coefficient, $\kappa'$ [not to be confused with the thermal diffusivity coefficient $\kappa$ in the right hand side of Eq. (3)], modified by noncontinuum gas-kinetic effects [11,12]. When computing the heat flux we also take into account ventilation effects due to finite Reynolds number, $\mathrm{Re}_n$, of the drop. The latent heat flux is proportional to the rate of change of drop mass and factor $L_v$ is the latent heat of vaporization [11].

Equation (10) describes the rate of change of the drop mass. The right hand side of Eq.(10) includes the modified diffusivity of water vapor, $D'$, the saturated vapor density, $\rho_{sat}^v$, and the difference between the surrounding relative humidity and relative humidity at the surface of the drop, $Q_n$ (cf. [11,12] for details).

The integration of Eqs. (1) - (4) is performed in curvilinear coordinates ($\xi, y, \eta$) which transform the lower wavy boundary, $z_s(x) = a \cos k\xi(x)$, into a plane boundary at $\eta = 0$ [8]. The instantaneous air-flow velocity, temperature and relative humidity at the location of $n$-th droplet, $U_i(r^n)$, $T_a(r^n)$ and $H(r^n)$, are evaluated by interpolation after mapping, Eqs. (17) and (18), the droplet Cartesian coordinates onto curvilinear coordinates.

Equations (1)-(4) are discretized in a rectangular domain with sizes $0 < x < 6\lambda$, $0 < y < 4\lambda$, and $0 < z < \lambda$ by employing a finite difference Adams-Bashforth method [13] of the second-order accuracy on a uniform staggered grid consisting of $360 \times 240 \times 180$ nodes. An additional mapping is employed to compress the grid in the vertical direction near the boundaries in order to resolve the viscous boundary layer [8].

The air-flow bulk Reynolds number in DNS is defined as:

$$\mathrm{Re} = \frac{U_0 \lambda}{\nu_a} \tag{13}$$

and set equal to Re = 15000. Two different wave-slopes, $ka$ = 0.1 and $ka$ = 0.2, are considered. We also prescribe the wave celerity to be sufficiently small, $c/U_0 = 0.05$, which corresponds to "slow" waves and strong wind-forcing conditions [8]. When prescribing the air temperature and fractional relative humidity at the waved surface and upper boundary we consider a situation typical of a tropical cyclone, and take $T_a = 27 + T_k$ (K) and $T_w = 28 + T_k$ (K) (where $T_k = 273.15$ K), and $H_w = 0.98$ and $H_a = 0.8$ [14,15]. Under these conditions, the effect of unstable temperature stratification on the air dynamics can be considered as negligible. Nevertheless, we take into account the buoyancy forces in the air momentum equation and prescribe a dimensionless gravity acceleration [analogous to the Richardson number $Ri = g(T_w - T_a)\lambda / (T_a U_0^2)$ ] to be equal to 0.01.

At the lower-plane boundary ($\eta = 0$) the no-slip (Dirichlet) condition for the air velocity, temperature, and relative humidity is prescribed. The air-flow velocity here coincides with the velocity of the water in the surface wave. At the upper boundary ($\eta / \lambda = 1$) the no-slip condition for the wind velocity is prescribed with respect to the plane moving with non-dimensional velocity ($1 - c$). The deviations of the air temperature and relative humidity form respective reference profiles $T_{ref}(z)$ and $H_{ref}(z)$, Eqs. (5) and (6), are put to zero at both the waved surface and the upper boundary. Periodical boundary conditions are prescribed for all fields at the side boundaries of the computational domain, namely, at $\xi/\lambda = 0, 6$ and $y/\lambda = 0, 4$.

Equations (7)-(10), are integrated in the Cartesian framework with the use of the second-order Adams method for the coordinate, Eq. (3), and Adams-Bashforth method for the velocity, Eq. (4). The inverse transform from the Cartesian to curvilinear droplets coordinates is preformed by an iterative Newton's method [8].

The feed-back contributions of each drop to the rate of change of air flow momentum, temperature and moisture, $f_{Ui}^n$ , $f_T^n$, and $f_H^n$, in Eqs. (1), (3) and (4) are modelled with the use of a point-force approximation. Thus the contributions by $n$-th droplet are evaluated by distributing them to the nearest eight grid nodes surrounding the droplet in the form [8]:

$$f_{Ui}^n = \frac{\pi d_n^3}{6} \frac{\rho_d^n}{\rho_a} \frac{1}{\tau_n} \left(V_i^n - U_i(r^n)\right)\left(1 + 0.15 \mathrm{Re}_n^{0.687}\right) \frac{w(r^n, r)}{\Omega_g}, \tag{14}$$

$$f_T^n = 2\pi\kappa' d_n \left(T_n - T_a(r^n)\right)\left(1 + 0.25 \mathrm{Re}_n^{0.25}\right) \frac{1}{\rho_a c_a} \frac{w(r^n, r)}{\Omega_g}, \tag{15}$$

$$f_H^n = -\frac{1}{\rho_{sat}^v} \frac{dm_n}{dt} \frac{w(r^n, r)}{\Omega_g}, \tag{16}$$

where $w(r^n, r)$ is a geometrical weighting factor inversely proportional to the distance between $n$-th droplet located at $r^n = (x^n, y^n, z^n)$ and the grid node located at $r = (x, y, z)$, and $\Omega_g$ ($r$) is the volume of the considered grid cell. Thus for each individual droplet, eight weighting factors are defined (for each of the surrounding grid nodes) and normalized so that the sum of the partial feedback contributions by the drop distributed to these nodes exactly equals the respective total feedback contribution. Therefore, there is no numerically-induced loss or gain of momentum, heat and moisture in the drops-air exchange processes.

We consider diameters of initially injected drops uniformly distributed in the range $100 \le d \le 300\,\mu m$. As is known from observations [1], for drop with diameter $d \approx 200\,\mu m$, the ratio of drop terminal settling velocity, $V_g = \tau_n g$ , to the product $\kappa u_*$ (where $\kappa$ = 0.4 is the Karman constant) is of the order of unity. This ratio ($V_g / \kappa u_*$)

indicates whether gravitational settling of the droplets is important as compared to the advection by turbulent eddies in the boundary layer [1]. In present DNS, ratio $V_g / \kappa u_* \approx 1$ for drop with diameter $d_n = 200\,\mu m$ and response time, Eq. (11), $\tau_n \lambda / U_0 = 0.5$. In order to investigate the impact of the drop mass fraction ($C_m$) and surface wave slope (*ka*) we also consider different cases of *ka* and $C_m$.

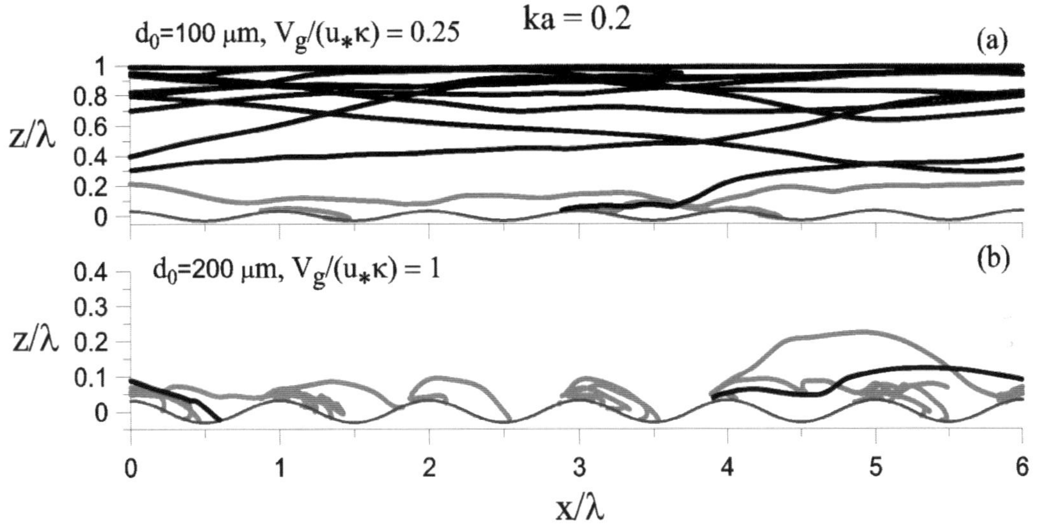


Fig. 2. The side view of the trajectories of drops injected with different diameters at injection, $d_0$: (a) $d_0 = 100\,\mu m$, $V_g/(\kappa u_*) = 0.25$; (b) $d_0 = 200\,\mu m$, $V_g/(\kappa u_*) = 1$. The trajectories are sampled at discrete time moments with increment $\Delta t = 0.2\lambda / U_0$ in DNS with wave slope *ka* = 0.2. The trajectories obtained during the total simulation time ($100 \le tU_0/\lambda \le 300$) are shown in grey color, whereas black color shows trajectories of drops from the moment of injection to the moment of their fall into the water.

## 3. Numerical results

Figure 2 presents a side view of the trajectories of drops with different diameters at injection obtained in DNS with wave slope *ka* = 0.2. [The trajectories obtained during the total simulation time ($100 \le tU_0/\lambda \le 300$) are shown in grey color.] The figure shows that a relatively small drop [case (a), $d_0 = 100\,\mu m$, ratio $V_g/(\kappa u_*)$ = 0.25], once injected, travels throughout the domain over many wavelengths without falling back into the water. Motion of larger drops [cases (b) $d_0 = 200\,\mu m$, $V_g/(\kappa u_*)$ = 1] is mostly confined to a near-water-surface layer at $z/\lambda$ < 0.1. In this case, drop injected above the up-wind side of the surface wave typically travel about one wavelength in the streamwise direction and falls back onto the water surface, either at the lee-side of the wave crest or in the vicinity of the wave trough.

Figures 3 and 4 show instantaneous heights of the drops above the water surface [$\eta_d = z_d - z_s(x_d)$] (a); drops diameters normalized by the diameter at injection ($d/d_0$) (b); the difference of the drops temperatures and the surrounding air temperature $(T_d - T_a)$ (normalized by the water-air temperature difference, $\Delta T$) (c); and the fluxes of momentum ($f_x^d$) (d), and sensible ($Q_S^d$) and latent ($Q_L^d$) heat (e) and the resulting enthalpy flux

($Q_S^d + Q_L^d$) from the drops to the surrounding air (normalized by the air heat capacity, density, and air-water temperature difference, $\Delta T$) (f). In terms of these fluxes, equations (8) and (9) describing the dynamics of the *x*-component of drop momentum and the drop temperature can be rewritten in the equivalent form:

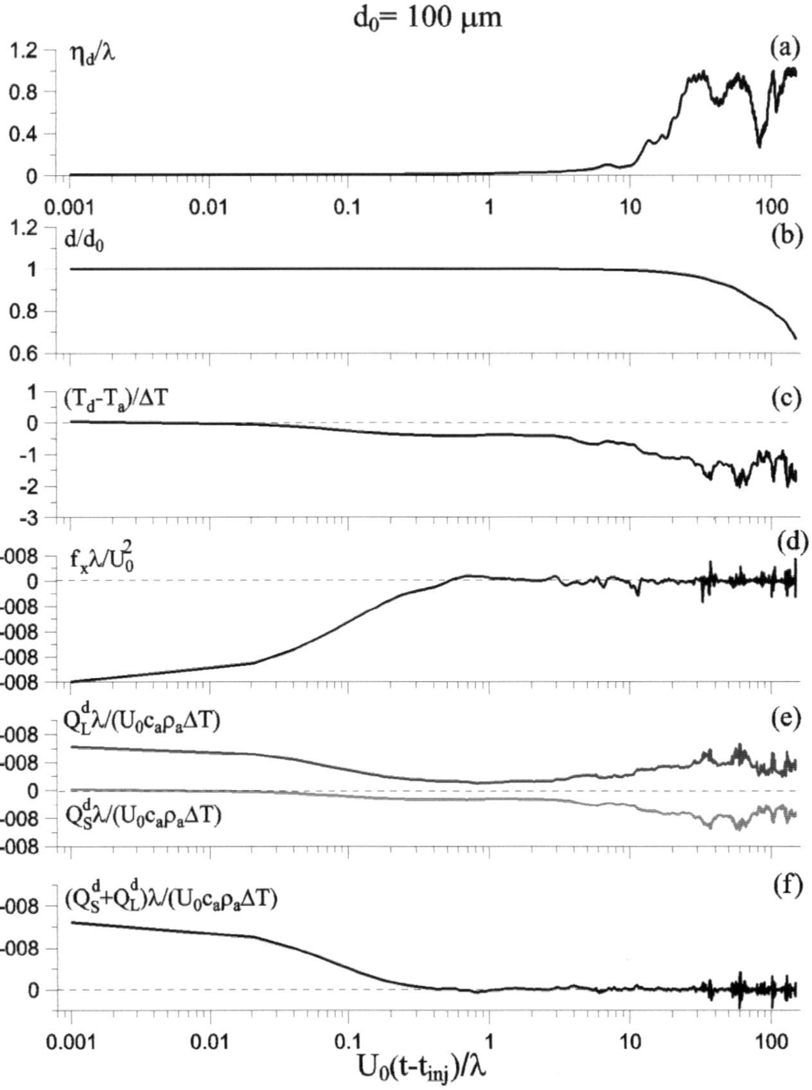


Fig. 3. Temporal development (along the drop Lagrangian trajectory) of the instantaneous drop height above the water surface, $\eta_d$; the difference between the drop temperature and the surrounding air temperatures, $T_d - T_a$; drop diameter, *d*; momentum flux from the drop to the surrounding air, $f_x$; sensible latent heat fluxes from the drop to the surrounding air, $Q_S$ and $Q_L$; and the enthalpy flux (the sum of $Q_S$ and $Q_L$). Wave slope $ka$ = 0.2; drop diameter at injection $d_0 = 100\,\mu m$.

$$m_d \frac{dV_x^d}{dt} = -\rho_a f_x^d, \tag{17}$$

$$m_d c_w \frac{dT_d}{dt} = -Q_S^d - Q_L^d, \tag{18}$$

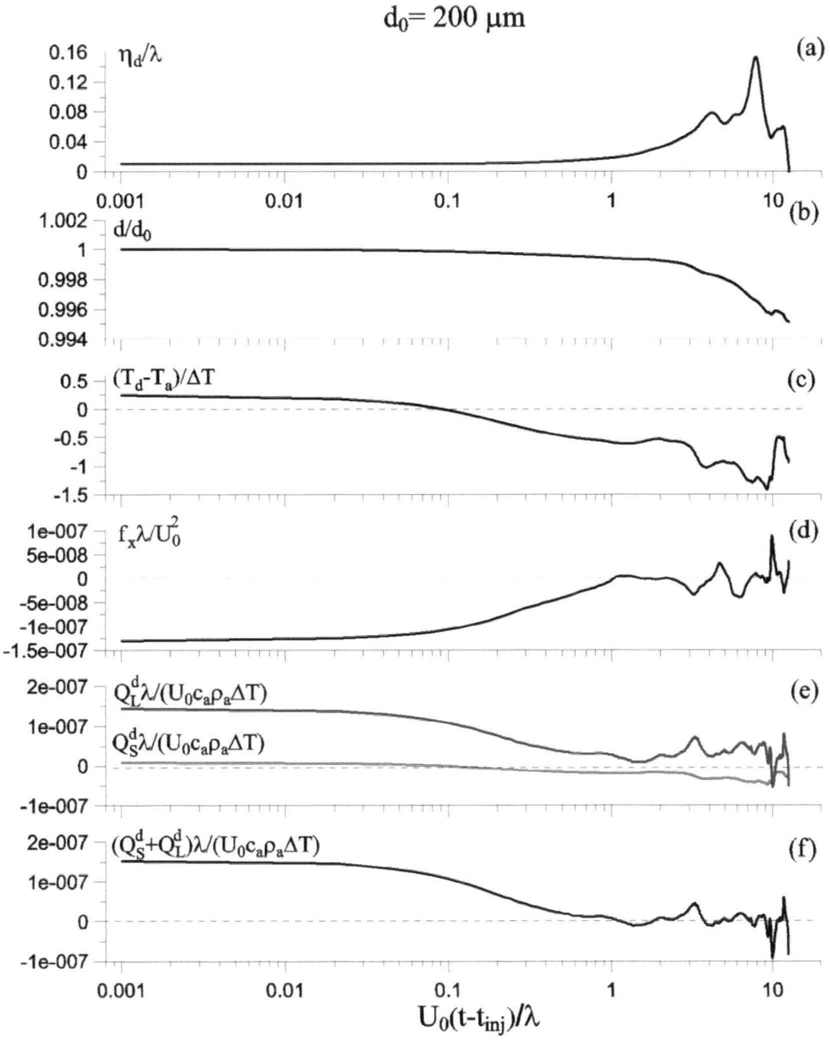


Fig. 4. The same as in Fig. 3 but for the drop diameter at injection $d_0 = 200\,\mu m$.

where

$$f_x^d = 3\pi d\nu\left(V_x^d - U_x(r^d)\right)\left(1 + 0.15\,\mathrm{Re}_d^{0.687}\right), \tag{19}$$

$$Q_S^d = 2\pi\kappa' d\left(T_d - T_a(r^d)\right)\left(1 + 0.25\,\mathrm{Re}_d^{0.25}\right), \tag{20}$$

$$Q_L^d = -L_v\frac{dm_d}{dt}, \tag{21}$$

where drop mass $m_d = \rho_d\pi d^3/6$, and $r^d$ and $\mathrm{Re}_d$ are the drop coordinate and Reynolds number. Other notations are the same as in Eqs. (8), (9). Similarly to the definitions of net fluxes in Eqs. (14) - (16), the signs of the $f_x^d$, $Q_S^d$ and $Q_L^d$ in Eqs. (19) - (21) are such that the fluxes are from the drop to the surrounding air.

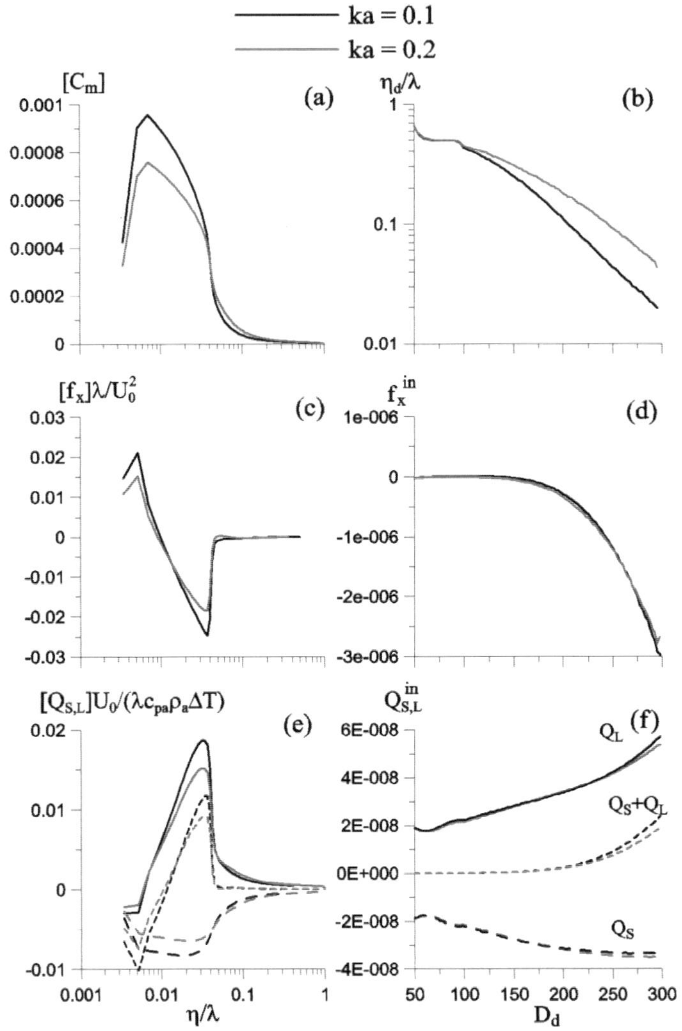


Fig. 5. Mean vertical profiles of drops concentration, $C$, (a), momentum flux, $f_x$, (c), and sensible and latent heat fluxes, $Q_{S,L}$ , (e), and distributions of $C, f_x$ and $Q_{S,L}$ over drop diameter (b,d, and e, respectively).

Figures 3b shows that diameter of a drop with $d_0 = 100\,\mu m$ monotonically decreases due to evaporation, especially at times $U_0(t-t_{inj})/\lambda > 1$ when the drop leaves the relatively moist near-surface layer ($\eta_d/\lambda < 0.1$, Fig. 3a) and travels far above the water surface where the air is more dry. The drop temperature is slightly above the water surface temperature at times $U_0(t-t_{inj})/\lambda < 0.01$ and decreases until time $U_0(t-t_{inj})/\lambda \approx 0.2$ due to the sink of heat due to evaporation, and remains approximately constant at times $0.2 \le U_0(t-t_{inj})/\lambda \approx 2$ (Fig. 3c). During this time interval, the drop resides near the water surface ($\eta_d/\lambda < 0.1$, Fig. 3a) and is in a state of "wet bulb", i.e. the latent heat flux, $Q_L^d$ , "consumed" by the drop evaporation, is almost exactly compensated by the diffusive heat flux from the surrounding air, $Q_S^d$ (Figs. 3d-

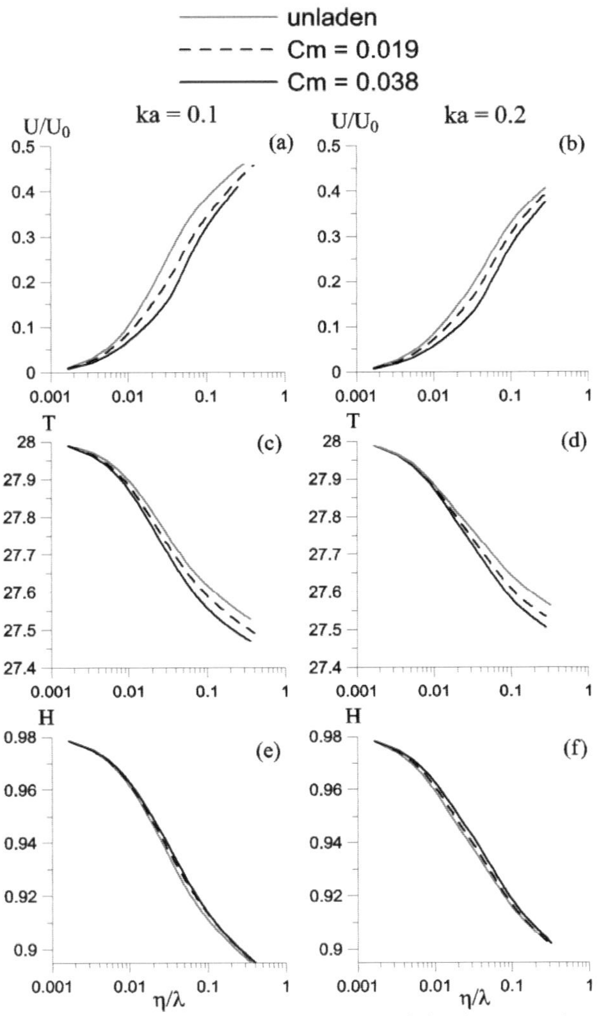


Fig. 6. Mean profiles of air velocity, temperature, and relative humidity for wave slope *ka* = 0.1 (a,c,e) and *ka* = 0.2 (b,d,f).

f) [14] and as a result their sum, which also equals the enthalpy flux from drop to air, vanishes:

$$Q_S^d + Q_L^d \approx 0\,. \tag{22}$$

As is also illustrated in Figs. 3d, the drop takes away momentum from the surrounding air (since being "torn off" from the water surface it is accelerated by the wind) at times $U_0(t-t_{inj})/\lambda < 0.4$. At later times, the exchange of momentum between the drop and air is relatively weak (near zero).

The dynamics of drop with $d_0 = 200\,\mu m$ (Fig. 4) is qualitatively similar to that of drop with $d_0 = 100\,\mu m$. However, in this case, the "wet bulb" state is not reached since the drops relatively quickly falls back to the water surface. The drop cools below the surrounding air temperature at time $U_0(t-t_{inj)}/\lambda \approx 0.1$, i.e. by the order of magnitude longer as compared to drop with $d_0 = 100\,\mu m$ (cf. Figs. 3c and 4c). Figure 4b also shows that the rate of change of drop diameter due to evaporation is much smaller as compared to case $d_0 = 100\,\mu m$.

Figure 5 presents mean vertical profiles of drops concentration, and momentum and sensible and latent heat fluxes (a, c, d) and distribution of average drops heights, and momentum and heat fluxes over drops diameter obtained in DNS for different wave slope (ka = 0.1 and 0.2). Figure 6 illustrates modifications of the air mean velocity, temperature, and humidity profiles by the drops, as compared to unladen flow.

Figures 5a and 5b show that drops concentration peaks at $\eta / \lambda \approx 0.01$, mostly due to drops with diameters about 300 $\mu m$. Momentum and sensible heat fluxes are negative, i.e. drops impose additional drag on the air and reduce wind velocity and also reduce the air temperature (Figs. 5b,d and 6a-d). On the other hand, drops provide positive latent heat flux (i.e., water wapor) to the air and increase relative humidity (Figs. 5e and 6e,f). As seen from Figures 5d,f, the drop contribution to momentum and sensible and latent heat fluxes with diameter.

## 4. Conclusions

We have performed direct numerical simulation (DNS) of turbulent atmospheric boundary layer over waved water surface and investigated the possible impact of sea spume drops on the momentum and sensible and latent heat exchange processes that are known to be of primary importance for parameterizations in large-scale prognostic models. DNS reveals many important details of both individual drops dynamics and the feedback effects by drops on the air flow. The results show that drops impose additional drag on the air, reduce air temperature and increase the relative humidity. The resulting enthalpy flux is found to be positive due to dominant contribution of latent heat (i.e. water vapor) by large drops.

**Acknowledgement:** This work was supported by RFBR (Nos. 16-55-52022, 16-05-00839, 17-05-00703 and 18-05-00292). Yu. Troitskaya and O. Druzhinin were partially supported by the Russian Science Foundation (Nos. 14-17-00667 and 15-17-20009, respectively).

## References

[1] Andreas E Jones K Fairall C 2010 *J. Geophys. Res.* **115** C12065, doi:10.1029/2010JC006458.
[2] Bortkovskii R 1987 *Air-sea exchange of heat and moisture during storms* (Dodrecht).
[3] Fairall C Kepert J Holland G *The Global Atmosphere and ocean system* **2** 121-142.
[4] Kudryavtsev V Makin V 2011 *Boundary-Layer Meteorol.* **140**, 383–410, doi: 10.1007/s10546-011-9624-2.
[5] Edson J Fairall C 1994 *J. Geophys. Res.* **99** 25295-25311.
[6] Mueller J Veron F 2014 *J. Phys. Oceanogr.* **44** 2817-2834, doi: 10.1175/JPO-D-13-0245.1.
[7] Troitskaya Yu Ezhova E Soustova I Zilitinkevich S 2016 *Ocean Dynamics* **66**, 659-669, doi:10.1007/s10236-016-0948-9.
[8] Druzhinin O Troitskaya Yu Zilitinkevish S 2017 J. Geophys. Res. Oceans **122** 1789-1807. doi: 10.1002/2016JCO12134.
[9] Monin A Yaglom A 1971 *Statistical Fluid Mechanics: Mechanics of Turbulence* **1** (Cambridge).
[10] Clift R Grace J Weber M 1978 *Bubbles, Drops, and Particles* (Academic Press).
[11] Pruppacher H Klett J 1978 *Microphysics of Clouds and Precipitation* (D. Riedel).
[12] Andreas E 1989 CRREL Rep. 89-11.
[13] Fletcher C 1991 *Computational Techniques for Fluid Dynamics* **2** (Springer).
[14] Andreas E Emanuel K 2001 *J. Atmos. Sci.* **58** 3741-3751.
[15] Andreas E 1995 *J. Atmos. Sci.* **52** 852-862.

# From particle condensation to polymer aggregation

**Wolfhard Janke**[1] **and Johannes Zierenberg**[1,2,3]

[1] Institut für Theoretische Physik, Universität Leipzig, Postfach 100 920, 04009 Leipzig, Germany

[2] Max Planck Institute for Dynamics and Self-Organization, Am Fassberg 17, 37077 Göttingen, Germany

[3] Bernstein Center for Computational Neuroscience, Am Fassberg 17, 37077 Göttingen, Germany

E-mail: wolfhard.janke@itp.uni-leipzig.de

**Abstract.** We draw an analogy between droplet formation in dilute particle and polymer systems. Our arguments are based on finite-size scaling results from studies of a two-dimensional lattice gas to three-dimensional bead-spring polymers. To set the results in perspective, we compare with in part rigorous theoretical scaling laws for canonical condensation in a supersaturated gas at fixed temperature, and derive corresponding scaling predictions for an undercooled gas at fixed density. The latter allows one to efficiently employ parallel multicanonical simulations and to reach previously not accessible scaling regimes. While the asymptotic scaling can not be observed for the comparably small polymer system sizes, they demonstrate an intermediate scaling regime also observable for particle condensation. Altogether, our extensive results from computer simulations provide clear evidence for the close analogy between particle condensation and polymer aggregation in dilute systems.

## 1. Introduction

Condensation and nucleation processes are underlying many phenomena from galaxy evolution over cloud formation to protein aggregation. For particles or colloids much is known from experiments, theory and computation [1, 2, 3]. For polymers, the situation is less settled. At first sight, particle condensation and polymer aggregation appear to be very different processes. While condensation is a phenomenon of one's every-day-life and the stereotypic example of a first-order phase transition, aggregation is usually associated with neurodegenerative diseases such as Altzheimer's, Parkinson's, and diabetes II [4]. However, if treated outside of complex environments such as the brain, proteins in dilute solution do show cluster formation [5] with a close resemblance to classical nucleation theory [6, 7, 8]. Consequently, concepts such as nucleation rates and barriers can be qualitatively carried over to more complex macromolecular solutions [9].

There exist a lot more qualitative similarities between particle and polymer systems. For example, the phase diagrams of particle and polymer solutions appear quite similar [9, 10, 11, 12, 13, 14]. In fact, already a single polymer shows similarities to a Lennard-Jones particle cluster in its low-temperature structure [15, 16, 17]. Also, the multi-step nucleation of a few constituents to a full cluster shows a comparable hierarchy by multiple backbending in a microcanonical analysis depending on the density [18, 19, 20, 21, 22, 23]. For an abstract protein model as unit-length stick on a periodic cubic lattice the formation of fibrils shows comparable scaling

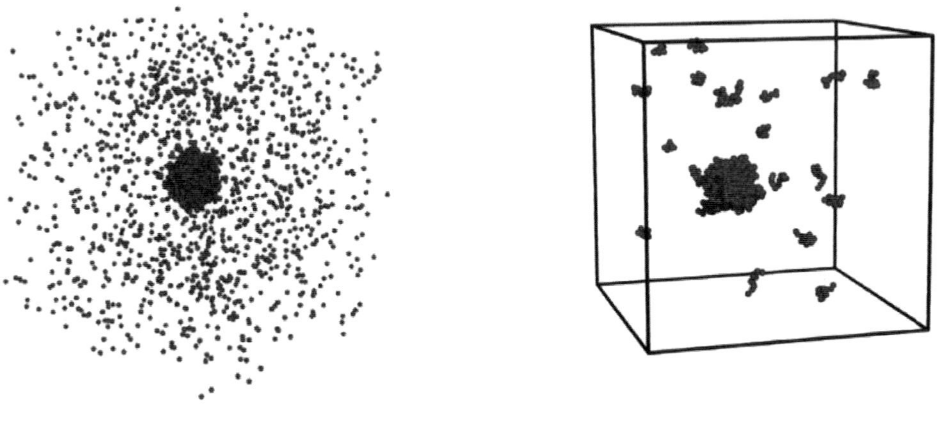

**Figure 1.** Illustration of droplet formation in a system of $N = 2000$ particles (left) and $N = 64$ polymers of length $P = 13$ (right).

behavior as droplet formation, which is expected since the model resembles a lattice gas with asymmetric interactions [24]. A natural question to ask is thus whether the cluster formation in polymer systems shows the same scaling behavior as in particle systems, despite their more complex structural properties [25, 26].

In this work, we want to recapitulate some evidence for an analogy between cluster formation in a particle gas and a polymer solution using a consistent model framework based on our own contributions in this field. By exploiting the similarity with particle systems, we will analyze the implications for the finite-size scaling behavior in dilute polymer systems and discuss how small system sizes may easily lead to misinterpretations.

## 2. Droplet formation in a particle gas

### *2.1. General arguments involving competition of energy and entropy at fixed temperature*

Let us consider the situation of a supersaturated gas in a finite system size at fixed temperature, i.e., in the canonical ($NVT$) ensemble [27, 28, 29, 30, 31]. Supersaturation refers to a particle excess $\delta N = N - N_0$ compared to the temperature-specific (grand-canonical) background gas density $\rho_{\rm g}(T) = N_0(T)/V$. Besides the constant free-energy contribution from the background density, the particle excess can contribute to the system's free energy either by an entropic part from the particle fluctuations $F_{\rm fluc}$, or by an energetic part $F_{\rm drop}$ related to the formation of a single macroscopic droplet of size $V_{\rm D}$. For an illustration see Fig. 1 (left). The fluctuations can be treated in a Gaussian approximation around the equilibrium background contribution and the condensate is treated as an ideally shaped droplet [28, 29]:

$$F_{\rm fluc} = \frac{(\delta N)^2}{2\hat{\kappa}V} \qquad \text{and} \qquad F_{\rm drop} = \tau_{\rm W} V_{\rm D}^{\frac{d-1}{d}}, \tag{1}$$

where $\hat{\kappa} = \beta\kappa = \beta\langle(N - \langle N\rangle)^2\rangle/V$ is the isothermal compressibility and $\tau_{\rm W}$ the surface free energy of a (Wulff shaped) droplet of unit volume.

This treatment differs from the general view on nucleation free-energy barriers, where interface free-energy loss competes with bulk free-energy gain because an infinite reservoir of attaching particles is assumed. This essentially focuses only on the droplet contribution

$\Delta F_{\text{drop}} = V_{\text{D}}\Delta f + \partial V_{\text{D}}\sigma$, where $\Delta f$ is the free-energy difference per unit volume, $\partial V_{\text{D}}$ denotes the surface of $V_{\text{D}}$, and $\sigma$ is the surface tension. This has a clear maximum, which is obtained by solving $d\Delta F_{\text{drop}}/dR = 0$ for the critical droplet radius $R_c$. If $R > R_c$ the droplet will grow infinitely due to the infinite reservoir of particles. In this case, $R_c$ is not system size dependent, such that all critical droplets have the same size. We, instead, consider the equilibrium situation of a finite particle number where the gas contribution needs to be considered explicitly. For a good overview of the historical development, see, e.g., Ref. [32].

We proceed our line of arguments by linking the droplet size to the particle excess inside the droplet, $\delta N_{\text{D}} = (\rho_{\text{l}} - \rho_{\text{g}})V_{\text{D}}$, where $\rho_{\text{l}}$ and $\rho_{\text{g}}$ are the background liquid and gas density, respectively. General arguments show that the probability of additional intermediate-sized droplets vanishes [28, 29], so that we can distribute the particle excess in a two-phase scenario between the excess inside the droplet $\delta N_{\text{D}}$ and the excess in the fluctuating phase $\delta N_{\text{F}}$. In terms of the scalar fraction

$$\lambda = \delta N_{\text{D}}/\delta N\,, \tag{2}$$

we obtain $\delta N_{\text{D}} = \lambda\delta N$ and $\delta N_{\text{F}} = (1-\lambda)\delta N$. Then, the total excess free energy $F = F_{\text{drop}} + F_{\text{fluc}}$ becomes

$$F = \tau_{\text{W}}\left(\frac{\lambda\delta N}{\rho_{\text{l}} - \rho_{\text{g}}}\right)^{\frac{d-1}{d}} + \frac{(1-\lambda)^2(\delta N)^2}{2\hat{\kappa}V} = \tau_{\text{W}}\left(\frac{\delta N}{\rho_{\text{l}} - \rho_{\text{g}}}\right)^{\frac{d-1}{d}}\left(\lambda^{\frac{d-1}{d}} + \Delta(1-\lambda)^2\right)\,, \tag{3}$$

with a dimensionless "density" parameter

$$\Delta = \frac{(\rho_{\text{l}} - \rho_{\text{g}})^{\frac{d-1}{d}}}{2\hat{\kappa}\tau_{\text{W}}}\frac{(\delta N)^{\frac{d+1}{d}}}{V} = \frac{(\rho_{\text{l}} - \rho_{\text{g}})^{\frac{d-1}{d}}}{2\hat{\kappa}\tau_{\text{W}}}(\rho - \rho_{\text{g}})^{\frac{d+1}{d}}\,V^{\frac{1}{d}}\,. \tag{4}$$

At fixed temperature, $\rho_{\text{l}}, \rho_{\text{g}}, \hat{\kappa}$, and $\tau_W$ are constants. By minimizing Eq. (3) with respect to $\lambda$ one obtains the fraction of particles inside the largest droplet $\lambda$ as a function of dimensionless density $\Delta$, as shown for a two-dimensional lattice gas in Fig. 2 by the black "analytic" line jumping at $\Delta_{\text{c}} = (1/2)(3/2)^{3/2} \approx 0.9186$ from $\lambda = 0$ to $\lambda = 2/3$. As derived in Refs. [28, 29], there exists in general a constant $\Delta_{\text{c}} = \frac{1}{d}\left(\frac{d+1}{2}\right)^{\frac{d+1}{d}}$ below which no droplet forms ($\lambda = 0$) and above which a single macroscopic droplet exists with non-trivial $\lambda > \lambda_{\text{c}} = \frac{2}{d+1}$. This threshold already includes the leading-order finite-size corrections, as can be seen if one rewrites Eq. (4) in terms of the density

$$\rho_{\text{c}} = \rho_{\text{g}} + \left(\frac{2\hat{\kappa}\tau_{\text{W}}\Delta_{\text{c}}}{(\rho_{\text{l}} - \rho_{\text{g}})^{\frac{d-1}{d}}}\right)^{\frac{d}{d+1}} V^{-\frac{1}{d+1}}\,. \tag{5}$$

It is less well visible that the critical droplet size $V_c = \lambda_c\delta N_c/(\rho_{\text{l}} - \rho_{\text{g}})$ scales non-trivially with system size, namely $V_c \propto R_c^d \propto V^{d/(d+1)}$ [27, 31, 43].

Figure 2 compares the results of simulations with the analytic prediction for a (square) lattice gas with nearest-neighbor interactions in 2D [33, 34, 35, 36]. In the lattice gas model, particles are modeled as non-empty lattice sites. This is equivalent to an Ising model, where particles correspond to either up or down spins while the vacuum corresponds to the other direction. For instance, identifying particles with up spins, the particle density $\rho$ is related to the magnetization of the Ising model by $\rho = N/V = (1+m)/2$ and the particle gas temperature is $T = T^{\text{Ising}}/4$. The data in Refs. [33, 34, 35, 36] was generated in this framework using mostly fixed-magnetization Metropolis [37] simulations (Kawasaki dynamics) at a given temperature $k_{\text{B}}T = 0.375$. In two dimensions, the equivalence to the Ising model additionally provides exact results or very precise estimates for the infinite-size quantities $\rho_{\text{g}}$, $\rho_{\text{l}} = 1 - \rho_{\text{g}}$, $\hat{\kappa}$, and $\tau_{\text{W}}$, necessary for the rescaling of the data according to Eq. (4) in order to compare with the theoretical (and, for the square lattice, rigorous [28, 29]) prediction [33, 34, 35, 36].

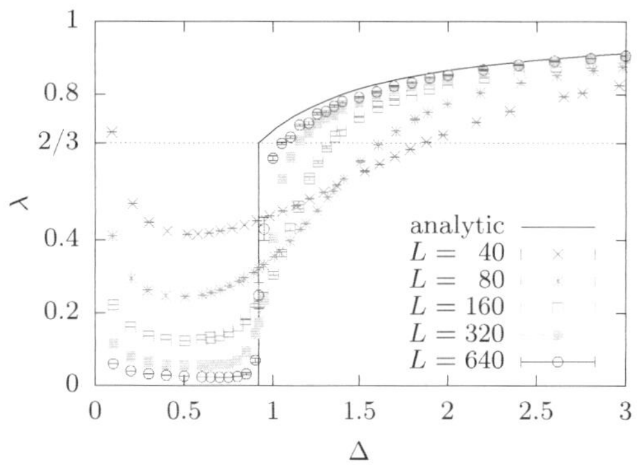


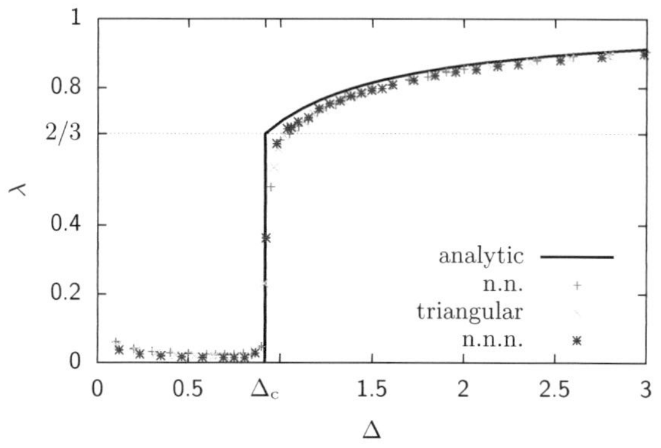


**Figure 2.** (left) Fraction of excess $\lambda$ that goes into the droplet as a function of dimensionless rescaled density $\Delta$ for 2D droplet formation in the Ising lattice gas (square lattice with nearest-neighbor interactions) at fixed temperature $k_{\mathrm{B}}T = 0.375$ (in lattice gas normalization). Note that $\Delta$ already accounts for the leading-order finite-size scaling behavior of the density. (right) Comparison of results for three different 2D lattice gas models (nearest-neighbor (n.n.) square lattice, n.n. triangular lattice, next-nearest-neighbor (n.n.n.) square lattice) at $T \approx (2/3)T_c$ for large system size $L = 640$. Adapted from Refs. [33, 34, 35, 36].

The data in Fig. 2 (left) clearly shows that for systems of finite size the analytic solution is overestimated for low densities and (slightly) underestimated for densities above the predicted transition density $\Delta_c$. The low-density finite-size effect is an unavoidable artefact due to the non-zero size of the largest cluster in the gas phase, which is at least $\mathcal{O}(1/N)$. The high-density deviations, on the other hand, can be attributed to physical effects such as non-ideal droplet shapes, including capillary wave distortions, and further non-trivial corrections to the theory [38]. As theoretically expected, these deviations rapidly vanish in the limit of large system size. In Refs. [34, 35] we have also checked that the analytic solution, which strictly speaking was derived for the square lattice with nearest-neighbor (n.n.) interactions in the infinite-size limit, describes other 2D lattice gases as well. This is demonstrated in Fig. 2 (right) where data for the next-neighbor triangular and next-nearest-neighbor (n.n.n.) square lattice gas are compared with the standard n.n. square lattice reference case.

Overall, Fig. 2 shows a good agreement with a surprising symmetry around $\Delta_c$. This may be caused by a judicious choice of the fixed temperature, far enough away from the critical point to avoid strong fluctuations but high enough to avoid slow dynamics. Results for $\lambda(\Delta)$ in 3D lattice models are less symmetric, see, e.g., Ref. [39], and are currently under closer investigation. The leading-order phenomenological scaling theory of droplet formation was in general found to be consistent with numerical results for both 3D Lennard-Jones [40] and lattice [41] models.

### *2.2. Changing perspective to fixed density*

The theory developed for fixed temperature quantitatively describes the formation and growth of a droplet with supersaturation, already including the leading-order finite-size corrections. Now, we consider instead a particle system at fixed density $\rho$ where the temperature is lowered until a droplet forms in an *undercooled* environment, again in the canonical ($NVT$) ensemble as in Refs. [42, 43, 44, 45]. This can be seen as an "orthogonal" perspective in the temperature-density phase diagram of the liquid-vapor transition [43]. Considering that the "density" parameter $\Delta$ already includes the dominant or leading-order finite-size scaling behavior, we can rewrite Eq. (4) as $V^{-1/(d+1)}\Delta^{d/(d+1)} = f(\rho, \beta)$, which admits a power-series expansion in $\beta$ around the infinite-size inverse temperature $\beta_0$ where $\rho_g(\beta_0) = \rho$. The

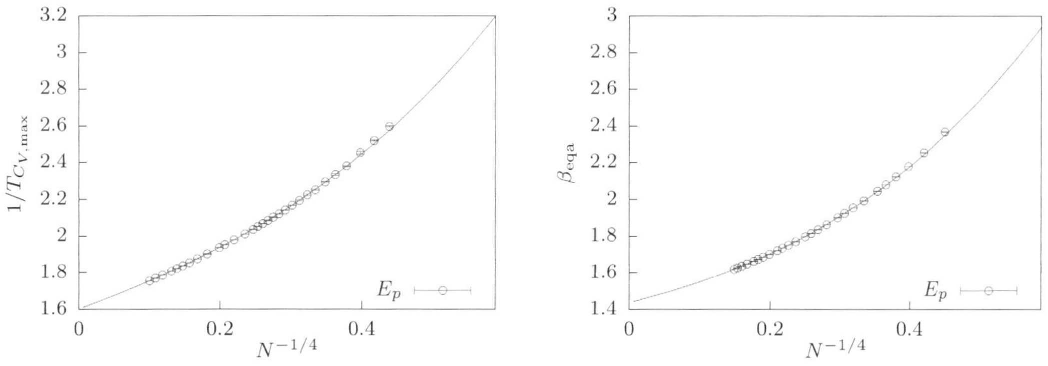


**Figure 3.** Finite-size scaling of the inverse transition temperature (fixed density $\rho = 10^{-2}$) for 3D droplet formation in (left) a lattice gas and (right) a Lennard-Jones gas. For the lattice gas the density is of course approximate due to discretization effects. The solid lines show fits to the ansatz (7). Adapted from Refs. [43, 45].

solution of $\Delta = \Delta_c$ then results in an asymptotic expansion in $V^{-1/(d+1)}$ for the droplet-formation point $\beta(N)$ of a finite number of particles $N$. Using the Taylor expansion $f(\rho,\beta) = f(\rho,\beta_0) + f'(\rho,\beta_0)(\beta-\beta_0) + \frac{1}{2}f''(\rho,\beta_0)(\beta-\beta_0)^2 + ...$, where $f(\rho,\beta_0) = 0$, one obtains

$$\beta(N) = \beta_0 + \frac{\Delta_c^{d/(d+1)}}{f'(\rho,\beta_0)}V^{-1/(d+1)} - \frac{f''(\rho,\beta_0)\Delta_c^{2d/(d+1)}}{2f'(\rho,\beta_0)^3}V^{-2/(d+1)} + \mathcal{O}\left(V^{-3/(d+1)}\right). \quad (6)$$

This of course neglects higher-order corrections to the fixed-temperature solution, especially expected non-trivial and logarithmic terms [38, 46, 47, 35, 48]. However, the scaling corrections correspond to powers of the critical droplet radius $R_c \propto V_D^{1/d} \propto V^{1/(d+1)}$ [31, 43] and we will show that fits to the higher-order terms from the Taylor expansion nicely describe the finite-size corrections in the data. Since the density is fixed, we can further replace $V \propto N$ and arrive at the ansatz for droplet formation in 3D

$$\beta(N) = \beta_0 + aN^{-1/4} + bN^{-1/2} + cN^{-3/4}. \quad (7)$$

Figure 3 shows data for droplet formation in a 3D lattice and Lennard-Jones gas based on data of Refs. [43, 45]. The lattice gas is directly treated as occupied sites on a cubic lattice with Ising-like nearest-neighbor interactions, while Lennard-Jones particles have a real-valued position in a cubic box with periodic boundary conditions interacting via the potential $V_{\rm LJ} = 4\epsilon\left[(\sigma/r)^{12} - (\sigma/r)^6\right]$ which exhibits a minimum at $r = 2^{1/6}\sigma$ of depth $-\epsilon$. Both data sets were generated using multicanonical simulations [49, 50, 51, 52] in a parallelized implementation [53, 54], which is a perfect choice for first-order-like phase transitions, especially upon variation of a continuously variable control parameter such as the temperature (for recent reviews, see Refs. [55, 56]). The parallel implementation shows linear speedup for particle condensation [39]. Canonical expectation values are obtained after a production run by standard reweighting techniques [57]. The finite-size transition temperatures $T_{C_V,\max}$ can be read off from the peak locations of the specific heat $C_V = \beta^2\left(\langle E_p^2\rangle - \langle E_p\rangle^2\right)/N$ (where $E_p$ is the potential energy; see below) in case of the lattice gas; and from an equal-area construction in the microcanonical inverse temperature $\beta(E_p)$ [58] in case of the Lennard-Jones gas, denoted in Fig. 3 (right) by $\beta_{\rm eqa}$. Since we have no analytical solution for the infinite-size quantities $\rho(T)$, $\kappa(T)$, or $\tau_W(T)$, we merely fit the amplitudes of the correction terms in Eq. (7). In both cases, the fit qualitatively

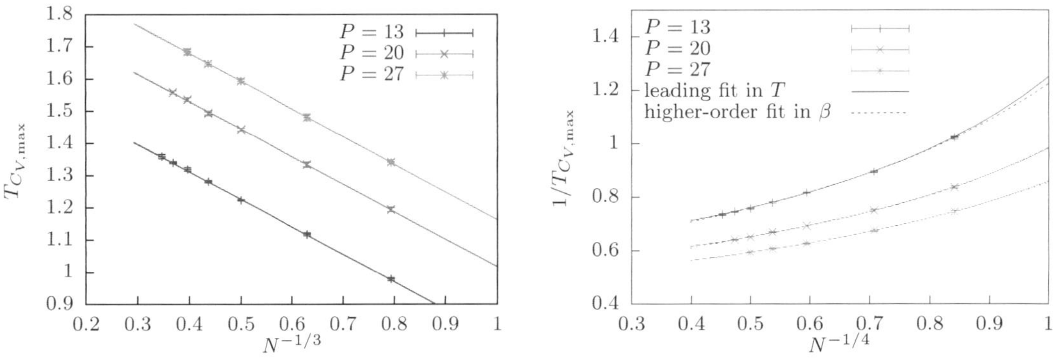


**Figure 4.** (left) Finite-size scaling of the transition temperature (fixed polymer density $N/V = 10^{-3}$) of cluster formation in a solution of bead-spring polymers of different length $P$ suggests a scaling as $T - T_0 \propto N^{-1/3}$ (solid lines). (right) If the inverse temperature is instead fitted with higher-order terms the leading behavior $\propto N^{-1/4}$ is consistent with Eq. (7) (dashed lines). Adapted from Ref. [60].

describes the data very well. For the lattice gas, the fit spans $N \in [106, 10\,000]$ with a goodness-of-fit parameter $Q \approx 0.002$, attributed to the merely approximate density that one can adjust at each value of $N$. For the Lennard-Jones gas, the fit spans $N \in [192, 2048]$ with $Q \approx 0.5$.

Overall, the data is in good agreement with the theory on droplet formation in an undercooled gas. It is worth noting that the comparably large finite-size corrections may enable a systematic study of finite-size scaling in experimental setups on the nanoscale. If one considers argon as a model system, our largest numerical Lennard-Jones setup with 2048 particles corresponds to a box of size $L' \approx 59\ r'_{\mathrm{min}} = 59 \times\ 2^{1/6}\ \sigma'$ and with $\sigma' \approx 3.4\mathring{A}$ we find $L' \approx 22.5$ nm. This is a scale that both experiments and simulations may reach.

## 3. Droplet formation in a solution of flexible polymers

Polymers are extended objects which bring an additional conformational entropy if isolated in the high-temperature phase [59]. Lowering the temperature results in the formation of a macroscopic aggregate surrounded by a gas of isolated chains [45, 60, 61], for an illustration see Fig. 1 (right). The shape of an aggregate of flexible polymers, however, shows strong structural similarities to the droplet in a gas [16, 60], with polymers actively unfolding when incorporated into the cluster [60, 61]. If one neglects the single-polymer free-energy contributions, one may consider polymers as extended particles with a more complex, potentially temperature dependent, interaction potential. Then, it would make sense to expect the same scaling behavior for the (inverse) transition temperature as predicted for particle condensation and outlined in Sec. 2.

When treating comparably small system sizes, we noticed, however, that the scaling of the aggregation transition temperature for flexible bead-spring polymers, obtained from parallel multicanonical simulations with up to $N = 24$ polymers of length $P = 13$ ("13mers"), 20 20mers or 16 27mers, shows an apparent $T - T_0 \propto N^{-1/3}$ scaling behavior [60], see Fig. 4 (left). This appears quite puzzling in the light of our above arguments. However, inspired by the findings for polymers, a close reexamination of particle condensation revealed a prominent *intermediate scaling regime* [43], consistent with the $N^{-1/d}$ scaling behavior observed for polymers in 3D [60]. Physically this means that for a small number of constituents either none or *all* of them assemble in the droplet – the surrounding gas phase cannot really develop. The problem is, that even for particles this regime extends up to $N = 10^3 - 10^4$, also depending on the dimension [43]. This is

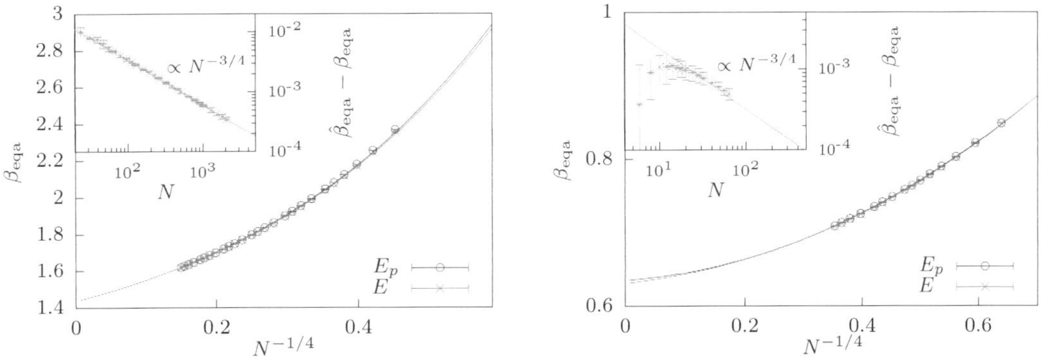


**Figure 5.** Finite-size scaling of the inverse transition temperatures (fixed monomer density $NP/V = 10^{-2}$) for 3D cluster formation in (left) a gas of Lennard-Jones particles and (right) a solution of flexible bead-spring polymers ($P = 13$ monomers each) reveals the same leading-order scaling as $\beta \sim N^{-1/4}$. The transition point is determined by the equal-area rule using the microcanonical inverse temperature as a function of potential energy $E_p$ or total energy $E$. Adapted from Ref. [45].

out of scope even of modern simulation tools for equilibrium estimates of polymer aggregation.

One may now argue that the observation of this intermediate regime is already a first hint on an analogy between cluster formation in particle and polymer systems. In addition, we observe that the empirical corrections from Eq. (7), as powers of the critical cluster size, describe the scaling of the inverse transition temperature quite well [Fig. 4 (right), dashed lines] and is almost indistinguishable from the leading-order fit to the transition temperature [Fig. 4 (right), solid lines].

Treating larger polymer systems with up to $N = 64$ polymers of length $P = 13$ in Ref. [45] further supports the strong similarity between particles and polymers, again obtained from parallel multicanonical simulations. The inverse temperatures plotted in Fig. 5 are obtained from the equal-area construction in the microcanonical inverse temperature as a function of energy (note that a *monomer* density of $10^{-2}$ corresponds for 13mers to a *polymer* density of $\approx 10^{-3}$ as used in Fig. 4). Here we also emphasize a conceptual point by using both the potential energy [$\beta(E_p)$] as used is most work using Monte Carlo simulations and the total energy [$\beta(E)$] as considered in any textbook on statistical physics. In the latter case we analytically included the kinetic energy $E_k$ what, technically, can be achieved by a convolution of the potential-energy probability distribution $P(E_p)$ with the Maxwell-Boltzmann distribution $P(E_k) = \frac{\beta^{3N/2}}{\Gamma(3N/2)} E_k^{(3N-2)/2} e^{-\beta E_k}$. For details see Ref. [45]. In Ref. [62] a complementary simulation method is discussed that samples directly the "real" microcanonical ensemble including the kinetic energy contribution [63, 64]. As expected both definitions show the same behavior with very small differences. It is interesting, however, that the difference between both definitions scales for large $N$ as $\hat{\beta} - \beta \propto N^{-3/4}$, i.e., with the volume of the critical cluster [43]. This systematic behavior can be anticipated also for the case of a polymer solution. In total, this supports our hypothesis that the formation of a cluster in a dilute solution of flexible polymers can, in leading order, be described by the generic arguments derived for particle systems.

As an outlook to future work it should be stressed that biological systems usually are comprised of more complex macromolecules that cannot be assumed to be flexible. A straightforward extension of the flexible bead-spring polymer model considered above is to include worm-like chain motivated bending stiffness, leading to semiflexible homopolymers. This

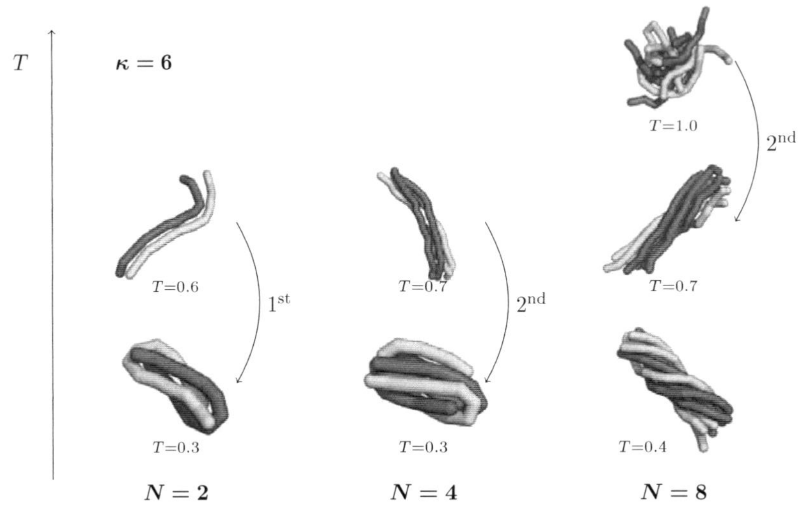


**Figure 6.** Typical aggregate morphologies of $N = 2, 4$, and 8 semiflexible polymers of length $P = 13$ in the intermediate-stiffness regime with bending-stiffness constant $\kappa = 6$, illustrating the size-dependence of the sub-aggregation re-ordering transitions within the aggregate: $N = 2$ polymers exhibit a first-order-like bundle-to-hairpin structural transition, whereas the finite-size transition for $N = 4$ polymers from elongated bundles to bundled hairpins is second-order-like. $N = 8$ polymers undergo a second-order-like amorphous-to-bundle transition followed by the formation of twisted bundles at even lower temperatures [65].

has an immediate impact on the structure of the low-temperature aggregate [65, 25, 26], cf. Fig. 6. While rather flexible polymers keep on forming spherical aggregates, sufficiently stiff polymers form (twisted) bundles [66, 67, 65, 25, 26] known from biopolymer systems, e.g., from amyloid protofibrils [68] or actin networks [69], or polymeric materials [70]. These bundles are not spherical but for finite diameter resemble a cylindrical structure. We do not suspect this to play an important role on the scaling of the cluster formation transition as long as the shape of the cluster does not become fractal. This is currently under active research.

## 4. Conclusions

In this paper we have compiled quite an extensive survey of computer simulation studies of particle condensation and polymer aggregation. By analyzing and interpreting the numerical data in a consistent theoretical framework, clear evidence for a close analogy between the cooperative behavior of particles and flexible polymers emerges.

To arrive at this main conclusion, we started with a lattice gas formulation at fixed temperature in 2D, for which we confirmed and extended in part rigorous theoretical scaling predictions in the infinite-size limit. We then proceeded to lattice and continuum Lennard-Jones particle systems in 3D, for which an "orthogonal" constant density approach was demonstrated to be more efficient. This enabled studies of large enough particle numbers to identify an intermediate scaling regime for smaller systems. Taking this crossover behavior into account we could finally show that the apparently more involved aggregation process of flexible polymers is governed by the same asymptotic scaling laws as for particles.

As an outlook we briefly mentioned on the technical side computer simulations in the "real" microcanonical ensemble that appear competitive to generalized-ensemble methods and on the physics side the more intricate aggregation behavior of semiflexible polymers. In general, our approach can be extended to elucidate nucleation mechanisms including nucleation barriers and rates in greater detail. These topics appear to be promising avenues for future work.

## Acknowledgments

This work has been partially supported by the Deutsche Forschungsgemeinschaft (DFG) under Grant No. JA 483/31-1 and SFB/TRR 102 (project B04), the Leipzig Graduate School of Natural Sciences "BuildMoNa", the Deutsch-Französische Hochschule DFH-UFA through the Doctoral College "$\mathbb{L}^4$" under Grant No. CDFA-02-07, and the EU Marie Curie IRSES network DIONICOS under Contract No. PIRSES-GA-2013-612707. JZ received financial support from the German Ministry of Education and Research (BMBF) via the Bernstein Center for Computational Neuroscience (BCCN) Göttingen under Grant No. 01GQ1005B.

## References

[1] Kelton K F and Frenkel D 2016 *J. Chem. Phys.* **145** 211501
[2] Binder K and Virnau P 2016 *J. Chem. Phys.* **145** 211701
[3] Herlach D M, Palberg T, Klassen I, Klein S and Kobold R 2016 *J. Chem. Phys.* **145** 211703
[4] Chiti F and Dobson C M 2006 *Annu. Rev. Biochem.* **75** 333
[5] Stradner A, Sedgwick H, Cardinaux F, Poon W C K, Egelhaaf S U and Schurtenberger P 2004 *Nature (London)* **432** 492
[6] Feder J, Russell K C, Lothe J and Pound G M 1966 *Adv. Phys.* **15** 111
[7] Oxtoby D W 1992 *J. Phys: Condens. Matter* **4** 7627
[8] Kashchiev D 2000 *Nucleation: Basic Theory with Applications* (Oxford: Butterworth-Heinemann)
[9] Sear R P 2007 *J. Phys.: Condens. Matter* **19** 033101
[10] Wilding N B 1995 *Phys. Rev.* E **52** 602
[11] Wilding N B 1997 *J. Phys.: Condens. Matter* **9** 585
[12] MacDowell L G, Virnau P, Müller M and Binder K 2002 *J. Chem. Phys.* **117** 6360
[13] Virnau P, Müller M, MacDowell L G and Binder K 2004 *New J. Phys.* **6** 7
[14] Virnau P, Müller M, MacDowell L G and Binder K 2004 *J. Chem. Phys.* **121** 2169
[15] Schnabel S, Vogel T, Bachmann M and Janke W 2009 *Chem. Phys. Lett.* **476** 201
[16] Schnabel S, Bachmann M and Janke W 2009 *J. Chem. Phys.* **131** 124904
[17] Schnabel S, Janke W and Bachmann M 2011 *J. Comp. Phys.* **230** 4454
[18] Junghans C, Bachmann M and Janke W 2006 *Phys. Rev. Lett.* **97** 218103
[19] Junghans C, Bachmann M and Janke W 2008 *J. Chem. Phys.* **128** 085103
[20] Junghans C, Bachmann M and Janke W 2009 *Europhys. Lett.* **87** 40002
[21] Junghans C, Bachmann M and Janke W 2011 *Comput. Phys. Commun.* **9** 1937
[22] Schierz P, Zierenberg J and Janke W 2015 *J. Chem. Phys.* **143** 134114
[23] Koci T and Bachmann M 2017 *Phys. Rev.* E **95** 032502
[24] Irbäck A and Wessén J 2015 *J. Chem. Phys.* **143** 105104
[25] Zierenberg J, Marenz M and Janke W 2016 *Polymers* **8** 33
[26] Janke W, Marenz M and Zierenberg J 2017 *Lobachevskii J. Math.* **38** 978
[27] Binder K and Kalos M H 1980 *J. Stat. Phys.* **22** 363
[28] Biskup M, Chayes L and Kotecký R 2002 *Europhys. Lett.* **60** 21
[29] Biskup M, Chayes L and Kotecký R 2003 *Commun. Math. Phys.* **242** 137
[30] Neuhaus T and Hager J S 2003 *J. Stat. Phys.* **113** 47
[31] Binder K 2003 *Physica* A **319** 99
[32] Schrader M, Virnau P and Binder K 2009 *Phys. Rev.* E **79** 061104
[33] Nußbaumer A, Bittner E, Neuhaus T and Janke W 2006 *Europhys. Lett.* **75** 716
[34] Nußbaumer A, Bittner E and Janke W 2008 *Phys. Rev.* E **77** 041109
[35] Nußbaumer A, Bittner E and Janke W 2010 *Prog. Theor. Phys. Suppl.* **184** 400
[36] Nußbaumer A, Zierenberg J, Bittner E and Janke W 2016 *J. Phys.: Conf. Series* **759** 012009
[37] Metropolis N, Rosenbluth A W, Rosenbluth M N, Teller A H and Teller E 1953 *J. Chem. Phys.* **21** 1087
[38] Kotecký R *Private communication*
[39] Zierenberg J, Wiedenmann M and Janke W 2014 *J. Phys.: Conf. Series* **510** 012017
[40] MacDowell L G, Virnau P, Müller M and Binder K 2004 *J. Chem. Phys.* **120** 5293
[41] Schmitz F, Virnau P and Binder K 2013 *Phys. Rev.* E **87** 053302

[42] Martinos S, Malakis A and Hadjiagapiou I 2007 *Physica* A **384** 368
[43] Zierenberg J and Janke W 2015 *Phys. Rev.* E **92** 012134
[44] Zierenberg J and Janke W 2016 *J. Phys.: Conf. Series* **750** 012017
[45] Zierenberg J, Schierz P and Janke W 2017 *Nat. Commun.* **8** 14546
[46] Langer J S 1967 *Ann. Phys. N.Y.* **41** 108
[47] Ryu S and Cai W 2010 *Phys. Rev.* E **81** 030601(R)
[48] Prestipino S, Laio A and Tosatti E 2012 *Phys. Rev. Lett.* **108** 225701
[49] Berg B A and Neuhaus T 1991 *Phys. Lett.* B **267** 249
[50] Berg B A and Neuhaus T 1992 *Phys. Rev. Lett.* **68** 9
[51] Janke W 1992 *Int. J. Mod. Phys.* C **3** 1137
[52] Janke W 1998 *Physica* A **254** 164
[53] Zierenberg J, Marenz M and Janke W 2013 *Comput. Phys. Commun.* **184** 1155
[54] Zierenberg J, Marenz M and Janke W 2014 *Physics Procedia* **53** 55
[55] Janke W and Paul W 2016 *Soft Matter* **12** 642
[56] Berg B A 2017 *Eur. Phys. J. Special Topics* **226** 551
[57] Janke W 2003 *Histograms and all that* in *Computer Simulations of Surfaces and Interfaces* NATO Science Series II. Mathematics, Physics and Chemistry – Vol. 114 ed by B Dünweg, D P Landau and A I Milchev (Dordrecht: Kluwer) pp 137–157
[58] Janke W 1998 *Nucl. Phys.* B (Proc. Suppl.) **63A-C** 631
[59] de Gennes P-G 1979 *Scaling Concepts in Polymer Physics* (Ithaca: Cornell University Press)
[60] Zierenberg J, Mueller M, Schierz P, Marenz M and Janke W 2014 *J. Chem. Phys.* **141** 114908
[61] Mueller M, Zierenberg J, Marenz M, Schierz P and Janke W 2015 *Physics Procedia* **68** 95
[62] Schierz P, Zierenberg J and Janke W 2016 *Phys. Rev.* E **94** 021301(R)
[63] Calvo F, Neirotti J P, Freeman D L and Doll J D 2000 *J. Chem. Phys.* **112** 10350
[64] Martín-Mayor V 2007 *Phys. Rev. Lett.* **98** 137207
[65] Zierenberg J and Janke W 2015 *Europhys. Lett.* **109** 28002
[66] Kierfeld J, Kühne T and Lipowsky R 2005 *Phys. Rev. Lett.* **95** 038102
[67] Heussinger C, Schüller F and Frey E 2010 *Phys. Rev.* E **81** 021904
[68] Giurleo J T, He X and Talaga D S 2008 *J. Mol. Biol.* **381** 1332
[69] Pandolfi R J, Edwards L, Johnston D, Becich P and Hirst L S 2014 *Phys. Rev.* E **89** 062602
[70] Kouwer P H, Koepf M, Le Sage V A, Jaspers M, van Buul A M, Eksteen-Akeroyd Z H, Woltinge T, Schwartz E, Kitto H J, Hoogenboom R, Picken S J, Nolte R J, Mendes E and Rowan A E 2013 *Nature (London)* **493** 651

IOP Conf. Series: Journal of Physics: Conf. Series **955** (2017) 012004 doi:10.1088/1742-6596/955/1/012004

# Manipulations with qubit states by short control pulses: the interpolation method for evolution operator and fidelity

**M V Denisenko[1,*], N V Klenov [2-4] and A M Satanin[1]**

[1] Lobachevsky State University of Nizhny Novgorod, Nizhny Novgorod, 603950, Russian Federation
[2] Skobeltsyn Institute of Nuclear Physics, Lomonosov Moscow State University, Moscow 119991, Russian Federation
[3] Moscow Technological University (MIREA), 119454, Moscow, Russian Federation
[4] Moscow Technical University of Communications and Informatics, 8a, Aviamotornaya St., Moscow 111024, Russia

*E-mail: mar.denisenko@gmail.com

**Abstract.** In this article the dynamics of the qubits states based on solution of the time-dependent Schrödinger equation is investigated. Using the Magnus method we obtain an explicit interpolation representation for the propagator, which allows to find wave function at an arbitrary time. To illustrate the effectiveness of the approach, the population of the levels a single and two coupled qubits have been calculated by applying the Magnus propagator and the result have been compared with the numerical solution of the Schrödinger equation. As a measure of the approximation of the wave function, we calculate fidelity, which indicates proximity when the exact and approximate evolution operator acts on the initial state. We discuss the possibility of extending the developed methods to multi-qubits system, when high-speed calculation methods of the operators of evolution is particularly relevant.

## 1. Introduction

Accurate and efficient solution of the time-dependent Schrödinger equation is important in the fields of the computational physics. In many situations, especially in the area of manipulating with quantum states of quantum bits (qubits), the evolution of the wavefunction occurs over multiple states and timescales, and a computationally efficient approach is a necessity for any progress. Simulation of qubits dynamics is important to obtain a fundamental understanding of "read-write" processes and to provide a theoretical base for the design and development of quantum gates for the optimal control over quantum registers [1, 2].

The numerical solution of the time-dependent Schrödinger equation for multilevel systems relies heavily on the discretization of the time-variable and to use standard methods of solutions of Cauchy problem for the differential equations with an initial value for wave function. The grid methods are usually used to find the evaluation operator (propagator) numerically, for instance, by applying the iterative methods, the method of fast-Fourier transformation, Crank-Nicolson methods and so on.

Another approach is based on composition of a suitable operator approximation when the propagator may be found explicitly at an arbitrary moment. In the case of a time-independent Hamiltonian the most popular long-time propagator is Chebyshev polynomial propagator [3]. It is accomplished as follows: the propagator at the first decomposed into Chebyshev polynomials, which are then computed using iterative procedures.

When the Hamiltonian is time-dependent, such as in a case where propagating ultra-short pulse acts on array of qubits [4-7] the efficiency of the evolving methods of the propagator is crucial for computer modeling of the quantum dynamics. In the present work, the Magnus expansion method for propagator will be presented. Using this method we will obtain a convenient representation for the evolution operator, which makes it possible to find an explicit expression for the propagator at any time. To illustrate the effectiveness of the approach, the populations of the levels a single and two coupled qubits at an arbitrary time will be calculated and compared with the numerical solution of the Schrödinger equation. As a measure of the approximation of the wave function, we calculate fidelity, a value that indicates proximity when the exact and approximate evolution operator acts on the initial state. In conclusion, we will discuss the possibility of extending the developed methodology to multi-qubit systems, where high-speed calculation methods are particularly relevant for evolution operators.

## 2.The model of the system and the basic equations

The Hamiltonian of the superconducting flux qubit [8] can be represented as

$$H(t)=\hbar\left(\Delta\sigma_z+\varepsilon(t)\sigma_x\right), \tag{1}$$

where $\hbar\Delta$ is the distance between qubit levels; $\varepsilon(t)$ is the control function (proportional to the magnetic flux); $\sigma_x,\sigma_z$ are Pauli matrices. When $\varepsilon=0$ the solution of the time-nondependent Schrödinger equation has the form: $\psi_0=\begin{pmatrix}0\\1\end{pmatrix}$, $E_0=-\frac{\hbar\Delta}{2}$ and $\psi_1=\begin{pmatrix}1\\0\end{pmatrix}$, $E_1=\frac{\hbar\Delta}{2}$, which corresponds to the choice of the two qubits levels with energies $E_0$ and $E_1$. We are interested in the processes of switching of the qubit from one basic state to another under the impact of an unipolar external pulse $\varepsilon(t)=\varepsilon_m\exp\left(-(t-t_0)^2/2\tau^2\right)$ (where $\varepsilon_m$ is the pulse amplitude; $\tau$ is the typical pulse duration and it is supposed to be $t_0>>\tau$ for a gaussian pulse). The time dependence of the two-component wave function of a qubit is determined by: $\psi(t)=U(t)\psi(0)$, where $\psi(0)=\begin{pmatrix}1\\0\end{pmatrix}$ is wave function of ground qubit state at the initial time, the propagator $U(t)\equiv U(t,0)$ (evolution operator, $U(0)=\mathrm{I}, U^{\dagger}=U^{-1}$) is the solution of the equation

$$i\hbar\frac{\partial U(t)}{\partial t}=H(t)U(t)\cdot \tag{2}$$

The solution of the Eq. (2) can be written in the form: $U(t)=\hat{\mathrm{P}}\exp\left(-\frac{i}{\hbar}\int_0^t dt_1 H(t_1)\right)$, where $\hat{\mathrm{P}}$ is chronological ordering operator. The extract solution of Eq. (2) is found by a numerical solution Eq. (2) of the system of equations obtained when the operator is parameterized in a certain basis.

In this paper we develop a technique for calculation of the operator Eq. (2) based on the construction of approximate representation of the evolution operator – Magnus expansion. The essence of the method is the following. At the first, the evolution operator can be represented in the form of: $U(t)=e^{A(t)}$, where $A(t)$ the anti-Hermitian operator. According to [9], the operator $A(t)$ is expanded in the series, where the terms are proportional to the corresponding degrees of the Hamiltonian operator: $A_M=A_1+A_2+A_3+\dots$. This decomposition is substituted in Eq. (2). The resulting equality for the corresponding components of the same order of smallness on the left and the right, which leads to the ratios:

$$A_1(t) = -\frac{i}{\hbar}\int_0^t dt_1 H(t_1), \qquad A_2(t) = \frac{1}{2\hbar^2}\int_0^t dt_2 \int_0^{t_2} dt_1 \left[H(t_1), H(t_2)\right],$$
$$A_3(t) = \frac{i}{4\hbar^3}\int_0^t dt_3 \int_0^{t_3} dt_2 \int_0^{t_2} dt_1 \left\{\left[\left[H(t_1), H(t_2)\right], H(t_3)\right] + \frac{1}{3}\left[H(t_1), \left[H(t_2), H(t_3)\right]\right]\right\} \text{ etc.} \tag{3}$$

If there is a small parameter in the system, we can keep only several terms in the series of the Magnus expansion.

However, the resulting expression for $U_M(t) = e^{A_M(t)}$ is difficult to use in practice, because we need to take action on the state of the operator in the exponent of the exponential. Further the idea that we can lead the term $A_M(t)$ in time $t$ to the expression $A_M(t) = V^{\dagger}(t)D(t)V(t)$ (where $D(t)$ is a diagonal matrix, $V^{\dagger}(t)V(t) = I$), and then to use the relation: $U_M(t) = e^{V^{\dagger}(t)D(t)V(t)} = V^{\dagger}(t)e^{D(t)}V(t)$.

For a two-level system – qubit – the Hermitian operator can be written in the form $\hat{A} = -i\hat{T}$ ($\hat{T}^{\dagger} = \hat{T} = T_x\sigma_x + T_y\sigma_y + T_z\sigma_z$). For the two-component wave function the evolution operator $U_M$ can be written as follows:

$$U_M = \cos T\begin{pmatrix} 1 & 0 \\ 0 & 1 \end{pmatrix} - \frac{i\sin T}{T}\begin{pmatrix} T_z & T_x + iT_y \\ T_x - iT_y & -T_z \end{pmatrix}, \tag{4}$$

where time-depended functions $T_x(t)$, $T_y(t)$ $T_z(t)$ are determined by the expressions:

$$T_x + iT_y = \int_0^t dt_1 \varepsilon(t_1) e^{-\Delta t_1}, \quad T_z = \frac{1}{2}\int_0^t dt_2 \varepsilon(t_2) \int_0^{t_2} \varepsilon(t_1)\sin(\Delta(t_1 - t_2)). \tag{5}$$

The eigenvalues of anti-Hermitian operator $\hat{T}$ are equal to $\pm T$, where $T = \sqrt{T_x^2 + T_y^2 + T_z^2}$ .

Thus, the qubit evolution procedure is reduced to constructing the matrix $U_M$ according to the Eq. (4) and further by calculating the wave function by using a simple rotation in the "pseudospin" space of the qubit as: $\psi_M(t) = U_M(t)\psi(0)$. As a criterion of proximity matrices one can take various measures matrix characterizing the proximity matrices $U(t)$ and $U_M(t)$. For the "write-read" quantum information tasks it's advisable to choose a measure (fidelity $F(t)$) defined by the wave functions $\psi(t)$ and $\psi_M(t)$ at the current time according to:

$$F(t) = \left| < \psi(0) | U_M^{\dagger}(t) U(t) | \psi(0) > \right|^2. \tag{6}$$

When $F(t) = 1$ during the evaluation the both the states are the same and we have a good approximation for propagator.

## 3. The populations and fidelity calculation for qubit

Now we shall demonstrate that the developed technique allows us to select analytically and numerically the optimal form of a single quantum pulse. For comparison of these two solutions, we follow the time changes in the probability of population of the qubit excited state under the action of the gaussian pulse with a given amplitude $\varepsilon_m$ and duration $\tau$. The qubit initialized (at the $t = 0$) to the ground state $\psi_0 = \begin{pmatrix} 0 \\ 1 \end{pmatrix}$, $E_0 = -\frac{\hbar\Delta}{2}$ and we are calculated the probability of state switching

$\psi_0 \to \psi_1$, respectively $p(t) = |\langle \psi_1 | \psi(t) \rangle|^2$, $p_M(t) = |\langle \psi_1 | \psi_M(t) \rangle|^2$. Analyzing the expression (4) it is possible to notice, that $\Delta\tau << 1$ is the criterion for good convergence of analytics and numerical calculation [1]. Figure 1 (a) demonstrates the behavior of the qubit ground state population at different tunneling energies Δ. Note that a solution based on the construction of approximate representation of the evolution operator – Magnus expansion (dashed red curve $\Delta\tau \square 0.15$) agrees well with the numerical calculations (solid red curve), while fidelity after unipolar pulse is $F = 0.99995$ (red curve in figure 1 b). The increase in the tunneling energy of the qubit leads to divergence of the solutions (see figure 1 b). When going beyond the conditions $\Delta\tau << 1$ for a more accurate analytical expression (4) it is necessary to consider more terms of the expansion in Eq. (3).

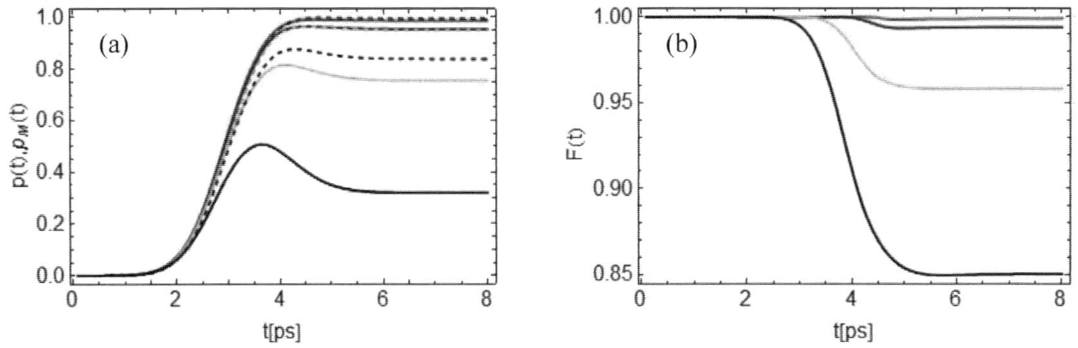


**Figure 1** (a) The results of numerical (solid curves) and analytical (dashed curves) calculations for the time-dependent population probabilities ( $p(t)$ and $p_M(t)$ ) of the qubit ground state and (b) the matching fidelity $F(t)$ of the numerical and analytical solutions at different tunneling energies of the qubit. Red curves – $\Delta = 0.05$ GHz, blue curves – $\Delta = 0.1$ GHz, green – $\Delta = 0.25$ GHz and black curves – $\Delta = 0.5$ GHz. The pulse parameters used here are: $\tau = 3$ ps, $t_0 = 1$ ps, $\varepsilon_m = 0.65$ GHz.

## 4. Multi-qubits systems

We have considered also the problem of controlling the populations of two qubits under the influence of a Gaussian pulse. The most interesting situation arises in the case when their parameters coincide, and control is carried out by the same impulse [10]. For identical qubits, the wave function can be decomposed into two orthogonal states: a singlet (scalar) and a triplet (vector).

$$H(t) = -\frac{\hbar}{2}\begin{pmatrix} 2\varepsilon(t)+J & \sqrt{2}\Delta & 0 \\ \sqrt{2}\Delta & -J & \sqrt{2}\Delta \\ 0 & \sqrt{2}\Delta & -2\varepsilon(t)+J \end{pmatrix}. \quad (7)$$

When the control parameter $\varepsilon = 0$, there are three levels of coupled symmetric qubits: $\left(-\hbar\sqrt{(J/2)^2+\Delta^2},\ -\hbar J/2,\ \hbar\sqrt{(J/2)^2+\Delta^2}\right)$. Time dependent external field will lead to transitions between them. For example, if a state with a lower energy is initially populated, then it is possible to raise the question of preparing the given populations of the remaining levels.

The interpolation evolution operator for a given system is constructed according to the scheme described above. We give an explicit expression for the case when the terms $\sim \Delta$ are taken into account in the exponent:

$$U_M = \frac{1}{T^2}\begin{pmatrix} e^{-iT}|\xi|^2 + |\eta|^2 & 0 & -2ie^{-iT/2}\xi\eta\sin(T/2) \\ 0 & T^2 e^{-iT} & 0 \\ -2ie^{-iT/2}\xi^*\eta^*\sin(T/2) & 0 & |\xi|^2 + e^{-iT}|\eta|^2 \end{pmatrix}, \quad (8)$$

where $\xi(t)$ and $\eta(t)$ are the functions of time:

$$\xi(t) = -\frac{\Delta}{\sqrt{2}}\int_0^t dt_1 \exp(iJt_1 + i\int_0^{t_1} dt_2\varepsilon(t_2)),\ \eta(t) = -\frac{\Delta}{\sqrt{2}}\int_0^t dt_1 \exp(-iJt_1 + i\int_0^{t_1} dt_2\varepsilon(t_2)),$$

and $T(t) = \sqrt{|\xi(t)|^2 + |\eta(t)|^2}$. The level populations for the selected values of the qubit parameters, calculated according to (8), for the chosen parameters only qualitatively agree with the exact numerical solution of the Schrodinger equation (see Fig. 2).

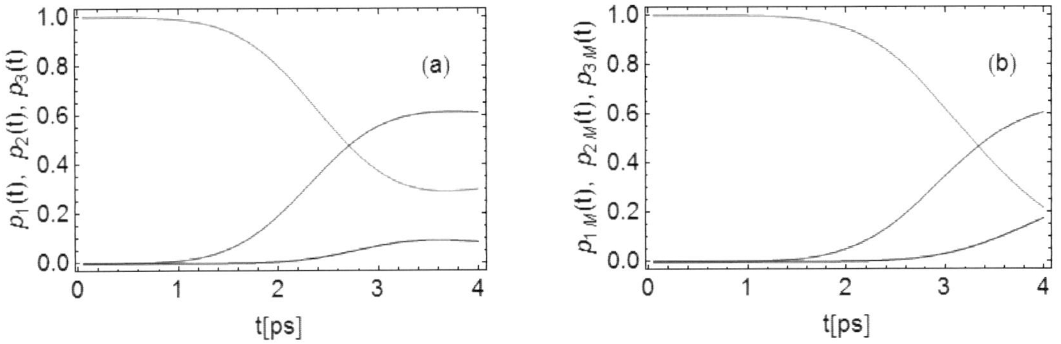


**Figure 2** The results of numerical $p_i(t)$ (a) and analytical $p_{iM}(t)$ (b) calculations for the time-dependent population probabilities of a pair of symmetric qubits: $J = 0.05$ GHz, $\Delta = 0.05$ GHz, $\varepsilon_m = 1$ GHz, $\tau = 3$ ps, $t_0 = 1$ ps. The red curve corresponds to population of ground state; the blue and black lines depict the populations of the second and the third levels.

Analyzing the data we arrive to the conclusion that for multi-qubits systems it may be necessary to take into account the neglected terms in the expansion (3).

## 5. Conclusion

We have developed here the interpolation method of approximation the evaluation operator of a multi-qubits system with a time-dependent Hamiltonian. To calculate the propagator system the Magnus expansion for the evolution operator has been applied, which makes it possible to find an explicit expression for the wave function at any time. The effectiveness of the approach has been illustrated by the calculation of the levels populations of a single and two coupled qubits and by comparing with the results with the numerical solution of the Schrödinger equation. We have investigated the fidelity which is playing role of the convenient measure of the approximation and that is indicated proximity when the exact and approximate evolution operator acts on the initial state. The carried out analytical and numerical simulations have demonstrated the possibility of accelerating the numerical solution of evolutionary problems by the developed method.

## Acknowledgments

This work was supported by the Ministry of Education of the Russian Federation (contract # 3.3026.2017) and by the RFBR (grants No. 17-12-01079, 16-57-51045, 16-29-09515, and 16-07-01012 NIF ).

## References

[1] N. V. Klenov, A. V. Kuznetsov, I. I. Soloviev, S. V. Bakurski, and O. V. Tikhonova, *Beilsten J. of Nanotechnology* **6**, 1946–1956 (2015).
[2] N. V. Klenov, A. V. Kuznetsov, I. I. Soloviev, S. V. Bakurskiy, M. V. Denisenko, and A. M. Satanin, *Low Temperature Physics* **43**(7), 789–798 (2017).
[3] H. Tal-Ezer and R. Kosloff, *J. Chem. Phys.* **81**, 3967 (1984).
[4] A. Fedorov, A. Shnirman, G. Schoen, A. Kidiyarova-Shevchenko, *Phys. Rev. B* **75**, 224504 (2007).
[5] I. I. Soloviev, N. V. Klenov, A. L. Pankratov, E. Il'ichev, and L. S. Kuzmin, *Phys. Rev. E* **87**, 060901(R) (2013).
[6] I. I. Soloviev, N. V. Klenov, S. V. Bakurskiy, A. L. Pankratov, L. S. Kuzmin, *Appl. Phys. Lett.* **105**, 202602 (2014).
[7] P. J. Liebermann and F. K. Wilhelm, Phys. Rev. Applied **6**, 024022 (2016).
[8] M. A. Nielsen, I. L. Chuang, Quantum Computation and Quantum Information (Cambridge University Press, 2001).
[9] W. Magnus, Commun. on Pure and Appl. Math., **7**, 649 (1954).
[10] A.M. Satanin, M.V. Denisenko, S. Ashhab, and F. Nori, *Phys. Rev. B* **85**, 184524 (2012).

# Towards on-the-fly Cryo-Electron Microscopy Data Processing by High Performance Data Analysis

**Evgeny Pichkur[1], Timur Baimukhametov[1], Anton Teslyuk[1,2], Anton Orekhov[1], Roman Kamyshinsky[1,2], Yury Chesnokov[1], Viacheslav Ilyin[1,2], Alexander Vasiliev[1] and Vasily Velikhov[1]**

[1] NRC Kurchatov Institute, Pl. Akademika Kurchatova, 1, Moskva, 123098, Russia
[2] Moscow Institute of Physics and Technology, 9 Institutskiy per., Dolgoprudny, Moscow Region, 141701, Russia

E-mail: anthony.teslyuk@grid.kiae.ru

**Abstract.** In this paper we present an approach to the on-the-fly Cryo-EM data acquisition and analysis using High Performance Data Analysis concept. An analysis of bottlenecks in the currently used data processing is presented, and the new iterative scheme for near real time implementation of the Cryo-EM data processing is discussed. The on-the-fly approach is proposed that will considerably reduce the time required for the protein (or other biological objects, e.g. viruses) 3D structure determination and allow eliminating manual interventions in the data processing.

## 1. Introduction

Last years, new trend, *High Performance Data Analysis (HPDA)*, is developing fast for the *compute-intensive* and *data-intensive* applications in different fields of science (modeling and simulations, mega science projects like Large Hadron Collider, humanitarian disciplines such as linguistics and archeology etc.), medicine, financial and commercial analytics, manufacturing, and many others. In short HPDA maintains applications where Big Data problems need High Performance Computing. Detailed discussion of the HPDA conception is presented in the book [1]. One of the notable HPDA example is structural biology where problem of 3D structure reconstruction of biomolecules and viruses is one of the most important methods for the design of new medicines. In this context reconstruction of the 3D structure of these biological objects is the serious computational problem that demands the high performance computing in the combination with big data treatment within the complicated workflows.

The Transmission Cryo Electron Microscopy (Cryo-EM) technique, in its modern development, becomes one of the most effective tools in the field of structural biology with the spatial resolution close to the traditional X-ray analysis [2, 3]. The progress in this method has been confirmed recently by the 2017 Nobel Prize award to J. Dubochet, R. Frank and R. Henderson. In contrary to the X-ray crystallography, Cryo-EM has good prospects to offer a unique option to investigate proteins and viruses in the native state with a possibility to reveal different conformations, which assume determination of the structure at different steps of biochemical reactions. Today the number of installations of modern Cryo-EM hardware is growing rapidly. Nowadays about 100 Hi-end cryogenic transmission electron microscopes are working in research centers around the world. The only installation in Russia is located in NRC Kurchatov Institute - Titan Krios TEM (FEI, USA) equipped with Falcon II electron detector and Cs-corrector.

Recent developments in the Cryo-EM data processing software allow to perform 3D reconstruction using the single particle analysis (SPA) method [4]. This method requires the acquisition of 105-106 projections of the individual researched objects (we will use the term particles for such projections). The images are acquired using highly sensitive and quick direct electron detecting (DED) cameras [5]. The typical size of such cameras is 4096 x 4096 pixels and the size of the single image is up to 32 MB for 16 bits pixel depth. The specialized software for the single particle data acquisition, e.g. EPU (FEI, USA), generates a stack of 7 images recorded during the exposure. Thus, the total data volume for one image stack is about 240 MB including metadata. Typical time between exposures may reach 1.5 min used for the setup of the next exposure, while duration of the typical microscope session is about a few days. The spatial resolution of the reconstructed model depends on the number of single particle images involved in the analysis. Therefore, for improving of the resolution and extracting fine details of the 3D model we need raw data up to few TB for each microscope session. Typical Cryo-EM lab performs up to several dozens of microscope sessions per year. Thus, the data volume per year could be of the PB scale for a Cryo-EM facility.One can note that manual work is required at many stages of the 3D reconstruction pipeline. As a result, the total time needed to complete the 3D reconstruction of the data generated during one microscope session can be about few months.

The paper is organized as follows.

In the next section, we describe the basic characteristics of input data coming from Cryo-EM.

In 3-rd section, typical scheme of the 3D reconstruction pipeline used by the Cryo-EM research groups at present time is discussed on example of the Cryo-EM lab and HPC facilities in NRC Kurchatov Institute.

In 4-th section, we formulate an approach to the on-the-fly implementation of the Cryo-EM data processing in fully automatic way following the HPDA concept. This approach will allow to eliminate bottlenecks in the state-of-the-art Cryo- EM data-processing, caused by extensive manual operations at various stages. Furthermore, we believe that this approach will also considerably reduce the time spent on the 3D reconstruction procedure.

In Conclusion, we discuss key advantages of the on-the-fly approach which might be valuable to Cryo-EM researchers.

## 2. Input data for 3D reconstruction

Below we discuss key details of the data acquisition process used in the 3D characterization of biological samples on Titan Krios TEM (FEI, USA), equipped with Falcon II electron detector, which is installed in NRC Kurchatov Institute. Sample preparation follows the standard procedures and details of our technique were described in [6, 7, 8]. The current data collecting rate is about 1000 image stacks per day [9]. In our current setup each image stack is transmitted to the NFS storage server over 1Gb Ethernet, which takes about 2 seconds. Next image stack acquisition starts 1.5 min later. During this time the acquisition of an image with lower magnification, centring and finding the foil hole, autofocusing procedure and timeouts for stage relaxation are taking place in the automatic mode. A DED Falcon II generates stacks of 4096x4096 images each. Up to 7 physical images per second are stored on the disk. Output image stacks are saved either as a stack of the 16 bits/pixel image or as a 32 bits/pixel stack after gain correction. A typical raw stack is about 240MB or up to 450MB after converting to 32bit and the motion correction.

Raw data volumes need to be stored. Dedicated disk storage is used of the total volume of 40 TB. The storage is accessible from computing nodes via NFS. For data processing two different

IOP Conf. Series: Journal of Physics: Conf. Series **955** (2017) 012005 doi:10.1088/1742-6596/955/1/012005

types of computing facilities are applied. For interactive operations as well as for debugging we use dedicated servers equipped with Tesla M2070 GPU cards. Each server has 2 Xeon X5660 CPUs and 24 GB RAM. For the intensive computing the parallel supercomputer installed in Kurchatov institute is utilized. It is equipped with 23 nodes each having Xeon Haswell (v3) CPU's, 128 GB RAM and three NVIDIA K80 GPU cards on board. 56Gb/s Infiniband fabric interconnects the computing nodes. The peak computing performance of this CPU/GPU cluster is 130 TFLOPs.

## 3. Current implementation of the image processing

Here we describe a current implementation of the data processing which includes a lot of manual operations. A representing scheme of this implementation is shown in Fig. 1.

Full dataset (about 4000 image stacks per microscope session) is collected and stored to the *Raw Image Repository*. Then full dataset is transferred to the *Processing Pipeline* session. It utilizes several image processing packages:

- MotionCor2 to correct sample motion, align image stacks and generate averaged output [10]
- Gctf to estimate the contrast transfer function and refine local defocus for each particle [11, 12]
- Gautomatch or EMAN2 for the particle picking [13, 14]
- RELION and CryoSPARC for the reference-free 2D classification, 3D classification and high-resolution 3D refinement [15, 16, 17]

Most of 2D classification strategies are based on the 2D multi-reference alignment [18], which is a straightforward implementation of iterative K-means clustering algorithm [19, 20]. There are different approaches to obtain high-quality 3D reconstruction: some of them rely on the 3D initial model (RELION), while others suggest stochastic gradient descent optimization scheme for *ab initio* structure determination and 3D classification without any prior models (CryoSPARC).

A typical 3D structure determination utilizes 4 nodes with peak performance of approximately 22 TFlops (double precision). This step may require up to 30 hours, depending on the size of the object of interest and amount of particles. The 2D/3D classification and 3D refinement are compute-intensive steps. A lot of manual operations are required to be performed by researchers. The particle selection is also a time consuming step and requires manual operations.

## 4. On-the-fly Cryo-EM 3D structure determination

Here we formulate an approach to the on-the-fly Cryo-EM 3D structure determination, corresponding flowchart is shown in Fig. 2. Key processing stages are as following.

### *4.1. Preprocessing*

Image stacks generated by Cryo-EM are stored in *Raw Image Repository* and transmitted to the dedicated GPU- accelerated workstation or node of the cluster for preprocessing, including motion correction and CTF estimation. The implementation of this step may be based on the existing system [9] or in-house software, which has to be developed. The cross-correlation of the simulated CTF with the observed experimental spectra (per image) is used for the automatic resolution cut-off after the setting up the required threshold. Also, the machine learning techniques could be applied for the identification of bad images in terms of presence of ice contaminations, large carbon film areas or, for example, very small particle concentration. All operations listed above have to be computed at near real time regime. The averaged images are stored in the fast *Averaged Images Local Storage*, which is connected to the computing nodes.

### *4.2. Processing Cycle*
This cycle is a core of the on-the-fly approach. Main steps of the *Processing Cycle* iterations are explained below. The processing cycle starts as soon as a chunk of images have been preprocessed and stored, yet the exact number depends on the sample and may vary. The chunk size is a parameter to be optimized as a trade-off between computing time and quality of analysis.

*4.2.1. Step 1* The particle picking is performed in a fully automatic manner using machine learning segmentation methods. Simulated re-projections of existing structures taken from EMDataBank [22] and Protein Data Bank [23] as well as public datasets with manually picked particles are used as a training dataset for the algorithm (software is to be developed). With 200 particles per image and the total amount of 2000 particles is a good estimate for the dataset size at the first iteration of the cycle. Unsupervised machine learning techniques like clustering methods or autoencoder neural networks [24] could be used to detect informative images. This is a subject for further research.

Nowadays this step requires manual operations, that is why automation of the picking is essential for the on-the-fly processing. Re-picking of the particles may be performed using embedded loop between *Step 1* and *Step 2* in the case of poor results. Machine learning tools may be used to check whether or not the picking routine is successful - if yes the re- picking loop is stopped. All available information after previous iterations, including the 2D class averages, the stacks of actual particles and irrelevant objects, re-projections of the 3D model as well as all averaged images stored in *Averaged Images Local Storage*, is used for re-picking. A predefined criterion should be designed to stop the re-picking loop as soon as the population and number of 2D classes are not changed significantly.

The re-picking loop is eliminated from the *Processing Cycle* by predefined criteria when coordinates of the actual particles don't change considerably between iterations.

*4.2.2. Step 2* After the fully automatic particle picking the particle stacks are stored in the *Iteration Cache* and the reference free 2D classification is performed. The purpose of the *Step 2* is to split the dataset of particles into a number of classes. Each class is formed by particles in close orientations. Particles are averaged inside each class to produce a class average image with significantly higher signal to noise ratio (SNR). The classification is used to select classes with actual particles and filter out the rest. The possibility of machine learning based classification of particle images in random orientations in Fourier space was recently studied [25].

*4.2.3. Step 3* After the 2D classification and elimination of the irrelevant images, containing contamination and noise, an *ab-initio* 3D reconstruction and classification [26] is performed. The 3D classification associates projections with 3D conformational classes using the expectation maximization (EM) algorithm. As soon as the 3D model is obtained, half-maps are used to calculate Fourier shell correlation (FSC) and determine the resolution of the model [27]. The decision to stop or continue the *Processing Cycle* (and, as a result, to finalize the microscope session) is based on the obtained resolution of the 3D model estimated in reciprocal or in real space [21].

All output data after *Steps 1, 2* and *3* is stored in the *Iteration Cache*, including particle stacks, coordinates of the particles, 2D class averages, 3D models, refined angles and offsets of the particles. One of the key aspects of the proposed approach is the usage of the 2D offsets and 3D rotations, determined on all previous iterations, for reconstruction of the 3D model in current iteration of the *Processing Cycle*. It will result in significant reducing of the computational time. All intermediate results obtained during *Processing Cycle* should be available for experts. This will allow a fast response to unexpected events during the experiment and reconstruction.

Next iteration (*Steps 1,2,3*) is started as soon as a new chunk of averaged images and corresponding metadata arrive after the *Preprocessing*.

It is worth to note that all decision-making criteria mentioned above will combine data-driven heuristics and expert knowledge. Further development of the software is required to bring the proposed concept to the Cryo-EM practice.

## 5. Conclusion

We propose the on-the-fly approach for cryo-EM structure determination, which relies on high level of automation and eliminates bottlenecks of the methods currently used. The use of machine learning techniques promises a significant improvement in particle picking and allows performing fast sorting of the good 2D and 3D class averages without human intervention. Once brought into practice, the proposed approach will incorporate all intermediate 3D rotations and 2D translation values, determined on previous iterations of the *Processing Cycle* significantly reducing computational time. The final model is expected to be ready in near real time when the data collection ends, thus reducing the whole duration of the experiment.

## Acknowledgments

This work has been carried out using computing resources of the federal collective usage center Complex for Simulation and Data Processing for Mega-Science Facilities at NRC Kurchatov Institute, http://ckp.nrcki.ru/.

The work was supported by the Russian Science Foundation grant No. 18-41-06001.

## References

[1] A. Osseyran and M. Giles, Industrial Applications of High Performance Computing, CRC Press, Taylor & Francis Group, ISBN 13:978-1- 4665-9681-8, 2015.I.S. Jacobs and C.P. Bean, *Fine particles, thin films and exchange anisotropy, in Magnetism*, vol. III, G.T. Rado and H. Suhl, Eds. New York: Academic, 1963, pp. 271-350

[2] Callaway, Ewen, The revolution will not be crystallized: A new method sweeps through structural biology, *Nature.* **525** (7568): 1724 (2015)

[3] Kuehlbrandt, Werner, Cryo-EM enters a new era, *eLife.* **3**: e03678 (2014)

[4] A. Doerr, Single-particle cryo-electron microscopy, *Nature Methods*, 13, 23 (2016)

[5] G. McMullan, et al, Comparison of Optimal Performance at 300 keV of Three Direct Electron Detectors for Use in Low Dose Electron Microscopy, *Ultramicroscopy* **147** (2014): 156163. PMC. Web. 20 Oct. 2017

[6] A. L. Kwiatkowski, et al, Impact of Salt Co- and Counterions on Rheological Properties and Structure of Wormlike Micellar Solutions, *J. Phys. Chem. B* 2016, **120**, 1254712556

[7] A. V. Shibaev, et al, Viscoelastic Synergy and Microstructure Formation in Aqueous Mixtures of Nonionic Hydrophilic Polymer and Charged Wormlike Surfactant Micelles, *Macromolecules* 2017, **50**, 339348

[8] T. Shtam, R. Samsonov, R. Kamyshinsky, R. Pantina, N. Verlov, A. Vasiliev, A. L. Konevega, and A. V. Malek, Exosomes: Some approaches to cancer diagnosis and therapy, *AIP Conference Proceedings 1882*, 020066 (2017)

[9] X. Li, S. Zheng, D. A. Agard and Y. Cheng, Asynchronous data acquisition and on-the-fly analysis of dose fractionated cryoEM images by UCSFImage, *J Struct Biol. 2015* November , **192**(2): 174178

[10] S. Q. Zheng, E. Palovcak, J.-P. Armache, K. A. Verba, Y. Cheng and D. A Agard, MotionCor2: anisotropic correction of beam-induced motion for improved cryo-electron microscopy, *Nature Methods* **14**, 331332 (2017)

[11] Zhang K., Gctf: real-time CTF determination and correction, Journal of structural, *Biology*, 2016; **193** (1), 1-12

[12] Wade R.H, A brief look at imaging and contrast transfer, *Ultramicroscopy*, 1992, **46**, pp. 145156

[13] http://www.mrc-lmb.cam.ac.uk/kzhang/

[14] G. Tang, L. Peng, P.R. Baldwin, D.S. Mann, W. Jiang, I. Rees & S.J. Ludtke, EMAN2: an extensible image processing suite for electron microscopy, *J Struct Biol.* **157**, pp. 38-46, (2017)

[15] Sjors H.W. Scheres, RELION: Implementation of a Bayesian approach to cryo-EM structure determination, *J Struct Biol.* 2012 Dec, **180** (3): 519530

[16] D. Kimanius, Bjorn O Forsberg, Sjors HW Scheres and Erik Lindahl, Accelerated cryo-EM structure determination with parallelisation using GPUs in RELION-2, *eLife* 2016, **5**: e18722

[17] Ali Punjani, John L Rubinstein, David J Fleet & Marcus A Brubaker, cryoSPARC: algorithms for rapid unsupervised cryo-EM structure determination, *Nature Methods* 2017 Mar; **14**(3):290-296
[18] Sjors HW Scheres, Mikel Valle, Rafael Nunez, Carlos OS Sorzano, Roberto Marabini, Gabor T Herman, Jose-Maria Carazo, Maximum-likelihood multi-reference refinement for electron microscopy images, *Journal of molecular biology* **348** (1), 139-149 2005
[19] Lloyd, S. P., Least squares quantization in PCM, *Technical Report RR-5497*, Bell Lab, September 1957
[20] MacQueen, J. B., Some methods for classification and analysis of multivariate observations, *Proceedings of the fifth Berkeley symposium on mathematical statistics and probability* (Vol. **1**, pp. 281297), 1967
[21] Alp Kucukelbir, Fred J Sigworth & Hemant D Tagare, Quantifying the local resolution of cryo-EM density maps, *Nature Methods* **11**, 6365 (2014)
[22] http://www.emdatabank.org/
[23] https://www.rcsb.org/pdb/home/home.do
[24] G. E. Hinton, R. R. Salakhutdinov, Reducing the Dimensionality of Data with Neural Networks, *Science*, JUL 2006, pp. 504-507
[25] S. A. Bobkov, A. B. Teslyuk, R. P. Kurta, O. Yu. Gorobtsov, O.M. Yefanov, V. A. Ilyin, R. A. Senin, and I. A. Vartanyants, Sorting algorithms for single-particle imaging experiments at X-ray free-electron lasers, *J. Synchrotron Rad.* **22**, 1345-1352 (2015)
[26] Scheres, Sjors H W, et al, Disentangling conformational states of macromolecules in 3D-EM through likelihood optimization, *Nature Methods*, vol. **4**, no. 1, 2007, p. 27
[27] Peter B Rosenthal, John L Rubinstein, Validating maps from single particle electron cryomicroscopy, *Current Opinion in Structural Biology*, Volume **34**, 2015, Pages 135-144

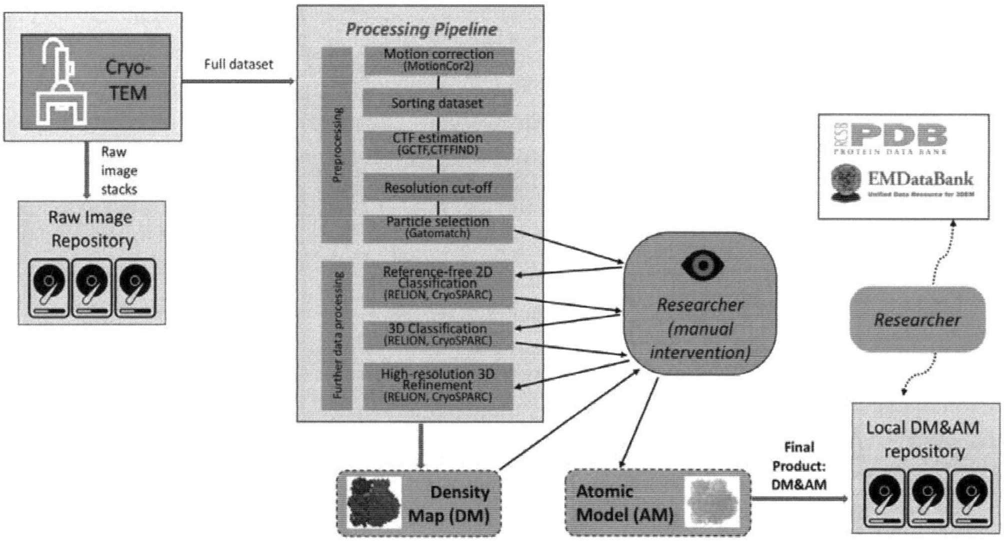


**Figure 1.** Current implementation of the image processing

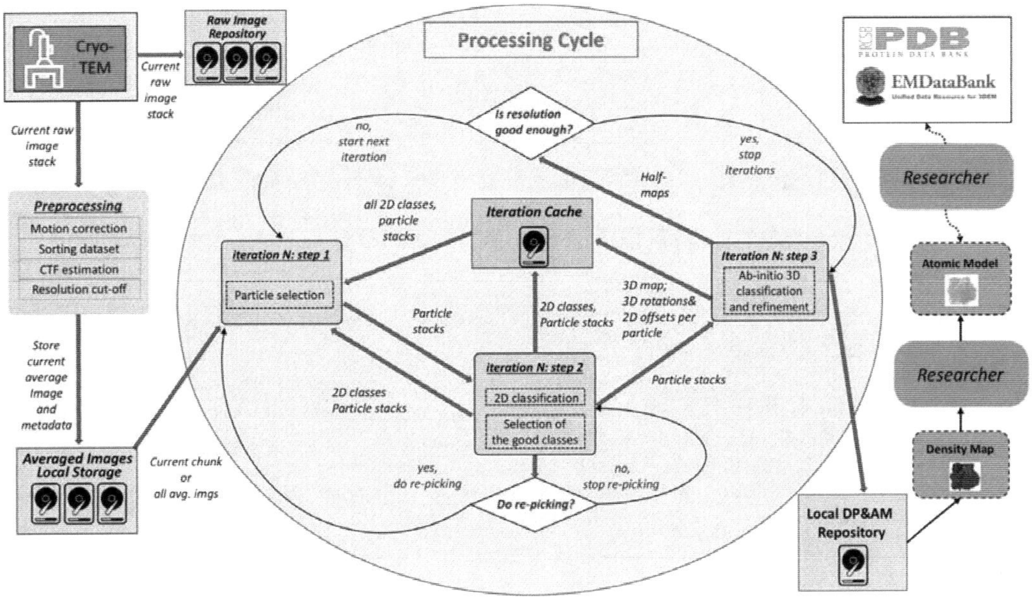


**Figure 2.** On-the-fly structure determination scheme

# Effect of aging on electrical conductivity of two-dimensional composite with rod-like fillers

**Yuri Yu Tarasevich[1], Valeri V Laptev[1,2], Andrei S Burmistrov[1], Nikolai I Lebovka[3,4]**

[1]Astrakhan State University, 20A Tatishchev Street, Astrakhan, 414056, Russia
[2]Astrakhan State Technical University, 16 Tatishchev Street, Astrakhan, 414025, Russia
[3]F. D. Ovcharenko Institute of Biocolloidal Chemistry, NAS of Ukraine, 42 Boulevard Vernadskogo, 03142 Kiev, Ukraine
[4]Taras Shevchenko Kiev National University, 64/13 Volodymyrska Street, 01601 Kiev, Ukraine

E-mail: `tarasevich@asu.edu.ru`

**Abstract.** Two-dimensional composite with rod-like fillers has been simulated using a lattice approach. The fillers were assumed to be conductors, whilst the substrate (host matrix) was assumed to be an insulator. Rod-like fillers were considered as linear $k$-mers of two mutually perpendicular orientations ($k_x$- and $k_y$-mers) on a square lattice with periodic boundary conditions applied along both directions of the lattice. A random sequential adsorption was used to produce an initial homogeneous and isotropic distribution of fillers. By means of the Monte Carlo (MC) technique, translational diffusion of the fillers was simulated as a random walk, while rotational diffusion was ignored. At each given MC step, the system under consideration was transformed into a random resistor network (RRN); the Frank–Lobb algorithm was applied to calculate the electrical conductivity of such RRNs. Diffusional relaxation of the system leads to a kind of aging. For example, when concentration of fillers exceeds the percolation threshold, the aging leads to decreasing of the electrical conductivity and, in some cases, the inhibition of percolation was observed.

## 1. Introduction

Among the alternatives for next-generation flexible, solution-processed transparent conductors, metal nanowire films attract a lot of attention as most promising (see, e.g., the reviews [1, 2] and the references therein). To accelerate design of such the systems and their adoption for different applications, the predictive tools are required. This is the reason why computational and analytical studies of electrical properties of two-dimensional (2D) random networks composed of rod-like particles are of increasing interest [3].

An important part of various technological processes, including production of conducting nanocomposites such as transparent flexible conductors is adsorption of large molecules or nanoparticles, e.g., nanotubes at the liquid-solid interface. This process can be simulated as random sequential adsorption (RSA) [4]. Numerous publications are devoted to 2D nanocomposites with rod-like conducting fillers (see, e.g., [1, 2, 5]).

The time changes in the electrical conductivity during the kinetic gelation [6] and aggregation of colloidal particles in aqueous solutions [7, 8] have been theoretically and experimentally studied. The data were interpreted in the framework of the percolation theory, fractal nature of the clusters, and cluster growth processes. The time changes in the electrical conductivity with composite aging have been observed in several experimental works. These changes were observed after transient shear during annealing of the polymer composites filled with carbon nanotubes (CNT) [9, 10]. The conductivity demonstrated noticeable increase with the time after the shear. The shearing can cause breaking and reformation of capacitors and resistors formed at the CNT–CNT contacts, and the conductivity recovery was explained by the reorganization of the conducting CNT network. The recovery of the conductivity was modelled using the electrical percolation approach combined with the cluster aggregation model [9, 11]. The time-dependent transition from insulating to conductive state was explained using the dynamic percolation concept [12]. The explanation was based on general effective medium model (GEM) coupled to a kinetic equation for CNT agglomeration. Thermal-induced dynamic percolation was observed in electrical conductivity behaviour in high-density polyethylene melt filled by carbon black [13, 14]. The dynamic percolation was modelled accounting for a first order aggregation model coupled with the classical percolation theory. Changes in electrical conductivity reflected the dynamic processes of destruction and re-formation of conductive filler networks in a low viscous CNT/epoxy suspension [15]. Dynamic percolation in highly oriented conductive networks formed with different carbon fillers (CNT and carbon black) was investigated during disorientation upon annealing [16]. The time-dependent changes of the electrical conductivity of carbon black filled polymethylmethacrylate were studied theoretically and experimentally [17]. The changes were attributed to the aggregation of freely dispersed aggregates into conductive pathways. The time dependence of the percolation in response to organic vapour was observed in water-borne polyurethane filled with carbon black [18].

Effects of fillers length, anisotropy and presence of defects on electrical conductivity of 2D systems were recently simulated [19, 20]. Although RSA with diffusion has been studied [21, 22, 23, 24, 25], effect of diffusion on electrical properties of 2D composites with rod-like fillers was not examined. In the present conference paper, we try to partially eliminate this gap. The rest of the paper is constructed as follows. In section 2, the technical details of the simulations are described. Section 3 presents our principal findings. Section 4 summarizes the main results.

## 2. Methods

The simulation was performed using a lattice approach. A low conducting substrate (host matrix) was approached as a square lattice of size $L \times L$ where $L = 256$. High conducting rod-like fillers were represented as linear $k$-mers (i.e., particles occupying $k$ adjacent lattice sites). The length of the $k$-mers (aspect ratio) was varied from 2 to 12.

An initial homogeneous and isotropic distribution of linear $k$-mers in a 2D film was produced using random sequential adsorption [4]. The $k$-mers were deposited randomly and sequentially onto the lattice, and their overlapping with previously placed $k$-mers was forbidden. The $k$-mers oriented along the $x$ and $y$ directions ($k_x$-mers and $k_y$-mers, respectively) were equiprobable in their deposition. To reduce finite-size effect, we used toroidal boundary conditions upon deposition, i.e., periodic boundary conditions (PBCs) along both the $x$ and $y$ axes. The concentration (packing density) of the particles varied in the range $p \in [0.1, p_j]$ where $p_j$ is the jamming concentration. However, the concentrations corresponding to a concentration slightly greater than the percolation threshold, $p_c$, were of our primary interest. Example of a system at initial state is presented in figure 1.

When the concentration of deposited particles reached the given value, the deposited particles were allowed to diffuse. The diffusion of $k$-mers was simulated using the kinetic MC procedure. In our simulation, only translational diffusion was taken into consideration. This is essentially the case for fairly dense systems in the jamming state [26], where rotational diffusion is impeded, especially for large values of $k$. Naturally, for dilute systems, rotational diffusion can occur,

**Figure 1.** Example of the initial pattern. $k = 10$, $p = 0.5$.

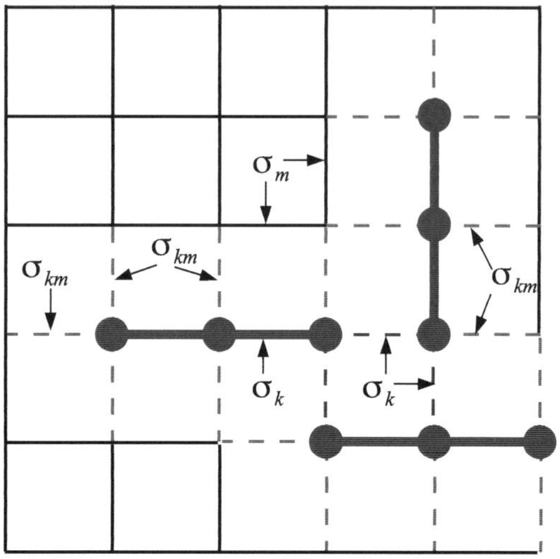


**Figure 2.** Representation of 3 deposited 3-mers as a RRN. Conductivities of the bonds between conducting sites are $\sigma_k$, between the insulating sites are $\sigma_m$, and between one insulating site and one conducting site are $\sigma_{mk}$.

however, this was ignored in our study. Although percolation threshold corresponds to a packing density less than jamming coverage, the packing density is high enough to essentially suppress rotations.

In our simulation, an arbitrary $k$-mer was randomly chosen and a translational shift by one lattice unit along either the longitudinal or the transverse axis of the $k$-mer was attempted. Equal probabilities to choose any of all the four possible directions to shift the $k$-mer were assumed. One time step of the MC computation, which corresponds to an attempted displacement of the total number of $k$-mers in the system, $N = pL^2/k$, was taken as the MC time unit. We started time counting from the value of $t_{\text{MC}} = 0$, being the initial moment (before diffusion). This model corresponds to [21, 23].

For characterization of the rearrangement of the system under consideration, the electrical conductivity, $\sigma$, were monitored at each given MC step. We used the model described in [27] to transform the system under consideration into a RRN. The torus was unrolled in a plane and two conducting buses were applied to its opposite sides. The electrical conductivity was calculated between these buses. The Frank–Lobb algorithm [28] was applied to calculate the electrical conductivity of such RRNs.

We considered an empty lattice as a regular resistor network. Each site is associated with four resistor. The conductivities of all resistors in the network are equal, $2\sigma_m$. Hence, the conductivity of a whole bond between two sites is $\sigma_m$. Each site of a $k$-mer is associated with four identical resistor, too. The conductivity of such the resistor is $2\sigma_k$. When a $k$-mer deposits onto the lattice, there are only three possible combinations of the resistors (figure 2).

**Two insulating** The conductivity of the entire bond is $\sigma_m$.

**Two conducting.** The conductivity of the entire bond is $\sigma_k$.

**Conducting insulating.** The conductivity of the entire bond is $\sigma_{km} = 2\sigma_m\sigma_k/(\sigma_m + \sigma_k)$.

In our study, we supposed $\sigma_m = 1$ a.u. and $\sigma_k = 10^6$ a.u., hence, $\sigma_{km} \approx 2$ a.u. Conductivities both along $x$ axis and along $y$ axis were taken into consideration.

Moreover, structure of percolation clusters was studied. The percolation clusters were identified using Hoshen–Kopelman algorithm [29]. After that, we extracted a backbone [30], i.e., a subset of the percolation cluster that carries an electrical current when a potential difference is applied between two its sites [31, 32, 33]. We used so-called 'burning algorithm' to find the backbone of the percolation cluster [34].

## 3. Results

We found that, for concentrations of fillers below the percolation threshold and close to the jamming concentration, the electrical conductivity of 2D composites with rod-like fillers is almost independent on time. In contrast, diffusion has a drastic impact on electrical conductivity when concentration of fillers slightly exceeds the percolation threshold, $p_c$, (figure 4). The latter case ($p \gtrsim p_c$) has attracted our special attention.

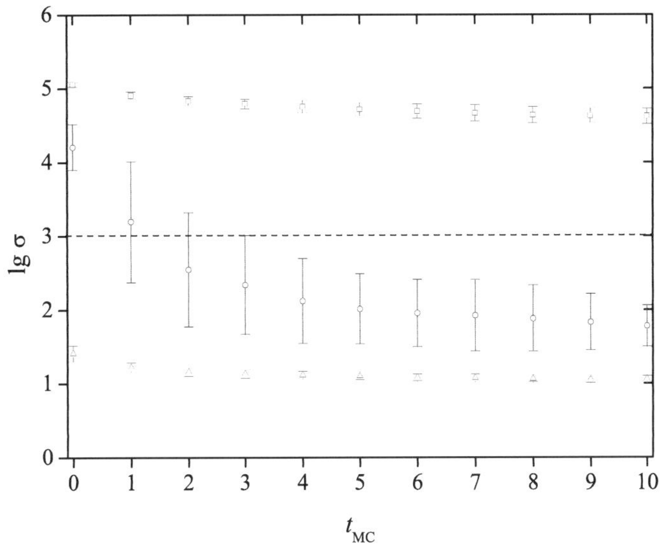


**Figure 3.** Examples of electrical conductivity (log scale), $\sigma$, vs MC steps, $t_{\mathrm{MC}}$, for different concentrations, $p$. $k = 10$, $L = 256$. $\triangle : p = 0.4$, $\circ : p = 0.5$, $\square : p = 0.6$. The results are averaged over 100 independent runs.

First of all, enormous error bars when $p \gtrsim p_c$ should be explained. These errors reflect permanent jumps of the electrical conductivity near the phase transition conductor-to-insulator. For the same MC step, some samples are conductors, whereas other samples are insulators. Thereby, a direct averaging of the electrical conductivity over all independent runs is meaningless. To obtain sensible behaviour, at each MC step, 100 independent runs were divided into two groups, viz., conducting samples and the insulating samples. The data inside each group should be averaged independently. In this case, the electrical conductivity vs MC steps is presented by two branches; the first one is corresponded to the conductor whereas the second one corresponds to insulator (figure 4).

Initially, the system is a conductor. At large times ($t_{\mathrm{MC}} \gtrsim 10^3$), the system approximately with equal probabilities may be an insulator or a conductor. During a transient mode ($10 \lesssim t_{\mathrm{MC}} \lesssim 10^2$), the system is more probably to be a conductor than an insulator. This behaviour of the electrical conductivity can be explained by the structure of percolation cluster. At initial sate, almost all occupied sites belong to backbone and make a contribution to conductivity. During a few dozens of MC steps, fairly dense initial backbone turn into a quasi-one-dimensional chain (figure 5).

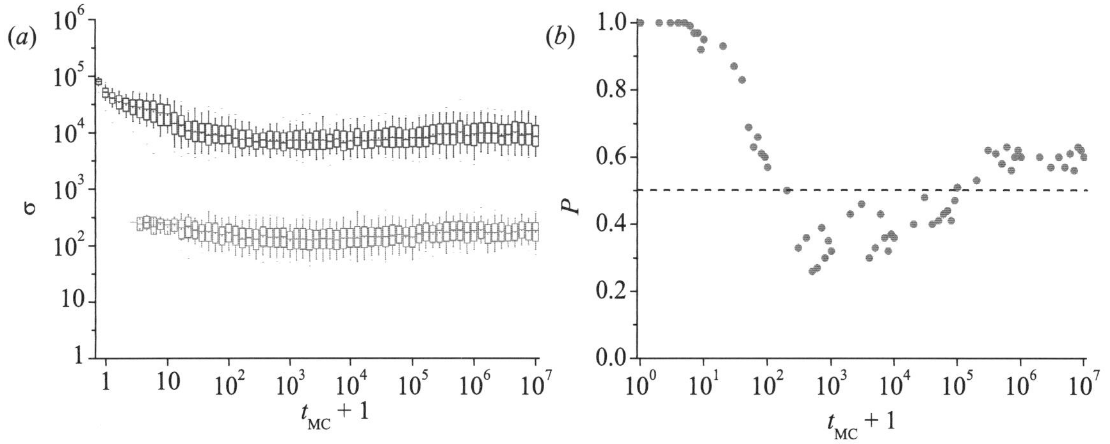


**Figure 4.** (*a*) Example of the electrical conductivity, $\sigma$, vs MC steps, $t_{\mathrm{MC}}$, for $k = 12$ and concentration of particles $p = 0.575$, box-and-whiskers plot in log-log scale. Upper branch corresponds to conducting states, whilst lower branch corresponds to insulating states. (*b*) Probability, $P$, that a conducting state occurs.

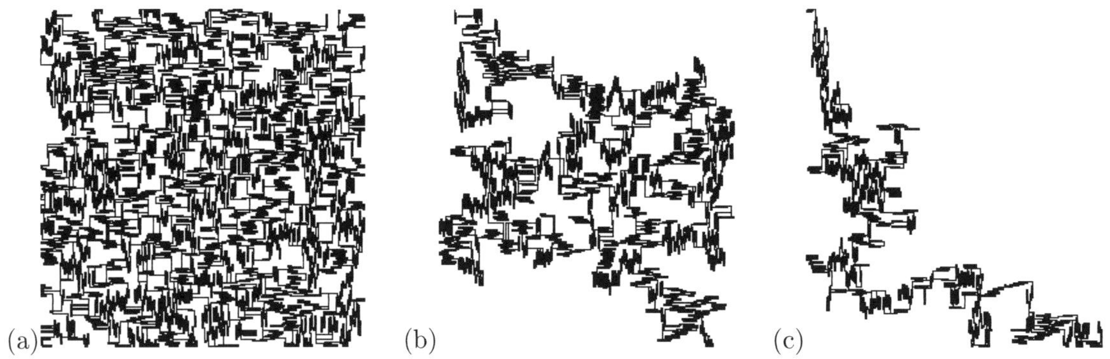


**Figure 5.** Example of temporal evolution of backbone of the percolation cluster. $k = 12$, $L = 256$, (a) $t_{\mathrm{MC}} = 0$,(b) $t_{\mathrm{MC}} = 10$, $d_f = 1.672$, (c) $t_{\mathrm{MC}} = 100$.

At the same time, number of sites belonging to the backbone decreases (figure 6). This has the drastic effect on electrical conductivity, because diffusion can easily both destroy such the weak backbone and recover it again. At $t_{\mathrm{MC}} \gtrsim 10^2$, a dynamical percolation occurs, i.e., percolation arises and fades. Whilst during several first MC steps, the monolayer is a conductor, after $t_{\mathrm{MC}} \sim 10^2$, the monolayer may be both conductor and insulator with different probabilities (figure 4).

## 4. Conclusion

In our study, a lattice approach has been applied to simulate effect of diffusion on electrical conductivity of a 2D composite with rod-like fillers. For concentrations of fillers above the percolation threshold but below the jamming concentration, we observed a kind of aging, i.e., decreasing of the electrical conductivity due to diffusion of fillers. This behaviour can be explained by evolution of the structure of the percolation cluster and its backbone. Diffusional relaxation leads to sparser percolation cluster with quasy-1D backbone. The diffusion of fillers

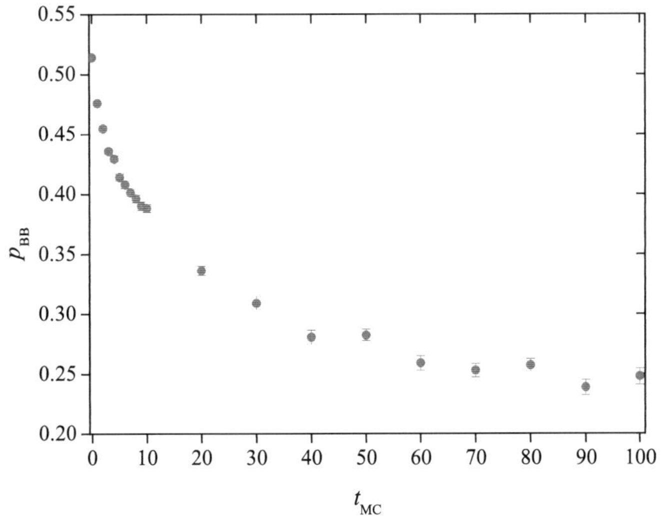


**Figure 6.** Example of fraction of lattice sites belonging to the backbone, $p_{BB}$ vs MC steps, $t_{MC}$. The results are averaged over 100 independent runs. $k = 12$, $L = 256$, $p = 0.6$.

can easily destroy this backbone and recover it again. As the result, electrical conductivity of the monolayer varies with time. Such the behaviour of electrical conductivity observed in our study can be treated as a kind of the dynamical percolation [35].

## Acknowledgments

The reported study was supported by the Ministry of Education and Science of the Russian Federation, Project No. 3.959.2017/4.6 (Y.Y.T., A.S.B., and V.V.L.), and the National Academy of Sciences of Ukraine, Project No. 43/17-H (N.I.L).

## References

[1] Mutiso R M, Sherrott M C, Rathmell A R, Wiley B J and Winey K I 2013 *ACS Nano* **7** 7654–7663 ISSN 1936-0851
[2] Mutiso R M and Winey K I 2015 *Prog. Polym. Sci.* **40** 63–84 ISSN 0079-6700 Progress in Polymer Hybrid Materials, Edited by Michael Bockstaller
[3] Mutiso R M and Winey K I 2013 *Phys. Rev. E* **88**(3) 032134 ISSN 2470-0045
[4] Evans J W 1993 *Rev. Mod. Phys.* **65**(4) 1281–1329 ISSN 0034-68611
[5] Du F, Fischer J E and Winey K I 2005 *Phys. Rev. B* **72**(12) 121404 ISSN 2469-9950
[6] Basta M, Picciarelli V and Stella R 1991 *Eur. J. Phys.* **12** 210–213 ISSN 1361-6404
[7] Cametti C, Codastefano P and Tartaglia P 1988 *EPL (Europhys. Lett.)* **5** 703–708 ISSN 1286-4854
[8] Bordi F, Cametti C, Codastefano P and Tartaglia P 1990 *Physica A* **164** 663–672 ISSN 0378-4371
[9] Alig I, Lellinger D, Dudkin S M and Pötschke P 2007 *Polymer* **48** 1020–1029 ISSN 0032-3861
[10] Alig I, Lellinger D, Engel M, Skipa T and Pötschke P 2008 *Polymer* **49** 1902–1909 ISSN 0032-3861
[11] Alig I, Skipa T, Lellinger D and Pötschke P 2008 *Polymer* **49** 3524–3532 ISSN 0032-3861
[12] Alig I, Skipa T, Lellinger D, Bierdel M and Meyer H 2008 *physica status solidi (b)* **245** 2264–2267 ISSN 1521-3951
[13] Cao Q, Song Y, Tan Y and Zheng Q 2009 *Polymer* **50** 6350–6356 ISSN 0032-3861
[14] Cao Q, Song Y, Tan Y and Zheng Q 2010 *Carbon* **48** 4268–4275 ISSN 0008-6223
[15] Schulz S C and Bauhofer W 2010 *Polymer* **51** 5500–5505 ISSN 0032-3861
[16] Zhang S, Lin L, Deng H, Gao X, Bilotti E, Peijs T, Zhang Q and Fu Q 2012 *Colloid Polym. Sci.* **290** 1393–1401 ISSN 1435-1536
[17] Krückel J and Schubert D 2014 *Eur. Polym. J.* **53** 50–57 ISSN 0014-3057
[18] Chen S G, Hu J W, Zhang M Q, Rong M Z and Zheng Q 2005 *J. Mater. Sci.* **40** 2065–2068 ISSN 1573-4803
[19] Tarasevich Y Y, Goltseva V A, Laptev V V and Lebovka N I 2016 *Phys. Rev. E* **94**(4) 042112 ISSN 2470-0045
[20] Tarasevich Y Y, Laptev V V, Goltseva V A and Lebovka N 2017 *Physica A* **477** 195–203 ISSN 0378-4371
[21] Grigera S A, Grigera T S and Grigera J R 1997 *Phys. Lett. A* **226** 124–126 ISSN 0375-9601
[22] Eisenberg E and Baram A 1998 *EPL (Europhys. Lett.)* **44** 168 ISSN 0295-5075

[23] Fusco C and Gallo P, Petri A and Rovere M 2001 *J. Chem. Phys.* **114** 7563–7569 ISSN 0021-9606
[24] Fusco C, Gallo P, Petri A and Rovere M 2002 *Philos. Mag. B* **82** 375–381 ISSN 1364-2812
[25] Fusco C, Fasolino A, Gallo P, Petri A and Rovere M 2002 *Phys. Rev. E* **66**(3) 031301 ISSN 2470-0045
[26] Torquato S and Stillinger F H 2010 *Rev. Mod. Phys.* **82**(3) 2633–2672 ISSN 0034-6861
[27] Lebovka N I, Tarasevich Y Y, Gigiberiya V A and Vygornitskii N V 2017 *Phys. Rev. E* **95**(5) 052130 ISSN 2470-0045
[28] Frank D J and Lobb C J 1988 *Phys. Rev. B* **37**(1) 302–307 ISSN 2469-9950
[29] Hoshen J and Kopelman R 1976 *Phys. Rev. B* **14** 3438–3445 ISSN 2469-9950
[30] Kirpatrick S 1978 The geometry of the percolation threshold *AIP Conference Proceedings* vol 40 ed Garland J C and Tanner D B (Columbus, OH, USA) pp 99–117 ISBN 0-88318-139-8
[31] Skal A and Shklovskii B 1975 *Sov. Phys. Semicond.* **8** 1029–1032 ISSN 0038-5700
[32] de Gennes P G 1976 *J. Physique Lett.* **37** 1–2 ISSN 0302-072X
[33] Stanley H E 1977 *J. Phys. A: Math. Gen.* **10** L211–L220 ISSN 1361-6447
[34] Herrmann H J, Hong D C and Stanley H E 1984 *J. Phys. A: Math. Gen.* **17** L261–L266 ISSN 1361-6447
[35] Lebovka N I, Tarafdar S and Vygornitskii N V 2006 *Phys. Rev. E* **73**(3) 031402 ISSN 2470-0045

IOP Conf. Series: Journal of Physics: Conf. Series **955** (2017) 012007 doi:10.1088/1742-6596/955/1/012007

# Modeling of monolayer charge-stabilized colloidal crystals with static hexagonal crystal lattice

**A N Nagatkin and P E Dyshlovenko**

Department of Applied Mathematics and Informatics, Ulyanovsk State Technical University, 32 Severny Venets Street, Ulyanovsk 432027, Russia

E-mail: p.dyshlovenko@mail.ru

**Abstract**. The mathematical model of monolayer colloidal crystals of charged hard spheres in liquid electrolyte is proposed. The particles in the monolayer are arranged into the two-dimensional hexagonal crystal lattice. The model enables finding elastic constants of the crystals from the stress-strain dependencies. The model is based on the nonlinear Poisson-Boltzmann differential equation. The Poisson-Boltzmann equation is solved numerically by the finite element method for any spatial configuration. The model has five geometrical and electrical parameters. The model is used to study the crystal with particles comparable in size with the Debye length of the electrolyte. The first- and second-order elastic constants are found for a broad range of densities. The model crystal turns out to be stable relative to small uniform stretching and shearing. It is also demonstrated that the Cauchy relation is not fulfilled in the crystal. This means that the pair effective interaction of any kind is not sufficient to proper model the elasticity of colloids within the one-component approach.

## 1. Introduction

Charge-stabilized colloidal systems are stable suspensions of electrically charged colloidal particles immersed in a liquid electrolyte [1]. In the colloidal crystals, particles are spatially ordered in such a way that they form a crystalline structure. These systems are of great interest as a basis for the photonic crystals and self-organizing structures [2]. Colloidal crystals are also interesting as model systems in the study of ordinary molecular crystals, as well as disordered colloidal systems. Monolayer colloidal systems contain one layer of particles usually near the substrate or at the interface. In the absence of displacement of particles in a direction perpendicular to the layer, such structures can be considered as two-dimensional.

Colloidal crystals of like-charged particles are stable only in the presence of an external pressure. Elastic properties of such systems have some specificity compared to conventional media [3, 4]. In particular, the series expansion of the stress with respect to strain includes a constant term corresponding to the stress in the absence of deformation.

The nonlinear Poisson-Boltzmann equation proved to be a good basis for theoretical study of the charge-stabilized colloids. It does not refer to any prescribed effective interaction between colloidal particles. In the present paper, we describe the numerical method based on the Poisson-Boltzmann equation to study elastic properties of the monolayer colloidal crystals. We then use this method to find elastic constants of the model crystals in the broad range of the crystals' densities.

## 2. Description of the Model

The model monolayer charge-stabilized colloidal crystal is shown schematically in figure 1. It is a system of identical absolutely rigid spherical particles of radius $R$ located between two parallel plates. The distance from the edge of the particles to the plates on both sides is the same and equal to $H$. The particles and plates are charged with a constant surface charge density $\sigma$ and $\Sigma$ respectively. The centers of colloidal particles are located at the nodes of a two-dimensional hexagonal crystal lattice with the lattice parameter $a$. The space between the plates and the particles is filled with electrolyte. The static lattice approximation is used, so that the thermal motion of the colloidal particles is neglected.

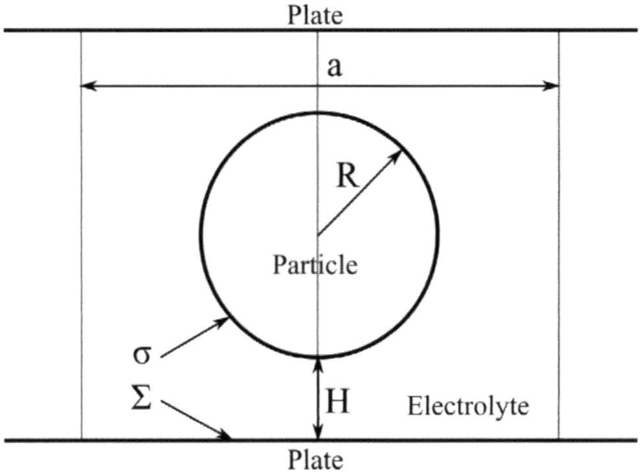


**Figure 1.** Elementary cell of a monolayer colloidal crystal. The description of the parameters is given in the text.

The electrostatic potential $\varphi$ in the electrolyte region obeys the nonlinear differential Poisson-Boltzmann equation [1]. We consider the case of a binary symmetric univalent electrolyte, or 1:1 electrolyte, which has two species with valences $z_1 = +1$ and $z_2 = -1$, while the volume concentrations of both species in the unperturbed electrolyte are equal to each other: $n_{01} = n_{02} = n_0$. The Poisson-Boltzmann equation for the electric potential $\varphi$ in the electrolyte region for the system under consideration has the form

$$\nabla^2\varphi = -\frac{q_e n_0}{\varepsilon_0 \varepsilon}\left[\exp\left(-\frac{q_e\varphi}{kT}\right) - \exp\left(\frac{q_e\varphi}{kT}\right)\right], \tag{1}$$

where $\varepsilon_0$ is the electric constant, $\varepsilon$ is the relative dielectric permittivity of the electrolyte, $q_e$ is the elementary charge, $k$ is the Boltzmann constant, $T$ is the absolute temperature. With the units $\kappa^{-1} = \left(2n_0 q_e^2 / \varepsilon_0 \varepsilon kT\right)^{-1/2}$, the Debye length for the 1:1 electrolyte, and $kT / q_e$ for the electric potential, the Poisson-Boltzmann equation (1) is reduced to the following dimensionless form:

$$\nabla^2\varphi = \sinh\varphi. \tag{2}$$

In what follows, the case of (infinitely) large dielectric permittivity of the electrolyte in comparison with the solid phase is considered, that is an appropriate limit case for aqua-based electrolytes. Then, the electric potential in the electrolyte region is independent of the potential inside the particles and plates and is determined only by the magnitude of the charge on their surfaces. This leads to the following Neumann boundary conditions on the surface of the particles:

$$-\nabla\varphi \cdot \mathbf{n} = \sigma, \tag{3}$$

and on the surface of the plates:

$$-\nabla\varphi \cdot \mathbf{n} = \Sigma, \tag{4}$$

where $\mathbf{n}$ is the unit normal vector directed into the interior of the electrolyte.

To find the electric potential, only one unit cell is needed due to the spatial periodicity of the crystal. The domain in the initial configuration is based on the Wigner-Seitz cell of one of the particles. This cell is a prism with a hexagonal base and height $2(H+R)$. The lateral size of the domain is characterized by the lattice parameter $a$ shown in the figure 1. The space of the colloidal particle in the center of the cell is subtracted from the domain. Only the lower half of the cell is sufficient due to the horizontal mirror symmetry plane passing through the centers of the particles. When a deformation is imposed, the domain is obtained by the corresponding transformation of the initial region.

The homogeneous Neumann boundary condition holds on the plane of mirror symmetry:

$$-\nabla\varphi\cdot\mathbf{n}=0 \tag{5}$$

On the side faces of the domain, the periodic boundary conditions hold for the potential:

$$\varphi(\mathbf{r})=\varphi\left(\mathbf{r}+\mathbf{r}^{(m)}\right),\qquad m=1,2,3, \tag{6}$$

and for the normal component of the potential gradient:

$$\nabla\varphi(\mathbf{r})\cdot\mathbf{n}^{(m)}=-\nabla\varphi\left(\mathbf{r}+\mathbf{r}^{(m)}\right)\cdot\mathbf{n}'^{(m)},\qquad m=1,2,3. \tag{7}$$

Here $m$, $m=1,2,3$, is the number of a pair of the opposing side faces, $\mathbf{n}^{(m)}$ and $\mathbf{n}'^{(m)}$ are the outer unit normals of the corresponding faces, and vectors $\mathbf{r}^{(m)}$, $m=1,2,3$, are the vectors of primitive translations that separate these opposite faces. The electric potential in any instantaneous crystal configuration, both initial and deformed, is determined by the solution of the boundary value problem for equation (2) in the domain mentioned above with the boundary conditions (3)–(7) on its boundaries.

## 3. Numerical Experiment

The elastic constants of the crystal are found from the stress-strain relations obtained in the numerical experiments. We consider only lateral deformations so that the colloidal particles are displaced in directions parallel to the plates with the centers remaining in the middle plane. Then, the system can be treated as a two-dimensional one. The symmetry analysis shows that such a system has three independent second-order elastic constants (elastic moduli), $B_{1111}$, $B_{1122}$, $B_{1212}$, and also one first-order elastic constant $B_{11}=-p$, where $p$ is an isotropic initial pressure. The remaining elastic constants are expressed in terms of the four indicated or identically equal to zero.

The first-order elastic constant is calculated as the stress at zero strain. To determine elastic constants of the second order, homogeneous deformation of two following types are used: stretching-compression along the $x$ axis of the form

$$\begin{pmatrix}\varepsilon & 0\\ 0 & 0\end{pmatrix} \tag{8}$$

and a shearing of the form

$$\begin{pmatrix}0 & \varepsilon\\ \varepsilon & 0\end{pmatrix}, \tag{9}$$

where $\varepsilon$ is the strain parameter. In this case, the stress-strain relations are the following:

$$T_{11}=B_{11}+B_{1111}\,\varepsilon+\ldots \tag{10}$$

$$T_{22}=B_{11}+B_{1122}\,\varepsilon+\ldots \tag{11}$$

for stretching-compression and

$$T_{12}=2B_{1212}\,\varepsilon+\ldots \tag{12}$$

for shearing [3]. Here dots denote terms of the orders higher than the first. Three expressions (10)–(12) are sufficient to find all elastic constants of the first and second order.

The stress tensor $T_{ij}$ in a two-dimensional crystal is expressed in terms of the usual osmotic stress tensor $S_{ij}$ (in three-dimensional space) as follows:

$$T_{ij} = 2(H+R)\left(S_{ij} - S_b\delta_{ij}\right), \tag{13}$$

where $2(H+R)$ is the height of the cell and $\delta_{ij}$ is the Kronecker delta symbol. Stress $S_b$ is the isotropic background stress, which is present in the very dilute system, when the particles are separated by sufficiently large distances and do not interact with each other.

The osmotic stress $S_{ij}$ is calculated using the fundamental stress tensor $\Pi_{ij}$ [7] by formula

$$S_{ik} = \frac{1}{V_c}\sum_m r_k^{(m)} \int_{S^{(m)}} \Pi_{ij} da_j . \tag{14}$$

Here $V_c$ is the volume of the unit cell, $\mathbf{r}^{(m)}$, $m = 1,2,3$, are the vectors of primitive translations that separate the equivalent points of each pair of the opposite lateral faces of the cell. Integration is carried out on only one of the two faces $S^{(m)}$ in each pair, namely on the face pointed by vector $\mathbf{r}^{(m)}$. The normals to the surfaces $S^{(m)}, m = 1,2,3$, are directed outward. In the initial configuration, $\mathbf{r}^{(1)} = \left(\frac{1}{2}, \frac{\sqrt{3}}{2}\right)a$ , $\mathbf{r}^{(2)} = (1,\ 0)a$ and $\mathbf{r}^{(3)} = \left(\frac{1}{2}, -\frac{\sqrt{3}}{2}\right)a$. For non-zero strain, the vectors $\mathbf{r}^{(m)}$, $m = 1,2,3$, as well as the volume $V_c$, are transformed in accordance with the strain. The background stress $S_b$ is also calculated by expression (14) but for an "empty" system, that is, a system with charged plates but without the particles.

The tensor $\Pi_{ij}$ for equation (2) has the form [1]

$$\Pi_{ij} = (\nabla\varphi)_i(\nabla\varphi)_j - \left(\frac{1}{2}|\nabla\varphi|^2 + \cosh\varphi - 1\right)\delta_{ij} . \tag{15}$$

Three relations $T_{11}(\varepsilon)$, $T_{22}(\varepsilon)$ and $T_{12}(\varepsilon)$ are obtained in two numerical experiments described above for every fixed set of the model's parameters $\sigma$, $\Sigma$, $R$, $H$ and $a$. Then, the polynomial approximation of these relations is carried out by the standard least square method. The coefficients of the approximation give the elastic constants in accordance with formulas (10)–(12). The numerical values of the strain parameter $\varepsilon$ vary in the range $-0.01 \div 0.01$ for stretching-compression and $0 \div 0.01$ for shearing in step of $0.001$. At each step of deformation, the new domain is constructed and the corresponding boundary value problem is solved numerically. The solution is carried out by the finite element method using gradient irregular meshes of tetrahedral elements. The calculations were partly supported by the Supercomputer Centre of Lomonosov Moscow State University [5].

## 4. Results and Discussion

Elastic constants were calculated for the crystal with the following model parameters: $\sigma = 2$, $\Sigma = 0$, $R = 1$, $H = 0.1$, while the lattice parameter $a$ varied in the range of $2.1 \div 6.0$ in steps of $0.1$. The results are shown in figure 2. The first-order elastic constant $B_{11}$, which is just the negative equilibrium pressure $p$, $B_{11} = -p$, tends monotonically to zero together with the grows of the lattice parameter. Behavior of the second-order elastic constants is also monotonic. While the lattice parameter decreases, the curves rise almost exponentially initially. Then, the rate of growth increases for smaller values of the lattice parameter. Positivity of all the three second order elastic constants means that the crystal is stable relative to the uniform stretch and shear strains within the present model.

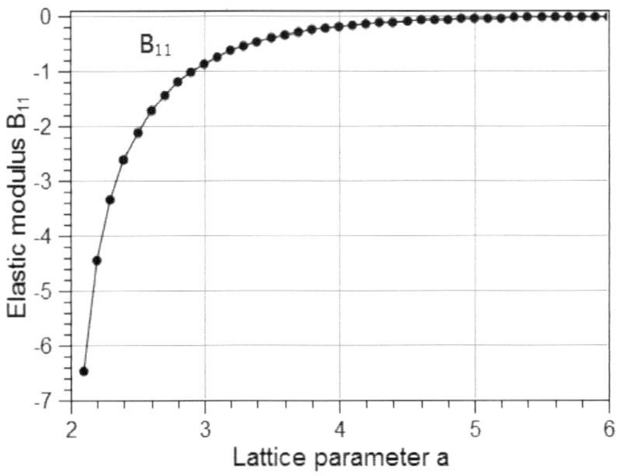


(a)

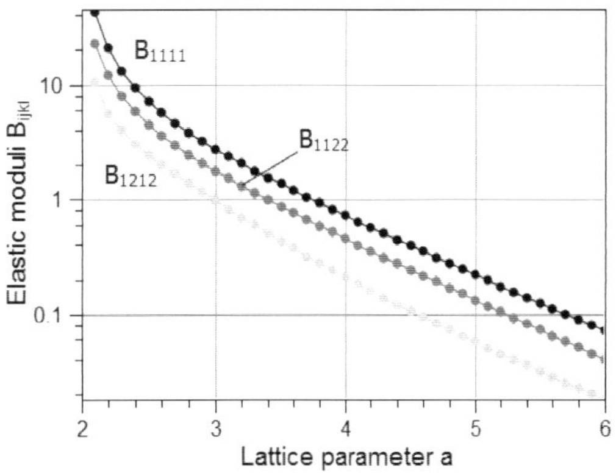


(b)

**Figure 2.** Elastic B-constants of the first (a) and second order (b).

Another set of elastic constants, *C*-constants, was also calculated. These constants are the coefficients in the series expansion of the second Piola-Kirchhoff stress tensor with respect to the components of the Lagrange strain tensor [6]. *B*-constants and *C*-constants are related to each other by the following expressions: $C_{11} = B_{11}$, $C_{1111} = B_{1111} - C_{11}$, $C_{1122} = B_{1122} + C_{11}$, $C_{1212} = B_{1212} - C_{11}$[3]. The results for the second-order elastic constants $C_{1111}$, $C_{1122}$, $C_{1212}$ are shown in figure 3. Behavior of these curves is similar to the one in figure 2b except that the rate of grows at smaller lattice parameters gets manifestly different for different constants.

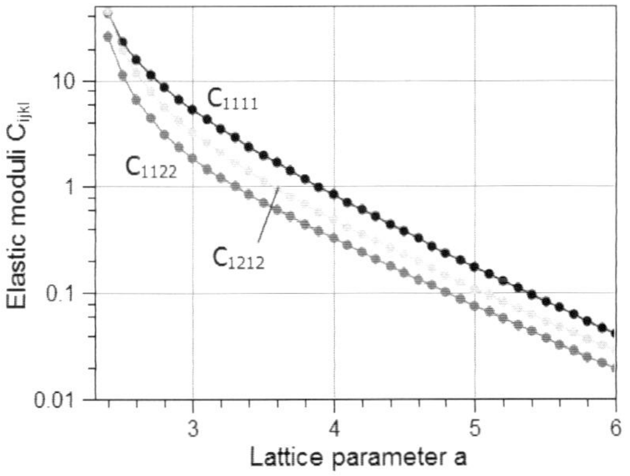


**Figure 3.** Elastic *C*-constants of the second order.

Instead of the theory based on the PB equation, the simpler one-component model [1] can be used to describe the colloidal systems. Within the one-component model, the complex interaction in the system is reduced to the effective interaction between the particles only. The effective potential is often chosen to be the central pair one. Having the C-constants, we can verify, whether the elastic properties of the particular crystal can be described by central pair effective potential of any kind. For pair effective interaction the Cauchy relation $C_{1122} = C_{1212}$ must hold for a static lattice. The ratio $C_{1122}/C_{1212}$ as function of the lattice parameter $a$ is shown in figure 4. The ratio $C_{1122}/C_{1212}$ definitely deviates from unity even for large lattice parameters where the pure pair interaction could be expected. This means that the effective interaction between colloids in the crystal is not purely pairwise, and three- and many-body effective interactions can play a role in the system.

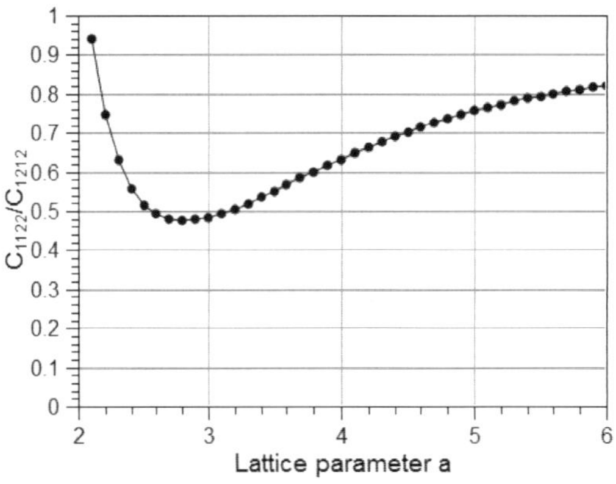


**Figure 4.** The ratio $C_{1122}/C_{1212}$ deviates from unity so that the Cauchy relation is not fulfilled.

## 5. Conclusion

The model of monolayer charge-stabilized colloidal crystals with two-dimensional monatomic hexagonal crystal lattice and constant surface charge density on the particles is proposed. The model is based on the Poisson-Boltzmann nonlinear differential equation. Numerical procedure for obtaining elastic constants from the stress-strain relations is proposed and applied to the crystals with different values of the lattice parameter. It was found that the crystals are stable relative to small uniform strains in the whole range of the lattice parameters studied. The breakdown of the Cauchy relation shows than any purely pair effective potential is not sufficient to proper simulate the elasticity of the colloidal crystals within the one-component model.

## References

[1] Belloni L 2000 *J. Phys.: Condens. Matter* **12** R549.
[2] Joannopoulos J D, Villeneuve P R and Fan S H 1997 *Nature* **386** 143.
[3] Barron T H K and Klein M L 1965 *Proc. Phys. Soc.* **85** 523.
[4] Wallace D C 1965 *Rev. Mod. Phys.* **37** 57.
[5] Sadovnichy V, Tikhonravov A, Voevodin Vl and Opanasenko V 2013 *"Lomonosov": Supercomputing at Moscow State University*. In Contemporary High Performance Computing: From Petascale toward Exascale (Chapman & Hall/CRC Computational Science), (Boca Raton, USA: CRC Press) pp 283-307.
[6] Holzapfel G A 2000 *Nonlinear Solid Mechanics: A Continuum Approach for Engineering* (Chichester: Wiley).
[7] Dyshlovenko P E 2010 *Colloid J.* **72** 627.

# Scaling laws during collapse of a homopolymer: Lattice versus off-lattice

**Suman Majumder, Henrik Christiansen and Wolfhard Janke**

Institut für Theoretische Physik, Universität Leipzig, Postfach 100 920, 04009 Leipzig, Germany

E-mail: suman.majumder@itp.uni-leipzig.de, henrik.christiansen@itp.uni-leipzig.de,wolfhard.janke@itp.uni-leipzig.de

**Abstract.** We present comparative results from simulations of a lattice and an off-lattice model of a homopolymer, in the context of kinetics of the collapse transition. Scaling laws related to the collapse time, cluster coarsening and aging behavior are compared. Although in both models the cluster growth is independent of temperature, the related exponents turn out to be different. Conversely, the aging and associated scaling properties are found to be universal, with the nonequilibrium autocorrelation exponent obeying a recently derived bound.

## 1. Introduction

Understanding of the nonequilibrium dynamics of coarsening of particle and spin systems is quite developed [1] and still is a topic of interest, especially in systems under special constraints [2]. Even though the pathways of collapse of a homopolymer, upon a transfer from a good to a poor solvent, bears resemblance to coarsening processes, the kinetics of the process has rarely been looked upon from that point of view until only recently [3, 4, 5, 6, 7]. The pioneer work on collapse kinetics goes back to the "sausage" model of de Gennes [8]. However, simulation results could rather be explained by the "pearl-necklace" picture of Halperin and Goldbart (HG) [9]. Those studies were motivated to understand the scaling of the collapse time, $\tau_c$, with the size of the polymer, $N$, using the form $\tau_c \sim N^z$. The exponent $z$ naturally depends on the dynamics of the simulations, and so far no agreement has been reached regarding its value.

According to HG, collapse occurs in three stages. The initial stage is the formation of nascent clusters of monomers. In the next stage, these clusters grow by pulling more monomers from the chain until they eventually coalesce with each other giving rise to a single cluster. In the final stage, monomers within the cluster reorganize to form a compact globule at equilibrium. Clearly, the second stage of the collapse, i.e., the cluster-growth stage can be identified with the usual coarsening processes. Consequently, it is found to be a scaling phenomenon where the average cluster size $C_s(t)$ at time $t$ follows a power law $C_s(t) \sim t^{\alpha_c}$, with $\alpha_c$ as the growth exponent.

Following the same analogy, it has been established [4] that one observes aging and related scaling during the collapse. To probe aging [10, 11] one defines a two-time autocorrelation function

$$C(t, t_w) = \langle O_i(t)O_i(t_w)\rangle - \langle O_i(t)\rangle\langle O_i(t_w)\rangle, \tag{1}$$

where $t$ and $t_w$ ($\ll t$) are the observation and waiting times, respectively, and $O_i$ is a parameter that reflects the spatio-temporal changes in the physical process, e.g., the time- and space-dependent order parameter for ferromagnetic ordering. Aging is manifested by the slower decay of $C(t, t_w)$ with increasing $t_w$ and the corresponding scaling of $C(t, t_w)$ is given as [10, 11]

$$C(t, t_w) = A_C x_c^{-\lambda_C}, \tag{2}$$

where $x_c = \ell/\ell_w$ is the ratio of growing length scales at time $t$ and $t_w$, as in ferromagnetic ordering [10, 11], and $\lambda_C$ is the nonequilibrium autocorrelation exponent. As for the collapse of a polymer [3, 6, 7] the relevant length scale is the cluster size $C_s(t) \equiv \ell(t)^d$ (where $d$ is the space dimension), $x_c = C_s(t)/C_s(t_w)$ is an obvious choice. In analogy with a bound on $\lambda_C$ popular in spin systems [12], for the collapse of a polymer, too, there exists a $d$-dependent bound [4]

$$(\nu d - 1) \leq \lambda_C \leq 2(\nu d - 1), \tag{3}$$

where $\nu$ is the critical exponent related to the size of the polymer in the extended state, i.e., $R_g \sim N^\nu$, with $R_g$ being the radius of gyration.

In this article, we present a comparative picture of the above scaling laws, viz., scaling of the collapse time, the cluster growth and the autocorrelation functions, from results involving an off-lattice model (OLM) and a lattice model (LM) via Monte Carlo (MC) simulations.

## 2. Models and Methods

For OLM, we opt for the bead-spring model of a flexible homopolymer in $d = 3$ where bonds between successive monomers are maintained via the standard finitely extensible non-linear elastic (FENE) potential

$$E_{\mathrm{FENE}}(r_{ii+1}) = -(K/2)R^2 \ln[1 - ((r_{ii+1} - r_0)/R)^2], \tag{4}$$

with $K = 40$, $r_0 = 0.7$ and $R = 0.3$. The nonbonded interaction energy is modeled by $E_{\mathrm{nb}}(r_{ij}) = E_{\mathrm{LJ}}(\min(r_{ij}, r_c)) - E_{\mathrm{LJ}}(r_c)$, where $E_{\mathrm{LJ}}(r) = 4\epsilon[(\sigma/r)^{12} - (\sigma/r)^6]$ is the standard Lennard-Jones (LJ) potential with $\sigma = r_0/2^{1/6}$ as the diameter of the monomers, $\epsilon(= 1)$ as the interaction strength and $r_c = 2.5\sigma$ as the cut-off radius.

For LM, we consider a variant of the interactive self-avoiding walk on a simple cubic lattice, where each lattice site can be occupied by a single monomer. The Hamiltonian is given by

$$H = -\frac{1}{2}\sum_{i \neq j, j\pm 1} w(r_{ij}), \quad \text{where } w(r_{ij}) = \begin{cases} J & r_{ij} = 1 \\ 0 & \text{else} \end{cases}. \tag{5}$$

In Eq. (5), $r_{ij}$ is the Euclidean distance between two nonbonded monomers $i$ and $j$, $w(r_{ij})$ is an interaction parameter that considers only nearest neighbors, and $J(= 1)$ is the interaction strength. We allow a fluctuation in the bond length by considering diagonal bonds, i.e., the possible bond lengths are 1, $\sqrt{2}$ and $\sqrt{3}$. The model has its origin in the bond-fluctuation model [13] and has been independently studied [14, 15] for equilibrium properties.

We introduce the dynamics in the models via Markov chain MC simulations [16]. For both models we apply local moves, i.e., after selecting a monomer randomly, for OLM, we shift it to a position randomly chosen within $[-\sigma/10 : \sigma/10]$ of its current position, and for LM, to another lattice site such that the bond-connectivity constraint and the excluded-volume criterion are preserved. For details on the allowed moves in LM we refer to Refs. [7, 14, 15]. A trial move is accepted or rejected following the Metropolis algorithm with Boltzmann criterion. One Monte Carlo step (MCS) consists of $N$ (where $N$ is the number of monomers in the chain) such attempted moves, effectively setting the time scale.

IOP Conf. Series: Journal of Physics: Conf. Series **955** (2017) 012008 doi:10.1088/1742-6596/955/1/012008

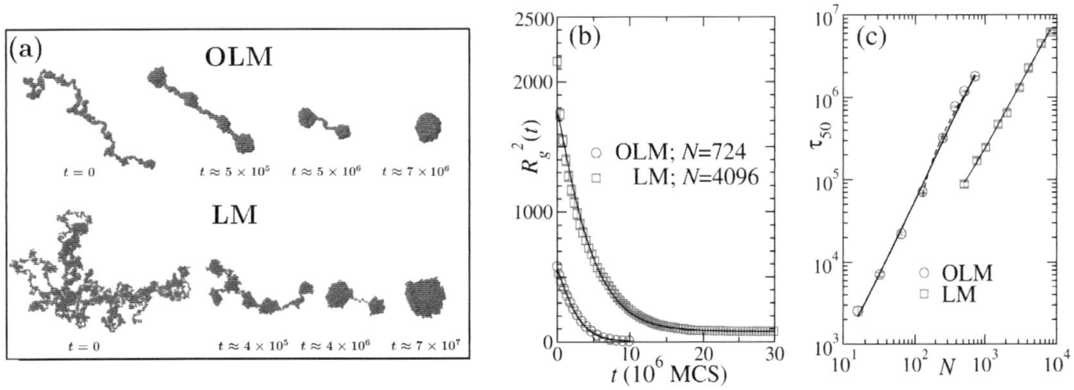


**Figure 1.** (a) Time evolution snapshots of the collapse of a homopolymer, after being quenched from an extended coil phase to a temperature, $T_q = 1$ for OLM, and $T_q = 2.5$ for LM, in the globular phase. (b) Decay of the squared radius of gyration, $R_g^2(t)$, with time. The solid lines are fits to a stretched exponential form described in the text. (c) Plot of collapse time, $\tau_{50}$, as function of $N$. The solid lines are fits to the form (6). The dashed line is a fit of the OLM data for $N \geq 128$, to the form (6) by fixing $z = 1$.

The collapse transition temperature is $T_\theta(N \to \infty) \approx 2.65\epsilon/k_B$ [6] and $\approx 4.0J/k_B$ [7], respectively, for OLM and LM. The unit of temperature is $\epsilon/k_B$ or $J/k_B$, where the Boltzmann factor $k_B$ is set to unity. We prepare initial conformations of the polymers at high temperatures $T_h \approx 1.5T_\theta$ that mimics an extended coil phase and then quench it to temperatures $T_q < T_\theta$. Since LM is computationally less expensive than OLM, for LM we simulate polymers up to $N = 4096$ whereas for OLM the longest polymer we simulate has $N = 724$. All the results presented (except the snapshots) are averaged over at least 300 different initial realizations.

## 3. Results

In Fig. 1 (a) we present snapshots showing the sequence of events during the collapse, for $N = 724$ and 4096, respectively, for OLM and LM. Both models provide the same phenomenological picture, i.e., initial formation of tiny clusters followed by coarsening of those clusters to form bigger ones and eventually a single cluster. This is in agreement with the "pearl-necklace" picture of HG. However, a careful observation reveals that in LM the coarsening occurs not only along the chain but also from lateral branches, which merely is an effect of using a much longer chain for LM.

As a first step to understand the kinetics, one observes the decay of the squared radius of gyration, $R_g^2 = \sum_{i,j}(r_i - r_j)^2/2N^2$, as shown in Fig. 1 (b). Although this does not provide any detailed information about the stepwise collapse of the polymer, one can extract a measure of the collapse time, $\tau_c$, by fitting the decay of $R_g^2(t)$ to a stretched exponential function $R_g^2 = a_0 + a_1 \exp[-(t/\tau_c)^\beta]$, represented by the solid lines in Fig. 1 (b). An elaborate description of such fitting can be found elsewhere [6, 7]. In addition, we also measure the collapse time, $\tau_{50}$, as the time when $R_g^2(t)$ has decayed to half of its total decay, i.e., $\left[R_g^2(0) - R_g^2(\infty)\right]/2$. In Fig. 1 (c), we show the variation of $\tau_{50}$ with $N$. Data for both the models show a power-law scaling, which can be quantified using the form

$$\tau_c = BN^z + \tau_0, \tag{6}$$

where $B$ is a nontrivial constant that depends on the quench temperature $T_q$, $z$ is the corresponding dynamic critical exponent, and the offset $\tau_0$ comes from finite-size corrections.

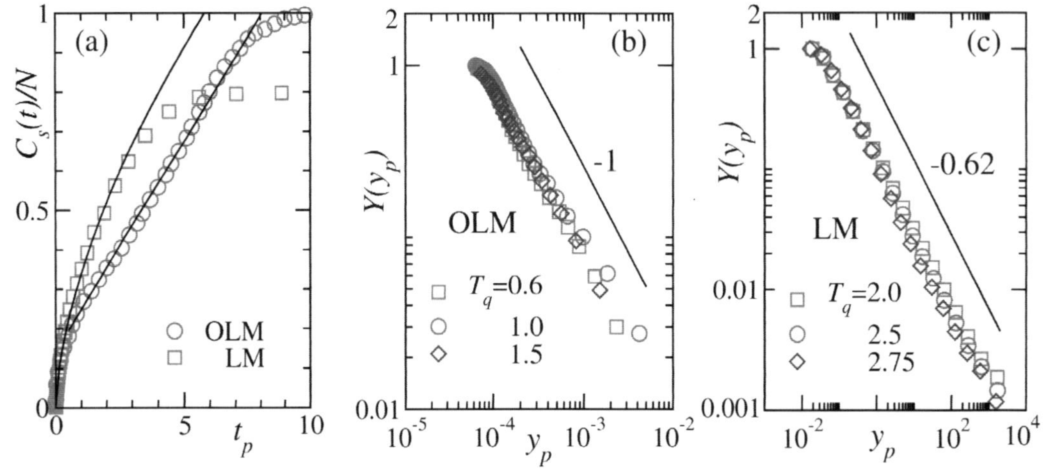


**Figure 2.** (a) Plots of the average cluster size $C_s(t)/N$ as function of time for the two models presented in Fig. 1 (a). To make both the data visible on the same plot, we divide the time axis by a factor $m$ to obtain $t_p = t/m$, where $m = 1 \times 10^6$ and $3.5 \times 10^6$, respectively, for OLM and LM. The solid lines there are fits to the form (7) with $\alpha_c = 0.98$ for OLM and $\alpha_c = 0.62$ for LM. The plots in (b) and (c) demonstrate the scaling exercise, respectively, for OLM with $\alpha_c = 1.0$ and LM with $\alpha_c = 0.62$, showing that data for $C_s(t)$ at different quench temperatures $T_q$ can be collapsed onto a master curve using a nonuniversal metric factor in the scaling variable. The solid lines there represent the corresponding $Y(y_p) \sim y_p^{-\alpha_c}$ behavior.

For LM a fitting [shown by the solid line in Fig. 1 (c)] with the form (6) yields $z = 1.61(5)$ and is almost insensitive to the chosen range. However, for OLM the fitting is sensitive to the chosen range. While using the whole range of data provides $z = 1.80(6)$ [shown by the solid line in Fig. 1 (c)], fitting only the data for $N \geq 128$ yields $z = 1.20(9)$. In this regard, a linear fit [$z = 1$ in (6)], shown by the dashed line, also cannot be ruled out [6].

We now move on to a comparative study of the scaling of the cluster growth between the two models. References [3, 6] provide details of the cluster identification method and subsequent calculation of $C_s(t)$ for OLM. On the other hand, for LM it is convenient to use $C_s(t) \equiv \ell(t)^3$, where $\ell(t)$ is obtained from the decay of the equal-time two-point density-density correlation function, for which we refer to Ref. [7]. However, here, for convenience, we use $C_s(t)$ as a notation for both the models. In Fig. 2 (a), we show the time dependence of $C_s(t)$ for OLM and LM. In coarsening kinetics of binary mixtures [17] such time dependence of the relevant length scale can be described correctly when one considers an off-set in the scaling ansatz. Similarly, it was later proved to be appropriate for the cluster growth during the collapse of a polymer [3]. Following that one writes down the scaling ansatz as

$$C_s(t) = C_0 + At^{\alpha_c}, \tag{7}$$

where $C_0$ corresponds to the cluster size after crossing over from the initial cluster formation stage, and $A$ is a temperature-dependent amplitude. The solid lines in Fig. 2 (a) are fits to the form (7) yielding $\alpha_c = 0.98(4)$ and $0.62(5)$, respectively, for OLM and LM.

As a step further, we verify the robustness of the growth by studying the dependence of cluster growth on the quench temperature $T_q$. For this we acquire data at different $T_q$ and perform scaling analyses based on nonequilibrium finite-size scaling (FSS) arguments [17]. An account of the FSS formulation in the present context can be found in Ref. [6]. In brief, one introduces in the growth ansatz (7) a scaling function $Y(y_p)$ as

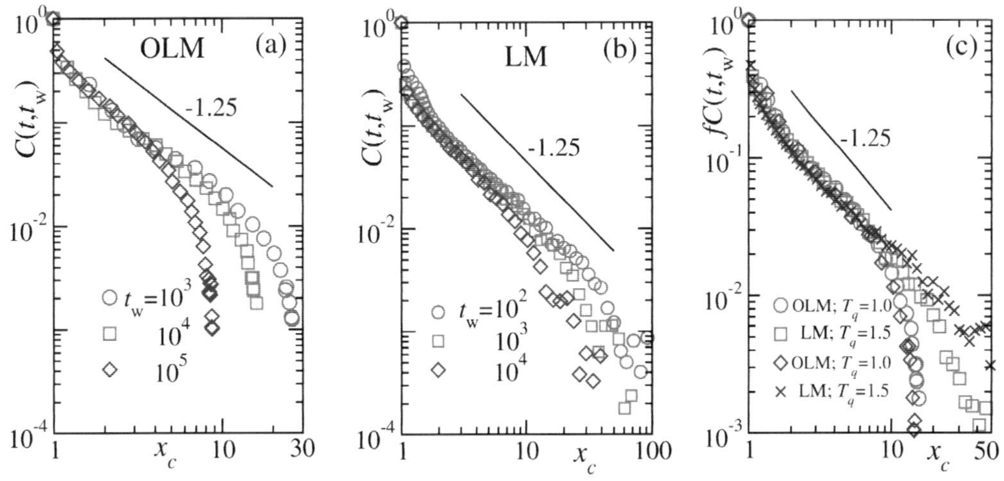


**Figure 3.** (a) Plot showing the scaling of $C(t, t_w)$ as function of $x_c = C_s(t)/C_s(t_w)$ in OLM, for three different waiting times $t_w$, at $T_q = 1$. The solid line shows the behavior (2) with $\lambda_C = 1.25$. (b) Same as (a) but for LM at $T_q = 1.5$. (c) Plot showing that aging scaling at different $T_q$ for the two models can be described by a single master-curve behavior. The solid line here also corresponds to Eq. (2) with $\lambda_C = 1.25$. Note that $C(t, t_w)$ is multiplied by a factor $f$ to make them collapse onto the same curve. For OLM $t_w = 10^4$ whereas for LM $t_w = 10^3$.

$$C_s(t) - C_0 = (C_{\max} - C_0)Y(y_p), \quad \text{i.e.,} \quad Y(y_p) = (C_s(t) - C_0)/(C_{\max} - C_0), \tag{8}$$

where $C_{\max} \sim N$ is the maximum cluster size a finite system can attain. In order to account for the temperature-dependent amplitude $A(T_q)$, one uses the scaling variable

$$y_p = f_s(N - C_0)^{1/\alpha_c}/(t - t_0), \quad \text{where} \quad f_s = [A(T_{q,0})/A(T_q)]^{1/\alpha_c}. \tag{9}$$

The metric factor $f_s$ is introduced for adjusting the nonuniversal amplitudes $A(T_q)$ at different $T_q$ [6]. Here, in addition to $C_0$ one also uses the crossover time $t_0$ from the initial cluster formation stage. A discussion of the estimation of $C_0$ and $t_0$ can be found in Refs. [6, 7]. While performing the exercise we tune the parameters $\alpha_c$ and $f_s$ to obtain a data collapse along with the $Y(y_p) \sim y_p^{-\alpha_c}$ behavior in the finite-size unaffected region. In Fig. 2 (b) and (c), we demonstrate such scaling exercises for OLM and LM with $\alpha_c = 1.0$ and 0.62, respectively. For $f_s$, we use the reference temperature $T_{q,0} = 1.0$ and 2.0, respectively, for OLM and LM. The collapse of data for different $T_q$ and consistency with the corresponding $y^{-\alpha_c}$ behavior in both plots suggest that the growth is indeed quite robust and can be described by a single finite-size scaling function with nonuniversal metric factor $f_s$ in the scaling variable. However, $\alpha_c$ in OLM is larger than for LM, a fact in concurrence with the values of $z$ estimated previously, and thus supporting the argument that $z \sim 1/\alpha_c$.

Next we compare the results from aging and related scaling during the collapse of the two model polymers. For OLM it is advised [4] to construct an autocorrelation function by assigning $O_i = \pm 1$ in (1), based on whether a monomer is inside a cluster or not, an analogy of the local density criterion. On the other hand, for LM we can define $O_i = \pm 1$ directly from the local density. For a description of the construction of such autocorrelation function we refer to Refs. [4, 6] and Ref. [7], respectively, for OLM and LM. In Fig. 3 (a) and (b), we demonstrate the

IOP Conf. Series: Journal of Physics: Conf. Series **955** (2017) 012008 doi:10.1088/1742-6596/955/1/012008

simple aging behavior by showing the scaling of $C(t, t_w)$ for three different $t_w$, with respect to the scaling variable $x_c$. Data for both models show consistency with Eq. (2), having an exponent $\lambda_C = 1.25$. The value indicates that $\lambda_C$ indeed follows the bound (3), which on transforming to numerical values [4] provides $0.762\,791 \leq \lambda_C \leq 1.525\,582$. For LM it has been shown [7] that scaling with respect to $t/t_w$ may lead to the misconclusion of the presence of sub-aging, which merely is the result of fitting with a complex variable, as pointed out previously in ordering of diluted ferromagnets [18]. Lastly, in Fig. 3 (c), we show that the data for the two models can be collapsed onto a single master-curve behavior, irrespective of $T_q$. The multiplier $f$ on the $y$-axis, [5, 6] adjusts different amplitudes, $A_C$, for different $T_q$ as well as models. Thus, unlike the growth exponent, the nonequilibrium autocorrelation exponent $\lambda_C$ is rather universal.

## 4. Conclusion

We have compared results from kinetics of the collapse transition in an off-lattice and a lattice homopolymer model. The smaller value of the exponent $z$, governing the scaling of collapse time, in the off-lattice model than in the lattice model suggests that the dynamics is faster in the former. This, perhaps, is controlled by the exponent $\alpha_c$ ($\alpha_c \sim 1/z$), characterizing the cluster-growth stage which seems to be the rate limiting stage of the overall collapse process. While the off-lattice model yields a linear growth ($\alpha_c \approx 1$), in the lattice model the growth is slower ($\alpha_c \approx 0.62$), which could be attributed to the topological constraints one experiences in a lattice model. On the other hand, surprisingly, both the models show evidence of simple aging scaling having the same autocorrelation exponent $\lambda_C \approx 1.25$, thus implying that the aging scaling is rather universal. This allowed us to demonstrate that scaling of the autocorrelation functions for the two models can be described by a single master curve.

## Acknowledgments

The work was funded by the Deutsche Forschungsgemeinschaft (DFG) under Grant Nos. JA 483/33-1 and SFB/TRR 102 (project B04), and further supported by the Leipzig Graduate School of Natural Sciences "BuildMoNa," the Deutsch-Französische Hochschule (DFH-UFA) through the Doctoral College "$\mathbb{L}^4$" under Grant No. CDFA-02-07, and the EU Marie Curie IRSES network DIONICOS under Contract No. PIRSES-GA-2013-612707.

## References

[1] Bray A J 2002 *Adv. Phys.* **51** 481
[2] Basu S, Majumder S, Das S K and Paul R 2017 *Europhys. Lett.* **116** 56003
[3] Majumder S and Janke W 2015 *Europhys. Lett.* **110** 58001
[4] Majumder S and Janke W 2016 *Phys. Rev.* E **93** 032506
[5] Majumder S and Janke W 2016 *J. Phys.: Conf. Ser.* **750** 012020
[6] Majumder S, Zierenberg J and Janke W 2017 *Soft Matter* **13** 1276
[7] Christiansen H, Majumder S and Janke W 2017 *J. Chem. Phys.* **147** 094902
[8] de Gennes P-G 1985 *J. Phys. (France) Lett.* **46** 639
[9] Halperin A and Goldbart A 2000 *Phys. Rev.* E **61** 565
[10] Zannetti M 2009 *Kinetics of Phase Transitions* ed S Puri and V Wadhawan (Boca Raton, FL: CRC Press) p 153
[11] Henkel M and Pleimling M 2010 *Non-Equilibrium Phase Transitions, Vol 2: Ageing and Dynamical Scaling Far from Equilibrium* (Heidelberg: Springer)
[12] Fisher D S and Huse D A 1988 *Phys. Rev.* B **38** 373
[13] Carmesin I and Kremer K 1988 *Macromolecules* **21** 2819
[14] Shaffer S 1994 *J. Chem. Phys.* **101** 4205
[15] Dotera T and Hatano A 1996 *J. Chem. Phys.* **105** 8413
[16] Landau D P and Binder K 2005 *A Guide to Monte Carlo Simulations in Statistical Physics* (Cambridge: Cambridge University Press)
[17] Majumder S and Das S K 2010 *Phys. Rev.* E **81** 050102
[18] Park H and Pleimling M 2010 *Phys. Rev.* B **82** 144406

# Influence of the random walk finite step on the first-passage probability

**Olga Klimenkova**[1,2], **Anton Menshutin**[2], **Lev Shchur**[1,2,3]

[1] National Research University Higher School of Economics, 101000 Moscow, Russia
[2] Science Center in Chernogolovka, 142432 Chernogolovka, Russia
[3] Landau Institute for Theoretical Physics, 142432 Chernogolovka, Russia

**Abstract.** A well known connection between first-passage probability of random walk and distribution of electrical potential described by Laplace equation is studied. We simulate random walk in the plane numerically as a discrete time process with fixed step length. We measure first-passage probability to touch the absorbing sphere of radius $R$ in 2D. We found a regular deviation of the first-passage probability from the exact function, which we attribute to the finiteness of the random walk step.

## 1. Introduction

Connection between stochastic processes and boundary value problems is well known. A Brownian motion is a classical example of continuous time stochastic process and it was shown by Einstein [1] that a collective motion of many Brownian particles is governed by diffusion equation. Let $B$ be a domain in the plane with boundary $\partial B$. Assume we have a Brownian particle that starts walking from some position $\boldsymbol{r}_0$, and the boundary $\partial B$ is absorbing which means that once a walker hits the boundary the walk is terminated. The time-dependent behavior of Brownian particle is described by diffusion equation:

$$\frac{\partial \rho(\boldsymbol{r},t)}{\partial t} = D\nabla^2 \rho(\boldsymbol{r},t) \tag{1}$$

with the boundary condition $\rho(\boldsymbol{r},t)=0$, $r\in\partial B$ and initial condition $\rho(\boldsymbol{r},t=0)=\delta(\boldsymbol{r}-\boldsymbol{r}_0)$.

If we are interested only in time-independent properties, e.g. first-passage probability, integration over time gives Laplace equation [2]

$$D\nabla^2\varphi(\boldsymbol{r}) = -\delta(\boldsymbol{r}-\boldsymbol{r}_0). \tag{2}$$

Equation (2) describes distribution of electrical potential $\varphi$ created by a point charge located at $\boldsymbol{r}_0$. The absorbing boundary condition in diffusion equation translates to the condition that a boundary $\partial B$ is grounded, $\varphi(\boldsymbol{r})=0$, $r\in\partial B$.

These connection allows one to analytically investigate Brownian motion by solving either diffusion equation or Laplace equation.

Vise-versa, these connection could be used to numerically solve some boundary value problems. It was shown by Kakutani [3] that solution of the equation

$$\nabla^2 u(x) = 0,\ x\in B \tag{3}$$

with the boundary condition

$$\lim_{y\to x} u(y) = g(x),\ x \in \partial B \tag{4}$$

at some point $x$ could be found as an expected value of $g(t)$, $t \in \partial B$ for a random walk starting at $x$, and $t$ being its first exit point.

Monte Carlo methods for boundary-value problems have several advantages. They are efficient for estimation of the function $u(x)$ at some point $x$. Monte Carlo simulations also have good performance for complex boundaries like fractals, and especially effective in high dimensional space. In addition, the parallel simulation of multiple random walks is straightforward and easy in implementation.

Special care should be taken while simulating random walks. In the study of the fractals formation by successive aggregation of random walks, e.g. diffusion limited aggregation (DLA) model [4], the domain $B$ is not bounded and diffusing particle is allowed to go infinitely far away. It is known that in 2D an escape probability is zero. In other words, all diffusing particles will be finally attached to DLA cluster after some time. This time can be infinitely large, and in practice one have to halt simulation if random walk goes at some large distance. For this purpose killing boundary of some big radius around the DLA cluster is used, and these may lead to the explosive growth of cluster in one direction. To avoid these unnecessary effect, the killing-free algorithm was proposed [5, 6], which allows to take into account the infinite boundary conditions exactly.

Discrete time random walk is easy to simulate but finite step of the random walk produces some bias. It was shown in [7] that particles undergoing discrete-time steps in three dimensions are captured with probability which is different from the probability generated with the infinitesimally short steps, and difference does depend on the root-mean square distribution of step length.

In the paper we present results of the simulations of the random-walk in the plane and propose regular form of the first correction to the first-passage probability which is due to the finite step of the random-walk.

## 2. Simulation algorithm

Let us consider particle at the point $(R_b, 0)$ in the plane at the distance $R_b$ from the center of absorbing circle of radius $R$ as illustrated in the Figure 1.

Particle at each time step jumps on the length $\delta$ in random direction. The random walk is symmetric thus directions of jumps are distributed uniformly. Particle position is continuous variable thereby such a process is the discrete time random walk in continuous space. If after making a move particle is inside the circle $R$ at position $(r\cos\phi, r\sin\phi)$, $r < R$ then it is assumed absorbed at position given by angle $\phi$. For simplicity we assume that $(R - r) \sim \delta$ and $\delta \ll 1$ so the difference in real absorption position and position after the last jump is negligible.

Probability $P_{exp}(\phi)$ for a particle to be absorbed at the angle $\phi$ is measured numerically. We compare it with analytic solution found by solving Laplace equation:

$$P(\phi) = \frac{1}{2\pi}\frac{x^2 - 1}{x^2 - 2x\cos\phi + 1} \tag{5}$$

where $x = R_b/R > 1$ (see for details Reference [5]).

Expression (5) can be interpreted as a first passage probability for a particle starting at the distance $R_b$ from the center of absorbing circle $R$. At the same time, it could be used to speed up simulation to deal with fly-away particles problem. The most trivial way to deal with particles that have gone far from the absorbing circle $R$ is to kill them. But this introduces an error in simulation and should be avoided. The correct way to solve problem, is to use expression (5) which allows us to return went away particles back to the simulation region. In our simulation

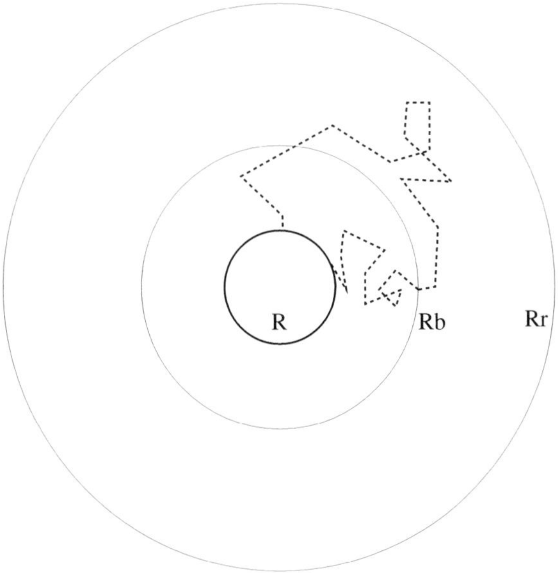


**Figure 1.** Illustration of the random walk in the plane. Random walk of the particle starts at the birth circle of radius $R_b$. Particle terminates while touching the absorbing circle of radius $R$. If particle cross returning circle $R_r$ it is returning back to the birth-circle at the corresponding angle (see Expr. 5) and discussion in the text.

if particle goes out of $R_r$, $R_r \gg R_b$, it is returned back to $R_b$ with probability (5) and angle is counted relative to the line connecting the particle position and the center of absorbing circle. In order to generate random variable $\phi$ with distribution (5) one can use the following mapping

$$\phi = f(u) = 2\arctan\left(\frac{x-1}{x+1}\tan u\frac{\pi}{2}\right), \tag{6}$$

where $u$ is a uniform random number, $u \in [-1, 1]$, and $x = R_b/R > 1$. See [9] for details. The killing-free algorithm effectively makes simulation region infinite which results in the correct account of the boundary condition.

Since we study the influence of finite jump length and we measure first-passage probability numerically and compare it with analytical solution given by Eq. (5), application of the same formula during the simulation should be done with care. To minimize this influence we must ensure that $(R_b - R) \gg \delta$ and $(R_r - R_b) \gg \delta$.

In the next section, we simulate random walk for different sets of parameters, estimate probability distribution $P_{exp}(\phi)$, and study its deviation from the exact result $P(\phi)$.

## 3. Simulation results

For calculation $P_{exp}(\phi)$, we divide interval of the possible values of $\phi$ $[-\pi : \pi]$ into 100 bins and count number of hits for each bin. Normalizing results over the total number of random walkers $N$ and over the bin size gives estimation of the hitting probability $P_{exp}(\phi)$.

Comparison of experimental result with analytical is presented in Figure 2. Data in the left panel is for the relatively small number of runs $N = 10^4$ while data in the right panel is for the larger number of runs $10^6$. Fluctuations become not visible on the scale of the figures for the large enough number of runs demonstrating quality of the data. At the same time, it comes clear, that there are some deviations of the measured in experiment probability function $P_{exp}(\phi)$ from the exact one.

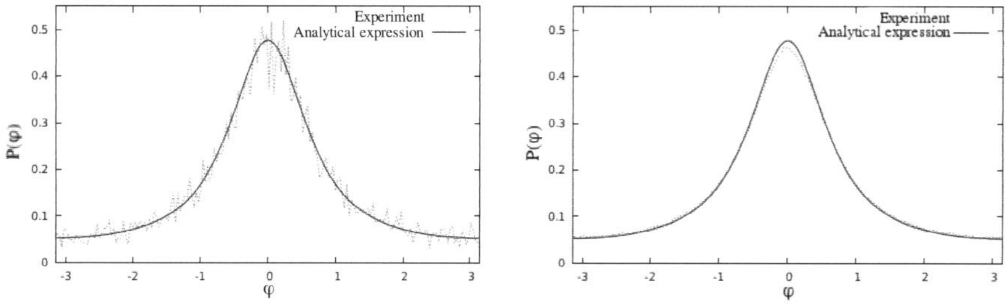


**Figure 2.** Comparison of the estimated probability $P_{exp}$ with the exact probability $P(\phi)$ in Expr. (5). Left: Parameters are $N = 10^4$, $R = 10$, $R_b = 20$, $R_d = 200$, $\delta = 1$. Right:Parameters are $N = 10^6$, $R = 10$, $R_b = 20$, $R_d = 200$, $\delta = 1$.

In order to make deviations more visibly pronounced we compute the relative deviation of the estimated probability $P_{exp}$ from the exact probability $P(\phi)$

$$f(\phi) = \frac{P_{exp}(\phi) - P_{exact}(\phi)}{P(\phi)}. \tag{7}$$

We estimate $f_i(\phi)$ from $N = 10^6$ runs, repeat this simulation $M$ times, and calculate an average

$$\langle f(\phi) \rangle = \frac{\sum_{i=1}^{M} f_i(\phi)}{M}. \tag{8}$$

Standard error of $f_i(\phi)$ is calculated accordingly

$$Ef(\phi) = \frac{\sqrt{\sum_{i=1}^{M} (f_i(\phi) - \langle f(\phi) \rangle)^2}}{M}. \tag{9}$$

Calculation of $\langle f(\phi) \rangle$ was done for $M = 100$ independent runs for $\delta = 1$, $\delta = 0.5$ and $\delta = 0.2$, and is shown in Fig. 3. Deviation is clearly depend on the size of the random walk step $\delta$, and decreases with the decreasing $\delta$.

$P_{exp}(\phi)$ has both random and systematic error. Function $< f(\phi) >$ shows systematic component while $Ef(\phi)$ shows random component. $Ef(\phi)$ depends on the total number of runs and decreases as $1/\sqrt{M}$ (see Fig. 4). Systematic deviation $< f(\phi) >$ depends mainly on the random walk step length $\delta$ and does not tend to 0 as $M$ grows.

For the reason of clarity we restrict our analysis to a single value $< f(0) >$ of $< f(\phi) >$ at $\phi = 0$ as a measure of deviation of $P_{exp}$ from $P_{exact}$. Fig. 5 shows $< f(0) >$ as a function of random walk step length $\delta$. This function shows decrease when $\delta$ decreases, but the way it converges to zero could not be derived from the available data. As $\delta$ goes to 0 various errors could become significant, e.g. rounding errors during calculation, and special care should be taken. We leave this question open for the future research.

## 4. Discussions

In this paper we have estimated numerically systematic deviation of the first-passage probability for the random walking particle to hit circle of radius $R$ in the plane. We may conclude from

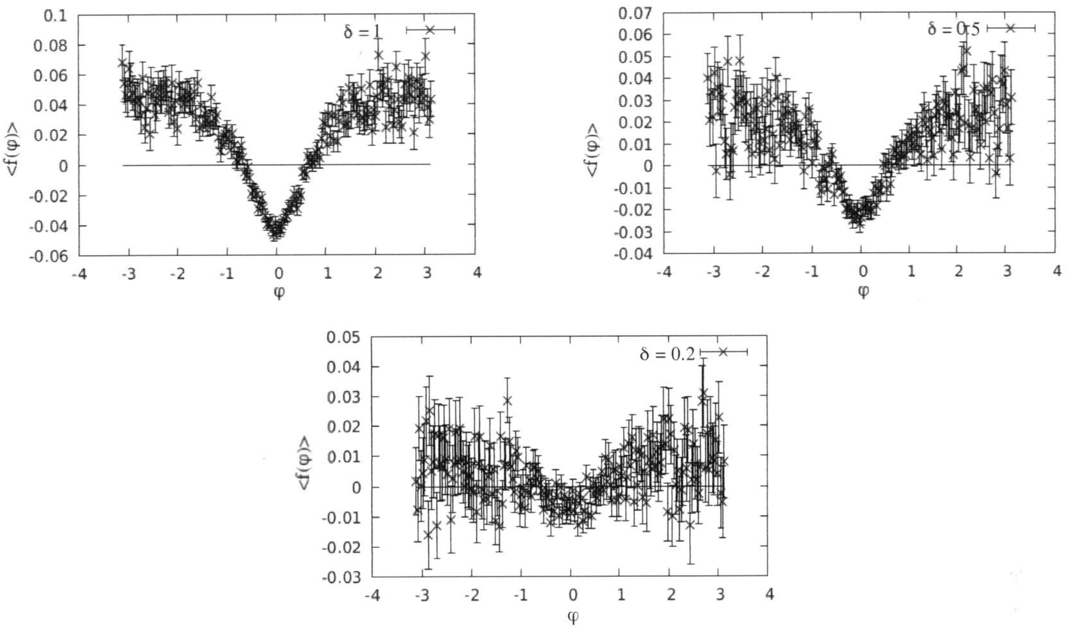


**Figure 3.** Deviation $\langle f(\phi)\rangle$ for the length of the random walk step $\delta = 1$, $\delta = 0.5$ and $\delta = 0.2$ (marked in the figures) averaged over $M = 100$ groups of runs, see Exprs. (8,9).

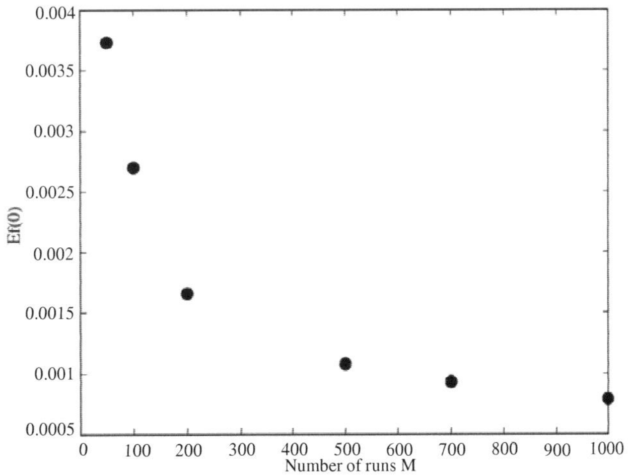


**Figure 4.** $Ef(0)$ dependence on the number of groups $M$ of the runs.

the results of our simulations, and especially from the analysis of the data presented in the Figures 3,5, that the following form of the correction may take place

$$P_{exp}(\phi) \approx P(\phi)\left(1 - \left(\frac{\delta}{R}\right)^{\alpha} cos(\phi)\right). \tag{10}$$

We propose this expression is the first-order correction to the exact result of the Laplace solution (5).

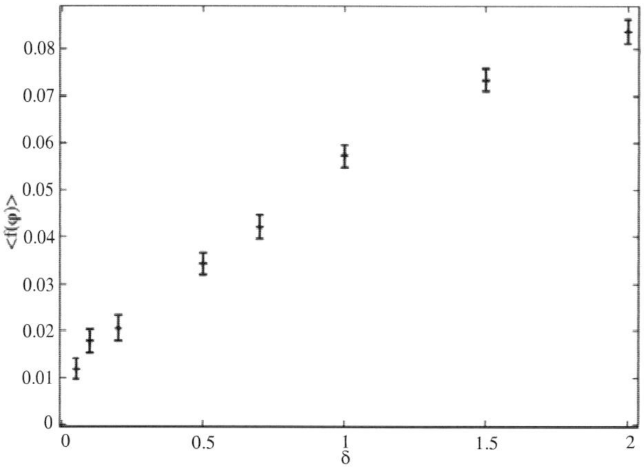


**Figure 5.** Dependence of $< f(0) >$ on the $\delta$ for $M = 100$ group of runs.

Future work, both simulations and analytical study, have to be done to check our proposal. Our findings can be important not only for the simulation of random walk in plane, but also in the study of boundary value problems, as well as in random growth fractals sumulations, e.g. in DLA problem [10].

## Acknowledgments
This work was supported by grant 14-21-00158 from the Russian Science Foundation.

## References
[1] A. Einstein, *Über die von der molekularkinetischen Theorie der Wärme geforderte Bewegung von in ruhenden Flüssigkeiten suspendierten Teilchen*, Annalen der Physik (in German). **322**, 549 (1905).
[2] reference on Redner paper
[3] S. Kakutani, *Two-dimensional Brownian motion and harmonic functions*, Proc. Imp. Acad. Tokyo **20**, 706 (1944).
[4] T. A. Witten, L. M. Sander, *Diffusion-Limited Aggregation, a Kinetic Critical Phenomenon*, Phys. Rev. Lett. **47**, 1400 (1981).
[5] A. Yu. Menshutin, L. N. Shchur, *Test of multiscaling in a diffusion-limited-aggregation model using an off-lattice killing-free algorithm*, Phys. Rev. E **73**, 011407 (2006).
[6] E. Sander, L. M. Sander, and R. M. Ziff, *Comput. Phys.* **8**, 420 (1994); L. M. Sander, *Contemp. Phys.* **41**, 203 (2000).
[7] R. M. Ziff, S. N. Majumdar, and A. Comtet, *Capture of particles undergoing discrete random walks*, J. Chem. Phys. **130**, 204104 (2009).
[8] J.F. Reynolds, *A Proof of the Random-Walk Method for Solving Laplace's Equation in 2-D*, The Mathematical Gazette **49**, 416 (1965).
[9] A. Yu. Menshutin. , L. N. Shchur, *Morphological diagram of diffusion driven aggregate growth in plane: Competition of anisotropy and adhesion*, Comp. Phys. Comm. **182** 1819 (2011).
[10] A. Yu. Menshutin, *Scaling in the Diffusion Limited Aggregation Model*, Phys.l Rev. Lett. **108**, 015501 (2012).

# Monte Carlo Simulation of Random-Anisotropy Amorphous Magnets

**A V Bondarev and I L Bataronov**

Voronezh State Technical University, 14 Moskovskii prospekt, Voronezh, 394026, Russian Federation

bondarev_a_v@mail.ru

**Abstract**. Using the Monte Carlo method, within the frame of the Heisenberg model, we studies the magnetic properties of amorphous Tb. The relaxation of magnetization of the model of amorphous Tb was studied. We stablished that the relaxation goes in two stages. On the first stage the magnetization sharply decreases by some amount $\Delta M_Z$, on the second stage the magnetization decreases with time according to the logarithmic law. The possible mechanisms of relaxation is discussed.

## 1. Introduction

Amorphous alloys based on rare-earth metals are of great interest due to their unique magnetic properties [1-3]. Particularly, the Re-Tb amorphous alloys transit into the spin-glass-like state at low temperatures. The nature of the spin-glass state in amorphous alloys is yet poorely understood. This ecourages us to construct and analyze computer models of these materials. In this paper we report on the Monte Carlo simulation of magnetization relaxation in pure amorphous terbium. Its properties have to be close to that of amorphous alloys with high concentrations of the terbium atoms.

## 2. Simulation technique

Using the molecular dynamics method, at $T = 100$ K we constructed the models of atomic structure of pure amorphous terbium containing 100 000 atoms in a cubic cell with periodic boundary conditions. The interatomic interaction was described by an empirical polynomial potential [4]. The parameters of the potential were estimated by the values of bulk modulus, atomization energy, atomic volume, and the interatomic distance for crystalline Tb. Application of this potential for the models of Re-Tb amorphous alloys revealed a good agreement of the model and experimental radial distribution functions and parameters of the topological short-range order [5].

Magnetic properties of these models were studied using the Monte Carlo method within the standard Metropolis algorithm [6, 7], in the framework of the Heisenberg model. For the model of amorphous Tb we used the Hamiltonian proposed by Harris, Plischke and Zuckermann for amorphous alloys containing rare-earth atoms with the non-zero orbital moment [8]:

$$H = -\frac{1}{2}\sum_{i,j} J_{ij}\left(\boldsymbol{S}_i \cdot \boldsymbol{S}_j\right) - D\sum_i \left(\boldsymbol{n}_i \cdot \boldsymbol{S}_i\right)^2 - \mu h \sum_i S_i^z , \quad (1)$$

where $J_{ij}$ is the integral of the exchange interaction between atoms $i$ and $j$; $D$ is random anisotropy; $\boldsymbol{S}_i$ is Heisenberg spin; $\boldsymbol{n}_i$ is unit vector determining the directions of the local anisotropy axes, which were chosen randomly in each site; $\mu = 9.72\mu_B$ is effective magnetic moment of the Tb atom; $h$ is external magnetic field strength.

Magnetic moments of the Tb atoms interact by means of the long-range RKKY exchange mechanism [9]. The exchange interaction integral rapidly decreases with the interatomic distance, therefore the positive (ferromagnetic) interaction between the nearest neighbours brings the maximal contribution to the exchange energy.

As an approximation of the RKKY interaction between the moments of the Tb atoms, we chose the linear dependence of the exchange integral on the interatomic distance $r$ [10]:

$$J(r) = \begin{cases} J_0 \dfrac{r_{\min} - r}{r_{\min} - r_1}, & r \le r_{\min} \\ 0, & r > r_{\min} \end{cases}, \tag{2}$$

where $J_0 = 19.26$ K is the average value of the exchange integral fitted such that the temperatures of the magnetic phase transitions $T_f$ would be close to the experimental values of $T_f$ for the Re-Tb amorphous alloys; $r_1$ is the first peak position on the partial pair radial distribution function $g_{Tb\text{-}Tb}(r)$; $r_{\min} = 0.446$ nm is the position of the first minimum of the $g_{Tb\text{-}Tb}(r)$ function. As the distribution of the interatomic distances in the first coordination sphere of an amorphous metal is close to normal law so the distribution of the exchange integrals $J_{ij}(r)$ is also close to normal law.

The value of the anisotropy $D$ in this paper varied in a wide range ($D/J_0 = 0-20$).

It is known from literature that for the crystalline terbium the anisotropy constant is $5.5 \cdot 10^8$ erg/cm$^3$ [9], or 127 K/at. When a material transforms from the crystalline state into the amorphous one the value of the anisotropy constant practically does not change [11]. So the basic value of the anisotropy constant for amorphous Tb was chosen as $D = 127$ K, or $D/J_0 = 127/19.26 \approx 6.6$.

Cooling of the models was conducted from the paramagnetic state in the temperature interval $T = 100-1$ K with steps $\Delta T = 2$, 5 and 10 K. During cooling, we calculated the temperature dependencies of the potential energy (1) and magnetization by the formula:

$$M = \frac{1}{N}\sum_i \sqrt{\langle S_i^x \rangle^2 + \langle S_i^y \rangle^2 + \langle S_i^z \rangle^2}, \tag{3}$$

where $N$ is number of spins in the system, $\langle \ldots \rangle$ is time averaging during $10^3$ MC-steps/spin.

Magnetic susceptibility $\chi$ was calculated by numerical differentiation of magnetization by external magnetic field strength:

$$\chi = \frac{\Delta M}{\Delta H}. \tag{4}$$

It was established that in the studied models in the fields of 0–5 kOe the magnetization depends on the field linearly. This allows us use the eq. (4) with the good accuracy and to assume $\Delta H = 5$ kOe.

## 3. Results and discussion

We studied the relaxation of the magnetization in the model of amorphous Tb at various values of the anisotropy ($D/J_0 = 0$, 2, 4, 6, 8, 10, 12, 14, 16, 18, 20). For every value of $D/J_0$ the calculation was carried out at the constant temperature $T = 1$ K. The external magnetic field with the strength 1000 kOe along the $Oz$ axis was switched on immediately, then the system was maintained in the field during 9–20 cycles of $10^3$ MC-steps/spin until the magnetization achieved saturation (from 0,999 at $D/J_0 = 0$ to 0,93 at $D/J_0 = 20$). Then the field was switched off and the time dependence of the projection of the magnetization on the $z$ axis $M(t)$ was studied.

At all studied values of $D/J_0 > 0$ the magnetization relaxation goes in two stages. On the first stage that proceeds immediately after the switching off the magnetic field, the magnetization sharply (during 1–2 cycles of $10^3$ MC-steps/spin) decreases by some amount $\Delta M_Z$. With increasing the $D/J_0$ ratio the value of the magnetization jump monotonically increases (Fig. 1) and probably comes to a constant value ~0,4 at

$D/J_0 \to \infty$. On this stage we observe the disorientation of magnetic moments of the Tb atoms from the state in which the spins are oriented in the direction of the external field to the state in which the orientation of the spins is determined mainly by the directions of the local anisotropy axes.

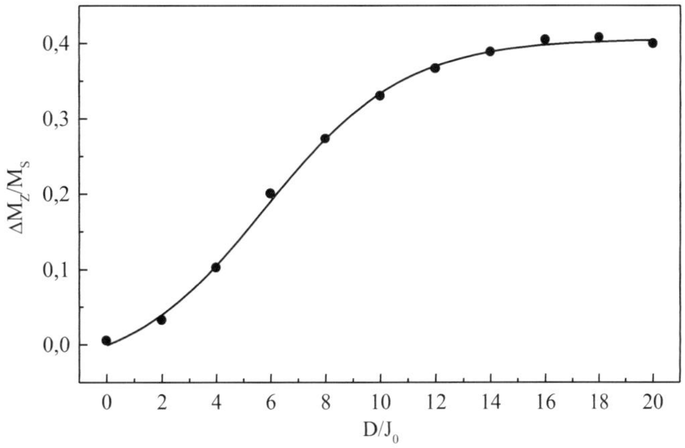


Fig. 1. Magnetization jump after switching off the external magnetic field for the model of amorphous Tb at $T = 1$ K

On the second stage of the relaxation, the magnetization very slowly decreases with time without any signs of achievement of an equilibrium value. On this stage the magnetic moments of the Tb atoms turn by the small angles, which is due to competition between exchange interaction and random anisotropy.

To verify this assumption, we calculated the distribution functions $F(\theta)$ of angles between directions of the magnetic moments of the Tb atoms and directions of the corresponding random anisotropy axes at $D/J_0 = 6{,}6$ and $T = 1$ K in the process of the magnetization relaxation (Fig. 2).

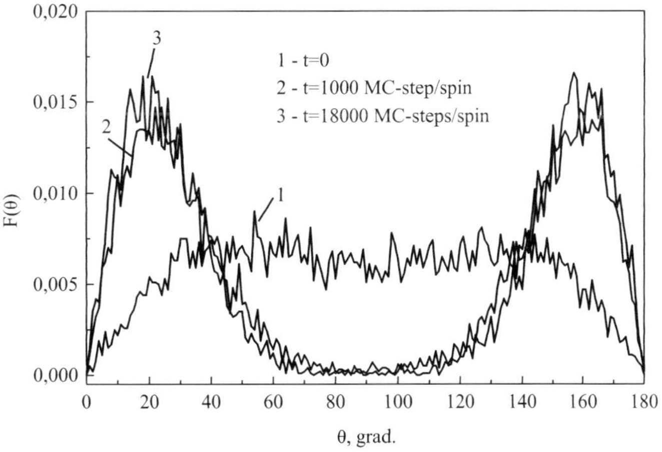


Fig. 2. Distribution function of the angles between directions of the spins and directions of the random anisotropy axes for the model of amorphous Tb at $D/J_0 = 6{,}6$ and $T = 1$ K

In the moment directly before the switching off the external magnetic field, the $F(\theta)$ function has the almost uniform distribution over the angles form ~30° to ~150°, since in the very strong fields ($H$ = 1000 kOe) the magnetization is totally determined by the magnitude of the external field and the influence of the random anisotropy is negligibly small. In one cycle of $10^3$ MC steps/spin, after the switching off the external field, the $F(\theta)$ function undergoes the essential changes: two high maxima at $\theta = 20°$ and $\theta = 160°$ and a broad minimum at $\theta = 70° - 110°$ appear on it. This confirms the assumption of that at the first stage of the relaxation the spin reorientation takes place from the directions determined by the external field towards the directions close to those of the random anisotropy axes. Further the form of the $F(\theta)$ curve almost does not change, only the maxima become a little higher. In Fig. 2 the curve $F(\theta)$ calculated after 18 cycles of $10^3$ MC-steps/spin after switching off the external field is also given. Therefore, on the second stage of the relaxation the spin rotations by small angles take place that is caused by competition of the exchange interaction and random anisotropy.

In Fig.3 we present the time dependencies of a projection of magnetization on the $Oz$ axis for the model of amorphous Tb at various values of the $D/J_0$ ratio. It is seen that at $D/J_0 = 0$ magnetization with time remains practically constant. At the other values of $D/J_0$, the dependencies $M(t)$ decrease monotonically.

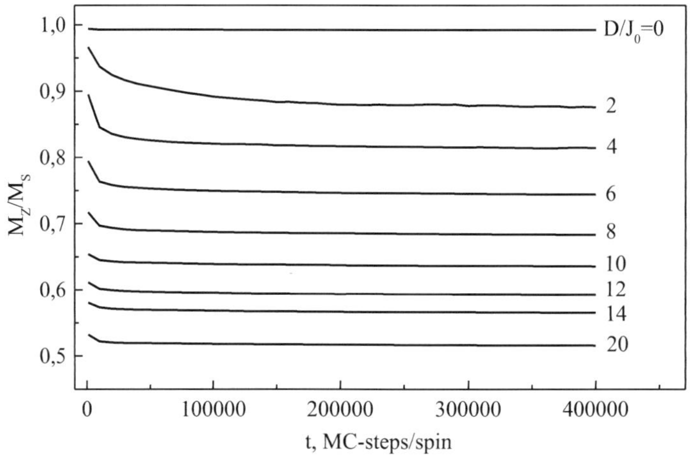


Fig. 3. Time dependence of the magnetization after switching off the external magnetic field for the model of amorphous Tb at $T = 1$ K

At all the values of the anisotropy constant, the dependence $M(t)$ was approximated by the least-square technique with a logarithmic function:

$$M(t) = a - b \cdot \ln t, \quad (5)$$

where $a$ and $b$ are fitting parameters.

It was confirmed experimentally that such a logarithmic relaxation of magnetization takes place in the real spin glasses [12-14]. With applying magnetic field to a sample, cooled to $T < T_f$ in zero field, magnetization first increases with a jump (in microscopic time), and then it slowly grows approaching to an equilibrium value according to the law close to logarithmic. If the sample is cooled in the field to $T < T_f$ then after the switching off the field the magnetization drops with a jump and then slowly decreases remaining finite even in time $10^5$ sec [14]. This feature proves that in spin glasses there is a wide spectrum of relaxation times that reaches the macroscopic times. The energy minima corresponding to the different sets of spin orientations are separated one from another by macroscopically high barriers, and thus the relaxation takes place in macroscopically large times.

For the model of amorphous Tb the magnetization relaxation was studied at $D/J_0 = 6{,}6$ and various temperatures. In Fig. 4 we present the time dependencies of the projections of magnetization on the $Oz$ axis after switching off the external magnetic field calculated at the temperatures 10–70 K with the step 10 K. At the temperatures $T = 10 - 50$ K corresponding to the spin-glass state, the magnetization decreases with time very slowly, according to the logarithmic law (6). At the temperature $T = 60$ K close to the temperature of the spin-glass transition $T_f$, the magnetization decreases rapidly and fluctuate around zero. At the temperatures $T > 60$ K corresponding to the paramagnetic phase, the magnetization drops to zero very rapidly.

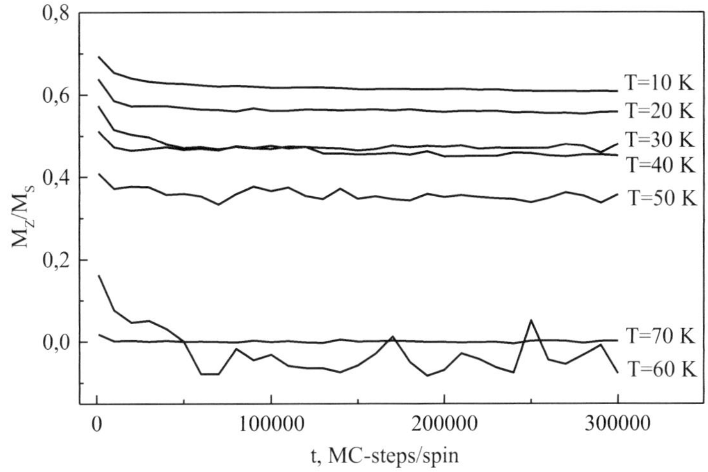


Fig. 4. Time dependence of the magnetization after switching off the external magnetic field for the model of amorphous Tb at $D/J_0 = 6{,}6$ and different temperatures

## 4. References

[1] Fernandez-Baca J A and Wai-Yim Ching (ed.) 1995 *The Magnetism of Anorphous Metals and Alloys* (Singapore: World Scientific)
[2] Fukamichi K, Hattori Y and Goto T 1996 *Sci. Rep. RITU. Ser. A* **42** 65
[3] Boucher B and Chieux P 1991 *J. Phys.: Condens. Matter* **3** 2207
[4] Bataronov I L, Bondarev A V and Barmin Yu V 2000 *Bull. Russ. Acad. Sci. Phys.* **64** 1329
[5] Bataronov I L, Bondarev A V, Urazov D V and Barmin Yu V 2005 *Bull. Russ. Acad. Sci. Phys.* **69** 1299
[6] Binder K 1997 *Rep. Prog. Phys.* **60** 487
[7] Binder K and Heermann D 1988 *Monte Carlo Simulation in Statistical Physics* (Berlin: Springer-Verlag)
[8] Harris R, Plischke M and Zuckermann M J 1973 *Phys. Rev. Lett.* **31** 160
[9] Taylor K N R and Darby M I 1972 *Physics of Rare-Earth Solids* (London: Chapman and Hall)
[10] Bondarev A V 2008 *Bull. Russ. Acad. Sci. Phys.* **72** 1251
[11] Chudnovsky E M 1988 *J. Appl. Phys.* **64** 5770
[12] Lundgren L, Nordblad P and Svendlindh P 1986 *Phys. Rev. B* **34** 8164
[13] Takayama H 2004 *J. Magn. Magn. Mater.* **272-276** 256
[14] Dotsenko V S 1993 *Phys.-Usp.* **36** 455

# Elastic properties of charge-stabilized colloidal crystals with static body-centered cubic lattice

**A A Batanova and P E Dyshlovenko**

Department of Applied Mathematics and Informatics, Ulyanovsk State Technical University, 32 Severny Venets Street, Ulyanovsk 432027, Russia

E-mail: p.dyshlovenko@mail.ru

**Abstract**. The model of charge-stabilized colloidal crystals with monatomic body-centered cubic crystal lattice is proposed to determine the first and second order elastic constants of these crystals. The crystals are described within the theory based on the Poisson-Boltzmann nonlinear differential equation. Electric behavior of the colloidal particles obeys the constant charge model. Elastic constants of the crystals are derived from the stress-strain relations obtained by means of the computer experiment. Elastic constants were obtained for a broad range of values of the lattice parameter. Stability of the crystals and the presence of the many-body effective interactions in them are briefly discussed.

## 1. Introduction

The charge stabilized colloidal crystals are spatially ordered systems of electrically charged submicron particles immersed into a liquid electrolyte. The centers of the particles are located in the nodes of some crystal lattice. Colloidal crystals are used in making photonic crystals and self-organizing structures [1]. These systems are interesting as model systems in the study of ordinary molecular crystals as well as disordered colloidal systems. They also can shed some light onto the more complex systems consisting of micelles, viruses or DNA molecules.

Within the model approach of the present work, colloidal crystals are treated as a medium with initial stress governed by the Poisson-Boltzmann (PB) nonlinear differential equation [2]. This approach does not require any prescribed interparticle potentials since the properties of the system are fully described by solution of the PB equation.

An external pressure is needed for the system of like-charged particles to be stable. For that reason, elastic properties of such systems have some specificity compared to conventional media [3]. In particular, the series expansion of the stress with respect to strain includes a constant term corresponding to the stress in the absence of deformation.

In the paper, we describe a numerical procedure and carry out computer calculations to obtain stress-strain relations for the colloidal crystal with a body-centered cubic (bcc) crystal lattice and constant surface charge density on the particles (cc-model). Elastic constants of the first and second order are then derived from these relations. Knowing the elastic constants we draw a conclusion about mechanical stability of the crystal. We also detect and estimate the many-body effective interaction in the system.

## 2. Description of the Model

The particles in the colloidal crystal are located in the nodes of the bcc crystal lattice with the lattice parameter $a$. The particles are charged with constant surface charge density $\sigma_0$. The whole array of the particles is embedded into the binary symmetrical univalent electrolyte (1:1 electrolyte).

Electric potential in the crystal obeys the PB equation for the case of the 1:1 electrolyte:

$$\nabla^2\varphi = \frac{2q_e n_0}{\varepsilon_0\varepsilon}\sinh\left(\frac{q_e\varphi}{kT}\right), \tag{1}$$

where $n_0$ is a bulk concentration of any of the two species of the electrolyte, $q_e$ is the elementary charge, $\varepsilon$ is a relative dielectric permittivity of the electrolyte, $\varepsilon_0$ is the electric constant, $k$ is the Boltzmann's constant, and $T$ is an absolute temperature. To convert equation (1) into the dimensionless form, appropriate for numerical solutions, Debye length $\kappa^{-1} = \left(2n_0 q_e^2/\varepsilon_0\varepsilon kT\right)^{-1/2}$ is used to normalize the distances and $kT/q_e$ to normalize the electric potential. Only dimensionless quantities are used hereafter in the paper. Then, equation (1) takes the following dimensionless form:

$$\nabla^2\varphi = \sinh\varphi . \tag{2}$$

The PB equation is solved within only a single unit cell due to the spatial periodicity of the crystal. The Wigner–Seitz cell of a bcc lattice is used as a domain for the crystal in equilibrium. In the case of the nonzero strain, the domain is a deformed initial Wigner-Seitz cell. The dielectric permittivity of the electrolyte is assumed to be much greater than the dielectric permittivity of the particle's material that is pertinent to the aqua solutions. Then, the interior of the particle is excluded from the domain due to the approximation of the infinite dielectric permittivity of the electrolyte. This leads to the following (dimensionless) boundary condition on the surface of the particle:

$$-\nabla\varphi\cdot\mathbf{n} = \sigma_0, \tag{3}$$

where $\mathbf{n}$ is an external unit normal vector of the particle.

A set of faces of the domain is resolved into seven pairs of oppositely located faces. The periodic boundary conditions for the electric potential and its gradient are

$$\varphi(\mathbf{r}) = \varphi\left(\mathbf{r}+\mathbf{r}^{(m)}\right), \quad m=1,\dots,7, \tag{4a}$$

$$\nabla\varphi(\mathbf{r})\cdot\mathbf{n}^{(m)} = -\nabla\varphi\left(\mathbf{r}+\mathbf{r}^{(m)}\right)\cdot\mathbf{n}'^{(m)}, \quad m=1,\dots,7. \tag{4b}$$

Here $m$ refers to a pair of opposite faces, $\mathbf{n}^{(m)}$ and $\mathbf{n}'^{(m)}$ are external unit normals to these faces for the $m$-th pair, and $\mathbf{r}^{(m)}$ is a vector of primitive translations separating the faces. In the equilibrium, $\mathbf{r}^{(1)} = (1,0,0)a$, $\mathbf{r}^{(2)} = (0,1,0)a$, $\mathbf{r}^{(3)} = (0,0,1)a$, $\mathbf{r}^{(4)} = \sqrt{3}/2(1,1,1)a$, $\mathbf{r}^{(5)} = \sqrt{3}/2(1,1,-1)a$, $\mathbf{r}^{(6)} = \sqrt{3}/2(1,-1,1)a$, $\mathbf{r}^{(7)} = \sqrt{3}/2(1,-1,-1)a$ in the Cartesian coordinate system. Under strain, vectors $\mathbf{r}^{(m)}$, $m=1,\dots,7$, are transformed accordingly.

Equation (2) and boundary conditions (3) and (4) constitute the boundary value problem for the PB equation on the unit cell. Solution of this problem fully describes the properties of the colloidal crystal within the adopted model in any particular spatial configuration, both equilibrium and deformed.

## 3. Numerical Experiment

The stress-strain relations for the crystals with initial stress are written as follows [3]:

$$T_{ij} = B_{ij}\delta_{ij} + B_{ijkl}\varepsilon_{kl} + \dots, \tag{5}$$

where $T_{ij}$ is the Cauchy stress tensor, $\varepsilon_{kl}$ is the infinitesimal strain tensor, $B_{ij}$ and $B_{ijkl}$ are tensors of elastic constants of the first and second order respectively, $\delta_{ij}$ is the Kronecker's delta-symbol and dots designate the quadratic and higher order terms omitted. The rule of summation over repeating subscripts is implied.

Due to the high symmetry, the bcc crystal has only one independent nonzero elastic constant of the first order, $B_{11}$, and three elastic constants of the second order, $B_{1111}$, $B_{1122}$ and $B_{1212}$. They can be found in two experiments in which the infinitesimal strain tensors have the forms

$$\begin{pmatrix} \varepsilon & 0 & 0 \\ 0 & 0 & 0 \\ 0 & 0 & 0 \end{pmatrix} \text{ and } \begin{pmatrix} 0 & \varepsilon & 0 \\ \varepsilon & 0 & 0 \\ 0 & 0 & 0 \end{pmatrix} \tag{6}$$

respectively. Here $\varepsilon$ is a strain parameter that varies during the experiment. The first experiment provides two stress-strain relations

$$T_{11} = B_{11} + B_{1111}\varepsilon + \dots, \tag{7a}$$

$$T_{22} = B_{11} + B_{1122}\varepsilon + \dots, \tag{7b}$$

while the second experiment gives

$$T_{12} = 2B_{1212}\varepsilon + \dots. \tag{7c}$$

Components of the Cauchy stress tensor $T_{ik}$ for charge stabilized colloids are calculated via the fundamental stress tensor $\Pi_{ij}$ associated with the Poisson-Boltzmann equation [4]:

$$T_{ik} = \frac{1}{V_c} \sum_m r_k^{(m)} \int_{S^{(m)}} \Pi_{ij} da_j \,, \tag{8}$$

where $r_k^{(m)}$ are components of vectors $\mathbf{r}^{(m)}$ described above, $V_c$ is a volume of a unit cell of the crystal and integration is carried out over the surface of the face $S^{(m)}$ indicated by the vector $\mathbf{r}^{(m)}$, $m = 1,\dots,7$, in each pair of opposite faces of the cell. Stress tensor $\Pi_{ij}$ in (8) is

$$\Pi = \nabla\varphi \otimes \nabla\varphi - (\tfrac{1}{2}|\nabla\varphi|^2 + \cosh\varphi - 1)I \,, \tag{9}$$

where $I$ is a unit tensor.

Elastic constants of the crystals with parameters $R = 1$ and $\sigma_0 = 2$ were calculated for the lattice parameter $a$ in range 2.4 to 6.0 in step of 0.1. The unit cell of the crystal was subjected to the two series of deformation described above and corresponding stress-strain relations (7) were obtained. Strain parameter $\varepsilon$ in (6) varied from –0.01 to +0.01 with step 0.001. An appropriate boundary value problem for PB equation was solved numerically for every given $\varepsilon$. The boundary value problems were solved numerically by the finite element method with gradient irregular meshes of tetrahedral elements. Calculations were partly supported by the Supercomputing Center of Lomonosov Moscow State University [5].

The experimental stress-strain curves were then fitted using standard least square method. The best fitting was obtained with cubic polynomials. Coefficients of these polynomials give needed elastic constants $B_{11}$ and $B_{1111}$, $B_{1122}$, $B_{1212}$.

## 4. Results and Discussion

Results for the first and second-order elastic constants $B_{11}$ are shown in figure 1. The first order elastic constant $B_{11}$ is just the negative equilibrium pressure $p$, $B_{11} = -p$. The pressure is shown in figure 1a. It decreases monotonically while the lattice parameter increases. The second-order elastic constants are shown in figure 1b. It is worth to notice that elastic constants $B_{1111}$ and $B_{1122}$ are almost equal to each other. The behavior of the second-order elastic constants is also monotonic. Starting from the larger lattice parameter $a$, the curves rise almost exponentially initially while the lattice parameter decreases. Then the rate of growth gets even larger for smaller lattice parameter, being the largest for the $B_{1212}$ elastic constant. Positivity of all the second order elastic constants means that the colloidal crystal with bcc lattice and cc model of the particles is stable relative to the uniform stretch and shear strains within the present model.

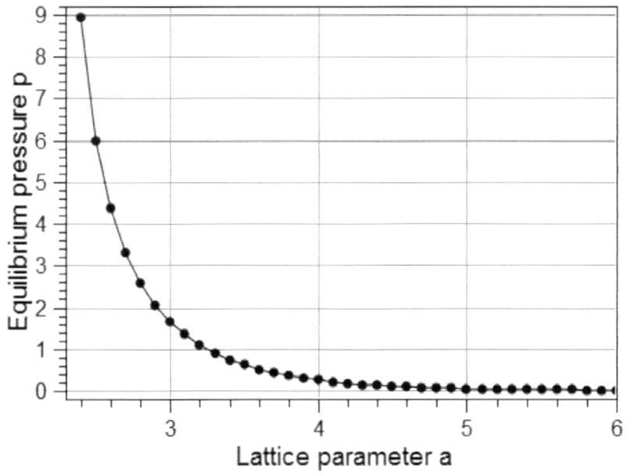


(a)

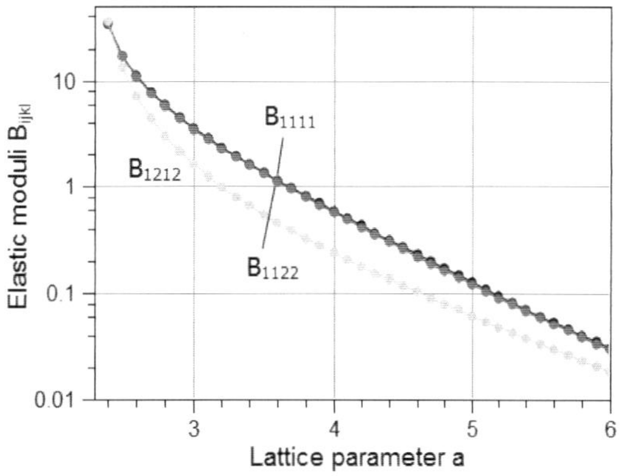


(b)

**Figure 1.** Equilibrium pressure (a) and elastic B-constants of the second order (b).

The coefficients in the series expansion of the second Piola-Kirchhoff stress tensor with respect to the components of the Lagrange strain tensor [6] provide another set of elastic constants. These *C*-constants are related to the B-constants in the following way: $C_{11} = B_{11}$, $C_{1111} = B_{1111} - C_{11}$, $C_{1122} = B_{1122} + C_{11}$, $C_{1212} = B_{1212} - C_{11}$ [3]. The results for the second-order elastic constants $C_{1111}$, $C_{1122}$, $C_{1212}$ are shown in figure 2.

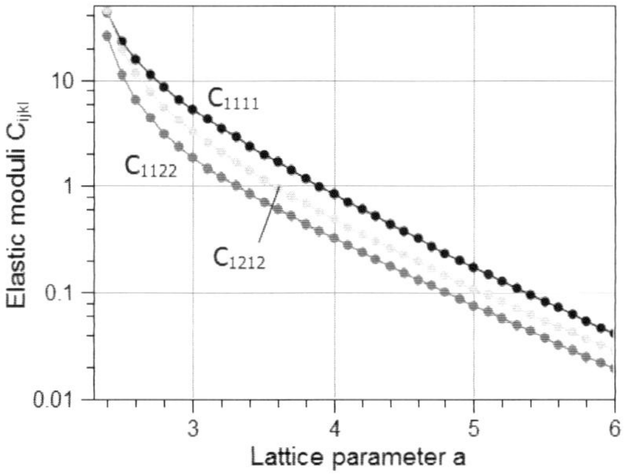


**Figure 2.** Elastic *C*-constants of the second order.

Within the one-component model [2], the complex interaction in the colloidal system is reduced to the effective interaction between the particles only. If the effective interaction is pair central, the Cauchy relation $C_{1122} = C_{1212}$ must hold in a static crystal lattice. In our case, the ratio $C_{1122}/C_{1212}$ as function of the lattice parameter $a$ is shown in figure 3. Some interesting oscillations of the ratio are observed for small lattice parameters. The $C_{1122}/C_{1212}$ ratio definitely deviates from the unity for all values of the lattice parameter and varies in the narrow range 0.55 to 0.7. It is worth to notice that the ratio $C_{1122}/C_{1212}$ remains much less than the unity even for big lattice parameters where the pure pair interaction could be expected. This means that the effective interaction between colloids in the crystal is not purely pairwise, and three- and many-body effective interactions can play a role in the system.

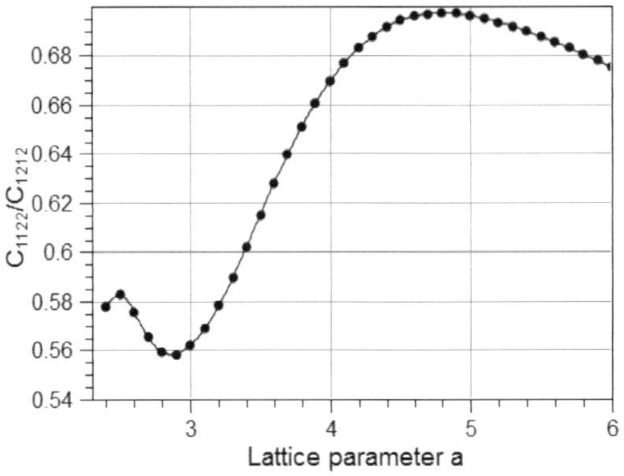


**Figure 3.** Breakdown of the Cauchy relation: the ratio $C_{1122}/C_{1212}$ strongly deviates from unity.

## 5. Conclusion

The model of charge stabilized colloidal crystals with monatomic body-centered cubic crystal lattice and constant surface charge density on the particles is proposed. The properties of the crystals are derived from the numerical solution to the Poisson-Boltzmann nonlinear differential equation. Numerical procedure for obtaining elastic constants from the stress-strain relations is proposed and applied to the crystals with different values of the lattice parameter. The crystal turns out to be stable relative to small uniform strains in the whole range of the crystal densities studied. The breakdown of the Cauchy relation means than many-body effective potential of some kind should be added to pair potential for proper simulation of the colloidal crystals within the one-component model. This can be of importance for disordered systems as well.

## References

[1] Joannopoulos J D, Villeneuve P R and Fan S H 1997 *Nature* **386** 143.
[2] Belloni L 2000 *J. Phys.: Condens. Matter* **12** R549.
[3] Barron T H K and Klein M L 1965 *Proc. Phys. Soc.* **85** 523.
[4] Dyshlovenko P E 2010 *Colloid J.* **72** 627.
[5] Sadovnichy V, Tikhonravov A, Voevodin Vl, Opanasenko V 2013 *"Lomonosov": Supercomputing at Moscow State University*. In Contemporary High Performance Computing: From Petascale toward Exascale (Chapman & Hall/CRC Computational Science), (Boca Raton, USA: CRC Press) pp 283-307.
[6] Holzapfel G A 2000 *Nonlinear Solid Mechanics: A Continuum Approach for Engineering* (Chichester: Wiley).

IOP Conf. Series: Journal of Physics: Conf. Series **955** (2017) 012012 doi:10.1088/1742-6596/955/1/012012

# The Different Types of Turbulence in Rotating Spherical Layers

**D Zhilenko[1], O Krivonosova[1] and M Gritsevich[2,3,4]**

[1] Moscow State University, GSP-1, Leninskie Gory, 119991,Moscow, Russia
[2] Department of Physics, University of Helsinki, Gustaf Hällströmin katu 2a, FI-00014 Helsinki, Finland
[3] Dorodnicyn Computing Center, RAS, Vavilova str. 40, 119333, Moscow, Russia
[4] Institute of Physics and Technology, Ural Federal University, Mira str. 19, 620002, Ekaterinburg, Russia

E-mail: jilenko@imec.msu.ru

**Abstract**. Turbulent flows of a viscous incompressible fluid in a layer between rotating concentric spheres under the action of the modulation of the velocity of one of the spheres have been studied experimentally and numerically. The form of spectra of turbulent pulsations of the azimuthal velocity depends on the sphere whose rotational velocity is modulated, as well as on the amplitude and frequency of modulation. The possibility of the formation of turbulence with spectra qualitatively similar to spectra obtained in measurements in the upper atmosphere is established: with the slope close to –3 at low frequencies and close to –5/3 at high frequencies and with the negative longitudinal velocity structure function of the third order. It has been shown that such spectra are formed in the regions of a flow that are strongly synchronized under the action of the modulation of the rotational velocity.

## 1. Introduction

Large scale flows in the atmosphere occur in the presence of fast rotation of the Earth, and their properties are usually explained within the concept of “two-dimensional” turbulence [1,2]. In two-dimensional turbulence, two inertial intervals are usually identified corresponding to energy transfer at low wave numbers and enstrophy transfer at high wave numbers [3]. Energy and enstrophy supply to a flow occurs owing to external forces with wave numbers between these intervals. The inertial interval of energy transfer from high to low wave numbers (inverse cascade) is described by the same Kolmogorov relation as in three-dimensional turbulence [4] for the dependence of the energy spectrum $E(k)$ on the wave number $k$: $E(k) \sim k^{-5/3}$. In the inertial interval of enstrophy transfer from low to high wave numbers (direct cascade), this dependence has the form $E(k) \sim k^{-3}$. The direction of the cascade is determined by the sign of the third order longitudinal velocity structure function [5], which is defined as $D_{LLL} = \langle [u(l) - u(l)]^{-3} \rangle$, where $u$ is the velocity at the spatially separated points $l$ and $l$ and angular brackets mean averaging over the ensemble of realizations. The negative and positive signs of $D_{LLL}$ correspond to the direct and inverse cascades, respectively. Conclusions of the theory of two-dimensional turbulence, in particular, the formation of spectra with a slope of –5/3 and –3 at large and small scales, respectively, were confirmed in numerous results reviewed in [6,7]. At the same time, several year measurements of the horizontal velocity of the wind in upper layers of the Earth’s atmosphere, performed during several years, revealed an anomalous location of spectral regions that is inconsistent with the theory of two-dimensional turbulence. In particular, spectra of turbulence with a slope of –3 begin at scales larger than 700 km and are limited by a strong peak at a scale of $10^4$ km. Spectra with a slope –5/3 were detected at scales smaller than 500 km [1]. Analysis of third order structure functions in [2] showed that only one of these regions with the slope of –3, corresponds to two-dimensional turbulence. This indicates the direct energy transfer cascade in both spectral regions under consideration. Despite the existing explanations [1, 8, 9]reasons for the inverse position of spectral regions, as well as the possibility of reproducing this phenomenon under laboratory conditions, are as yet unclear. Both viscous dissipation [10]and vertical motions that are components of large scale circulation [6] prevent two-dimensional turbulent flows in the atmosphere. Large scale circulation also exists in turbulent flows induced by the rotation of the boundaries of the spherical layer, which is responsible for the motion of viscous incompressible fluid sandwiched between them [11]. It is exactly why the model spherical Couette flow is studied in this work for the qualitative simulation of processes in the atmosphere. By analogy with [12], where transitions between two- and three-dimensional turbulence were studied in the presence of the azimuthal jet counter to the direction of the rotation of cylinders, we chose the case of the counter rotation of spheres. Under stationary boundary conditions, oppositely directed vortices with an interface between them are formed in the meridional plane of such a flow (see figure 1, which is similar to figure 1 in [13]). A similar circulation can be observed in the case of the rotation of only the inner sphere in the presence of altitude-inhomogeneous external heating [14], typical of the atmosphere. In spherical layers the formation of turbulence with a high correlation dimension occurs by the increase in the rotation velocity of one of the boundaries [11, 15] as well as by their modulation [16]. The spectrum of developed turbulence in the latter case depends on the parameters of force action [17].

## 2. Method of calculation

An isothermal flow of a viscous incompressible fluid is described by the Navier–Stokes and continuity equations:

$$\frac{\partial U}{\partial t} = U \times rotU - grad\left(\frac{p}{\rho} + \frac{U^2}{2}\right) - \nu rotrotU, \; divU = 0,$$

where $U$, $p$, ν, and ρ are the velocity, pressure, viscosity, and density of the fluid, respectively. These equations were numerically solved in a spherical coordinate system where the impermeability and no-slip conditions for the azimuthal $u_\varphi$, radial $u_r$, and polar $u_\theta$ components of the velocity have the form $u_\varphi(r = r_{1,2}) = \Omega_{1,2}(t)r_{1,2}\sin\theta$, $u_r(r = r_{1,2}) = 0$, $u_\theta(r = r_{1,2}) = 0$, where subscripts 1 and 2 correspond to the inner and outer spheres, respectively. We used an algorithm of numerical solution [18] based on a conservative finite difference scheme of the discretization of the Navier–Stokes equations in space and semi-implicit Runge–Kutta scheme of the third order integration accuracy in time. Discretization in space was performed on grids nonuniform in $r$ and θ directions with concentration near the boundaries and equatorial plane and the total number of nodes $5.76 \times 10^5$. The sensitivity of the results to the parameters of grid was analyzed in detail in [13], [19]. This algorithm was used for calculations with both stationary [11] and periodic [19] boundary conditions. Agreement was shown to be between the experimental and calculated results, including the integral properties of turbulent flows. $S-$ spectra of pulsations of the square of the azimuthal velocity component $u_\Phi$ (minus the average value determined for the entire sample) were calculated at points *1*–*7* shown in figure 1 ($\theta$ and $\varphi$ are constant and only $r$ is varied). To this end, $u_\varphi$ time series with a length of no less than 72000 points were written with a time step $\Delta t = 0.015$–$0.025$ s. $D_{LLL}$ was obtained using the dependence of $u_\varphi$ on the azimuth angle $\varphi$ during 16 rotation periods ($0 \leq \varphi \leq 32\pi$). All calculations were performed for the initial and boundary conditions corresponding to the experimental conditions.

## 3. Experimental setup

The experimental setup consisted of two coaxial spheres with inner and outer radii $r_1 = 0.075$ m and $r_2 = 0.150$ m. The space between the spheres was filled with silicone oil with viscosity $\nu = 50 \times 10^{-6}$ m$^2$/s at the temperature of the working fluid of 22°C to which aluminum powder was added for visualization of flows. The rotation velocity was periodically varied by the law $\Omega_k(t) = \Omega_{k0}[1 + A_k \sin(2\pi f_k t + \Phi_k)]$ with an accuracy of no worse than 0.5% (where $A_k$ and $f_k$ are the amplitude and frequency of modulation; $\Omega_{k0}$ is average angular velocity of rotation; initial phase $\Phi_k$ is arbitrary). The modulation frequencies $f_1 = 0.01 - 0.1$ Hz and $f_2 = 0.01 - 0.02$ Hz were no higher than the average rotation frequencies of the spheres ($\Omega_{10}/2\pi = 0.59$ Hz, $\Omega_{20}/2\pi = 0.32$ Hz). The measurements of $u_\varphi$ were performed by a SDS 01.11 laser anemometer with the allowable velocity interval of 0.005–1 m/s and a data sample rate of 20.16 Hz. The measurement point was located near point *7* in figure 1. The experiments were performed at Reynolds numbers $\mathrm{Re}_1 = \Omega_{10} r_1^2/\nu = 412.5 \pm 0.5$ and $\mathrm{Re}_2 = \Omega_{20} r_2^2/\nu = -900 \pm 1$. At these Reynolds numbers in the absence of modulation, a periodic flow with the frequency $f_0 = 0.0376$ Hz is formed in the layer; this flow referred to as initial is a result of mutual synchronization of individual linear modes [13]. The initial flow has the form of traveling azimuthal waves with the wave number $m = 3$. The modulation of the rotation velocity of one of the boundaries leads to the flow induced synchronization. With an increase in the amplitude of modulation at a fixed frequency, the initial flow is destroyed. Turbulence appears at the transition from mutual synchronization to induced synchronization [16]. More detailed description of the setup and experimental technique can be found in [16].

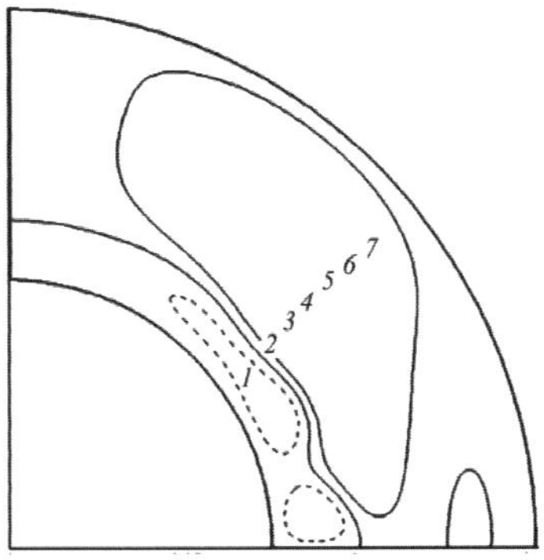


**Figure 1.** Calculated stream functions $\psi$ (in cubic meters per second) (see Zhilenko and Krivonosova [13]) in the meridional plane of the axisymmetric steady state flow at $\mathrm{Re}_2 = -900$, $\mathrm{Re}_1 = 414$: $\psi_{max} = 6 \times 10^{-6}$, $\psi_{min} = -6 \times 10^{-6}$, and $\Delta\psi_{max} = 6 \times 10^{-6}$. Dashed lines are negative value contours. Points *1*–*7* are located at the relative distance $l = (r - r_1)/(r_2 - r_1) = 0.135$, 0.246, 0.359, 0.484, 0.611, 0.7, and 0.803 from the inner sphere (where $r_1$ and $r_2$ are the radii of the inner and outer spheres) with a deviation of $0.206\pi$ from the equatorial plane.

## 4. Results

The processing of the results of measurements of $u_\varphi$ shows that the form of spectra $S$ near the threshold of formation of turbulence is independent of the modulation frequency. In this case, the spectra have a constant slope in the frequency range limited from below by the largest of the quantities $f_0$ and $f_k$. In the case of the modulation of the rotation velocity of the inner sphere, the slope of the spectra is in the range between –5/3 and –3 and approaches –5/3 with an increase in the amplitude. With a further increase in the modulation amplitude, the spectra can be transformed to the form characteristic of two-dimensional turbulence. For modulation of $\Omega_1(t)$ at $f_1 \leq f_0$ (figure 2a), the spectra obtained both from measurements and numerically exhibit a pronounced segment with a slope of –5/3 at low frequencies (0.06–0.27 Hz) and a segment with a slope of –3 at high frequencies (0.27–0.8 Hz).

IOP Conf. Series: Journal of Physics: Conf. Series **955** (2017) 012012 doi:10.1088/1742-6596/955/1/012012

With an increase in $f_1 \geq f_0$, the form of the spectrum is transformed (figure 2b) and only the segment with a slope of –3 remains between the modulation frequency $f_1$ and the end of the inertial interval.

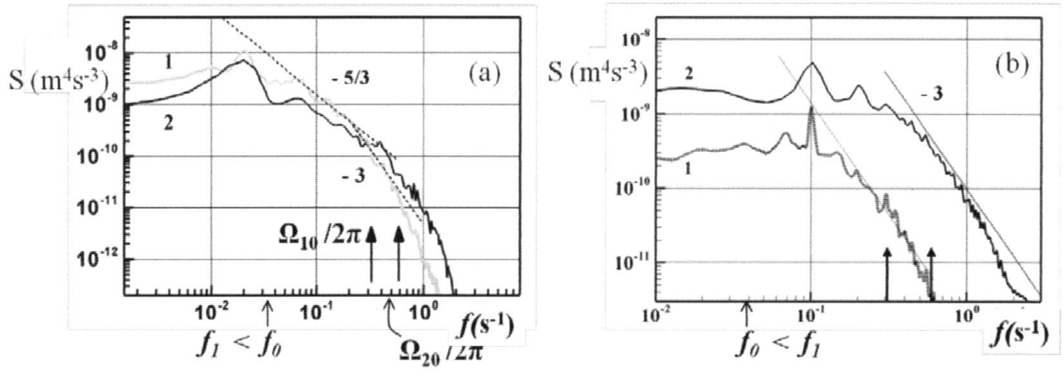


Figure 2. Inner sphere modulation. Spectra at point *7* (see figure 1). (a) - $f_1$ = 0.01 Hz, $A_1$ = 0.163; (b) - $f_1$ = 0.1 Hz, $A_1$ = 0.217. 1 – experiment and 2 – calculation.

In the case of the modulation of $\Omega_2(t)$ near the transition to turbulence, spectra with a slope in the range between –5/3 and –3 are observed. At an increase in the amplitude, the spectra are modified to the form qualitatively corresponding to the spectra of atmospheric turbulence [1] with a slope –3 at frequencies below 0.1 Hz and –5/3 at higher frequencies (0.1–0.31 Hz) (figure 3a). A further increase in the amplitude can result in spectra with a constant slope between –5/3 and –3. Under the condition $f_k \leq f_0$, the form of the spectrum depends on the position of the point at which the azimuthal velocity is calculated. In the single studied case $f_1 \geq f_0$, the spectra were spatially uniform. The most characteristic differences in the form of the spectra at points *1–7* (figure 1) are observed in the case of $\Omega_2(t)$ modulation. In particular, near the outer sphere and at a certain distance from it (points *7–3*), the observed spectra are typical to atmospheric turbulence, whereas the spectrum observed near the inner sphere (point *1*) has a constant slope of -5/3 and is typical to three-dimensional turbulence (figure 3b). We tried to determine the direction of the energy cascade in the cases corresponding to the spectra shown in figure 2 and 3 from the sign of the third order longitudinal velocity structure function.

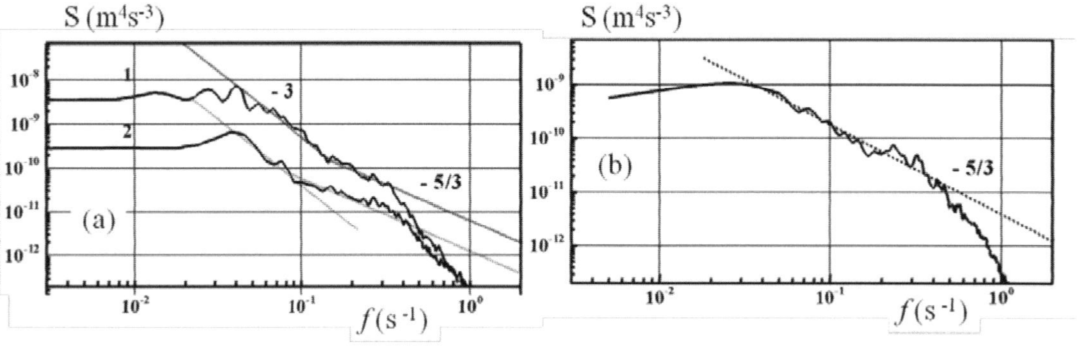


Figure 3. Outer sphere modulation with $f_2$ = 0.02 Hz, $A_2$ = 0.2. Spectra: (a) – at point *7* (1- experiment and 2- calculation); (b) – at point *1* (see figure 1).

The sign of the quantity under consideration alternates with a period of $2\pi/3$, because large scale coherent structures [11] characteristic of the initial flow are held in the turbulent flow. Similar large scale coherent structures in the upper layers of the atmospheres of planets (e.g., Venus) were assumingly interpreted as Rossby waves [20]. For this reason, to determine the sign of $D_{LLL}$, the results of the calculation were approximated by sixth order polynomials. figure 5 shows the dependence of $D_{LLL}$ on the frequency $f$ given by the expression [21] $f = \langle u_\Phi \rangle / l$, where $0 < l < 32\pi r$ and $\langle u_\Phi \rangle$ is the average velocity at a distance of $r\sin\theta$ from the axis. We first consider flows for which the observed spectra were typical to two-dimensional (figure. 2a) and three-dimensional (figure 3b) turbulence. In the former case (figure 4, line *1*), transition from positive $D_{LLL}$ values to negative is observed at $f = 0.2$ Hz. At the same frequency, transition from a slope of –3 to a slope of –5/3 is observed in the experiment (figure 2, line *1*). In the latter case (figure 4, line *2*) $D_{LLL} < 0$. Both of these cases confirm the correctness of the estimate of the sign of $D_{LLL}$. In the case of atmospheric turbulence (figure 4, line *3*), $D_{LLL} < 0$ in the frequency range corresponding to the segments of the spectrum with both slopes of –3 and –5/3. Thus, the direct energy transfer cascade is observed in both segments of the inertial intervals in complete agreement with the results of processing of natural measurements in [2].

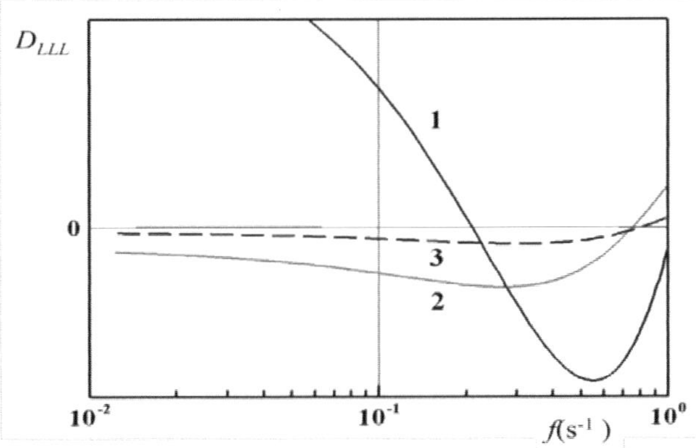


Figure 4. Approximation of the third-order longitudinal velocity structure function $D_{LLL}$ in arbitrary units for (*1*) $f_1 = 0.01$ Hz, $A_1 = 0.163$ and (*2*, *3*) $f_2 = 0.02$ Hz, $A_2 = 0.2$ at points *7*(*1*, *3*) and *1* (*2*).

The results obtained for the model flow under consideration imply that the form of spectra of turbulence in the upper layers of the atmosphere is explained by the induced synchronization of the periodic part of atmospheric flows (e.g., Rossby waves) by an external periodic action with a longer period. Since the main source of the energy for all atmospheric processes is solar heat, seasonal variations of this quantity can be considered as such a periodic external action on the atmosphere.

## 5. Conclusions

The results of the performed experimental and numerical studies have shown that a decrease in the modulation frequency is accompanied by an increase in differences in the behaviors of the azimuthal and meridional components of the kinetic energy of the flow. The former component remains periodic, whereas the latter component changes the periodic behavior to chaotic. The suppression of turbulence of the azimuthal kinetic energy of the flow promotes the formation of quasi-two-dimensional turbulence. The form of spectra of turbulent pulsations of the azimuthal velocity depends on the sphere whose rotation velocity is modulated, as well as on the amplitude and frequency of modulation. Spectra characteristic of two-dimensional turbulence with a constant slope of –5/3 and an inverse cascade ($D_{LLL} > 0$) at low frequencies and with a slope –3 and a direct cascade ($D_{LLL} < 0$) at high frequencies have been observed in the case of the modulation of the inner sphere velocity. At a modulation frequency below the frequency of the initial periodic flow, the form of the spectra is spatially nonuniform. In the case of the modulation of the outer sphere velocity, spectra with the qualitative form characteristic to turbulence in the upper layers of the atmosphere with a constant slopes of –3 and –5/3 at low and high frequencies, respectively, are observed in the region of circulation induced by the outer sphere. For both segments of the inertial interval $D_{LLL} < 0$. The form of the spectrum near the inner sphere is characteristic of three-dimensional turbulence: the segment with a constant slope of –5/3 presents and $D_{LLL} < 0$. It has been found that the level of synchronization between the rotation velocity of the boundary and the velocity of the flow is different in all flows considered above. The lowest and highest levels of synchronization are observed where spectra are similar to spectra of three-dimensional and atmospheric turbulence, respectively.

## Acknowledgments

This work was supported by the Russian Foundation for Basic Research project No. 160500004. MG acknowledges support from the ERC Advanced Grant No. 320773, and the Russian Foundation for Basic Research, project No. 16-07-01072. Research at the Ural Federal University is supported by the Act 211 of the Government of the Russian Federation, agreement No 02.A03.21.0006.

## References

[1] Gage K S and Nastrom G D 1986 *J. Atmos. Sci.* **43** 729
[2] Lindborg E 1999 *J. Fluid Mech.* **388** 259
[3] Kraichnan R 1967 *Phys. Fluids* **10** 1417
[4] Kolmogorov A N 1941 *Dokl. Akad. Nauk SSSR* **32** 19
[5] Frisch U 1995 *Turbulence: The Legacy of A.N. Kolmogorov* (Cambridge: Cambridge Univ. Press) p 510
[6] Danilov S D and Gurarii D 2000 *Phys. Usp.* **43** 863
[7] Boffetta G and Ecke R E 2012 *Ann. Rev. Fluid Mech.* **44** 427
[8] Xia H, Burne D, Falkovich G, and Shats M 2011 *Nature Phys.* **7** 321
[9] Vallgren A, Deusebio E, and Lindborg E 2011 *Phys. Rev. Lett.* **107** 268501
[10] Gledzer A E, Gledzer E B, Khapaev A A and Chkhetiani O G 2011 *J. Exp. Theor. Phys.* **113** 516
[11] Zhilenko D Yu and Krivonosova O E 2010 *Tech. Phys.* **55**, 449
[12] Baroud C N and Swinney H L 2003 *Physica D* **184** 21
[13] Zhilenko D Yu and Krivonosova O E and Nikitin N V 2007 *Fluid Dyn.* **42** 886
[14] Yavorskaya I M 1975 *Izv. Akad. Nauk SSSR, Mekh. Zhidk. Gaza* **2** 15
[15] Belyaev Yu N, Monakhov A A, Shcherbakov S A and Yavorskaya I M 1978 *JETP Lett.* **29** 295
[16] Zhilenko D Yu and Krivonosova O E 2013 *Fluid Dyn.* **47** 452
[17] Gordienko S N and Moiseev S S 1999 J. *Exp. Theor. Phys.* **89** 880
[18] Nikitin N 2006 *J. Comp. Phys.* **217** 759
[19] Zhilenko D Yu and Krivonosova O E 2013 *Tech. Phys. Lett.* **39** 84
[20] Monin A S 1999 *Hydrodynamics of Atmosphere, Ocean and Earth's Depths* (St.Petersburg: Gidrometeoizdat) p 674
[21] Golitsyn G S 2007 *Dokl. Phys.* **52** 376

IOP Conf. Series: Journal of Physics: Conf. Series **955** (2017) 012013 doi:10.1088/1742-6596/955/1/012013

# An integral model of a convective jet with a pressure force and forms of vertical fluxes in the atmospheric surface layer

**A N Vulfson**[1,2], **P V Nikolaev**[3]

[1]Institute of Oil and Gas Problems, Russian Academy of Sciences, ul. Gubkina 3, 119333 Moscow, Russian Federation
[2]National Research University Higher School of Economics, ul. Myasnitskaya 20, 101000 Moscow, Russian Federation
[3]National University of Science and Technology MISIS, Leninsky pr. 4, 119991 Moscow, Russiam Federation

vulfson@ipng.ru

**Abstract**. In the article, it is considered a modification of an integral model of an unsteady turbulent jet with a pressure force. Stationary solutions of the presented model are compared with well-known analytical results of classical models. It is shown that the inclusion of the pressure forces changes the dynamic parameters of a jet by about 12%. Analytical solutions of a steady forced buoyant jet of the atmospheric convective boundary layer and a spontaneous jet of a surface layer are presented. The simplest model of an ensemble of spontaneous jets of convective surface layer is constructed. It is shown that an ensemble of spontaneous jets forms a dependence of the turbulent moments and heat eddy diffusivity on the altitude within the convective boundary layer.

## 1. Introduction

Integral models of convective jets (plumes) are widely used for a description of volcanic activity and turbulent mixing within the boundary layers of the atmosphere and ocean.

Most of the well-known integral models of convective jets (e.g.: [1-7]) use the approximation of the vertical boundary layer. It enables the pressure force in the equation of motion to be neglected.

The present paper discusses modifications of the integral model of a convective jet by including the pressure force. The modifications have exact solutions that can be compared against the analytical solutions of classical models. It allows us to estimate the effect of a pressure forced, that changes the vertical velocity and the buoyancy of a jet by about 15%.

In geophysical environments, we can observe both isolated convective jets [8] and ensemble of buoyant jets [9]. An ensemble of convective thermals forms in the atmosphere over a heated surface of the land or ocean. A steady spontaneous jet and its analytical solution may be considered as a simplest model of an isolated element of the ensemble of thermals. It is shown that an ensemble of the dynamically identical spontaneous jets forms the vertical profile of turbulent eddy diffusivity and the profiles of the turbulent moments of velocity up to fourth order.

## 2. Integral model of a convective jet in a neutrally stratified atmosphere

Suppose that a convective jet forms over a point source of heat and buoyancy. Experimental measurements in a neutrally stratified environment show that the jet has a near-conical shape with the radius $R_w$. The shape of the cone is defined by $\partial R_w / \partial z = \alpha_R$, where $\alpha_R \approx 0.1$ is a constant coefficient. The hypothesis about a conical shape of a jet is a part of the integral model [4].

The vertical velocity and buoyancy in each horizontal cross section of the jet are assumed to have stepwise $\Pi$-shaped profiles with amplitudes $\hat{w}$ and $g\hat{\theta}$. Amplitude equations of the model [3, 4] in the vertical boundary-layer approximation with the hypothesis about a conical shape of a jet are

$$\begin{cases} \dfrac{\partial}{\partial t}\hat{w}R_w^2 + \dfrac{\partial}{\partial z}\hat{w}^2 R_w^2 = g\hat{\theta}R_w^2, \\ \dfrac{\partial}{\partial t}\hat{\theta}R_w^2 + \dfrac{\partial}{\partial z}\hat{\theta}\hat{w}R_w^2 = 0, \quad \dfrac{\partial}{\partial z}R_w = \alpha_R \end{cases} \tag{1}$$

The boundary conditions for the point sources of heat and momentum are

$$\lim_{z\to 0}\hat{w}^2R_w^2 = P_*(t), \quad \lim_{z\to 0}\hat{\theta}\hat{w}R_w^2 = S_*(t) \tag{2}$$

Here, $P_* = P_*(t) > 0$ and $S_* = S_*(t) > 0$ are the time dependent strengths of momentum and heat fluxes, $[P_*] = m^4/s^2$, $[S_*] = m^3/s$.

A comparison between the numerical calculation by the model [3, 4] and other known integral models [1-5] was made in [5]. Those results are shown in Fig. 1. Here we use normalized vertical velocity $\tilde{w}$, buoyancy $g\tilde{\theta}$ and altitude $\tilde{z}$ which description provides in models [2-5].

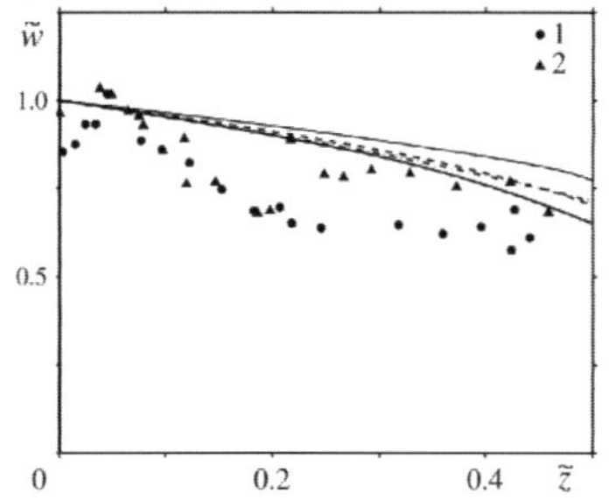


**Figure 1:** A comparison between integral models [1-5] and experimental data [2]. The normalized velocity $\tilde{w}$ field. Circles and triangular dots are experimental data. Dash-and-dot line – model [1]; solid line – model [2]; thin solid line – model [3,4]; dotted line – model [5].

Let us consider an alternative way of establishing the amplitude equations of unsteady convective jet that does not use the approximation of the vertical boundary layer.

According to [10], in the structure of a turbulent jet, we can distinguish a coherent system consisting of a sequence of moving eddies that can be stylized as a chain of spherical thermals (Fig. 2).

Let us consider an equation of the isolated thermal in the chain. Suppose that $V_w = V_w(t)$ is a volume of the spherical thermal with radius $R_w = R_w(t)$; $\bar{w} = \bar{w}(t)$, $\bar{\theta} = \bar{\theta}(t)$ are an averaged mean velocity and dimensionless potential-temperature fluctuation and $\bar{z} = \bar{z}(t)$ is the height of the centre of mass of a thermal. The equation of the motion of a chain spherical thermal was suggested in [10]

$$\frac{d}{dt}\bar{w} = \frac{2}{3}g\bar{\theta} - \frac{\alpha_R}{R_w}\bar{w}^2, \quad \frac{d}{dt}\bar{w} = \frac{2}{3}g\bar{\theta} - \frac{1}{\bar{z}}\bar{w}^2, \tag{3}$$

The factor $2/3$ in (3) is related to introduction of a pressure force, that is parameterized by the added-mass, (see [11]).

The concept of a jet as a chain of thermals allows us to construct an alternative model of a convective jet. Instead of (1) we consider a modified system that include a parametric form of the pressure force

$$\begin{cases} \dfrac{\partial}{\partial t}\hat{w}R_w^2 + \dfrac{1}{2}\dfrac{\partial}{\partial z}\hat{w}^2R_w^2 = \dfrac{2}{3}g\hat{\theta}R_w^2, \quad \dfrac{\partial}{\partial t}\hat{\theta}R_w^2 + \dfrac{\partial}{\partial z}\hat{\theta}\hat{w}R_w^2 = 0, \quad \dfrac{\partial}{\partial z}R_w = \alpha_R \\ \lim\limits_{z\to 0}\hat{w}^2R_w^2 = P_*(t), \quad \lim\limits_{z\to 0}\hat{\theta}\hat{w}R_w^2 = S_*(t) \end{cases} \tag{4}$$

Suppose that $d/dt = \partial/\partial t + \hat{w}\partial/\partial z$ is an individual time derivative. When $\bar{w} = \hat{w}$ and $g\bar{\theta} = g\hat{\theta}$, equation of impulse (4) turns into the form (3).

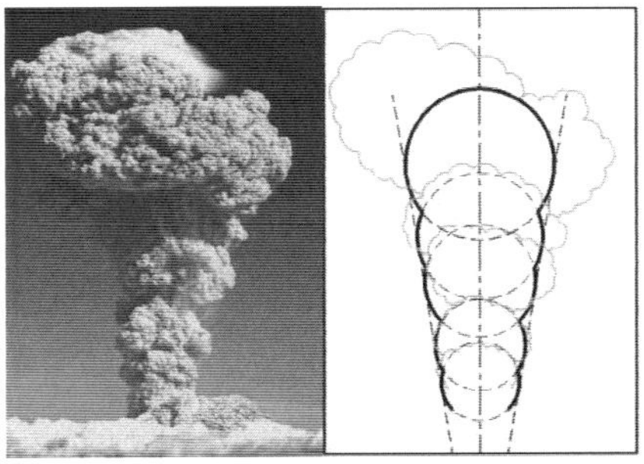

**Figure 2.** Internal coherent structure of a turbulent jet and its geometric representation. (a) a convective turbulent ash plume from the eruption of Shiveluch volcano (Kamchatka), September 24, 2014; (b) geometric representation of a convective plume as a chain of spherical thermals.

The equation (4) includes an approximation of pressure force and therefore cannot be derived in the vertical boundary-layer approximation.

The Euler form of the modify stationary equations (4) with the boundary conditions (2) is

$$\begin{cases} \frac{1}{2}\frac{d}{dz}\hat{w}^2R_w^2 = \frac{2}{3}g\hat{\theta}R_w^2, \quad \frac{d}{dz}g\hat{\theta}\hat{w}R_w^2 = 0, \quad R_w = \alpha_R z, \\ \lim\limits_{z\to 0}\hat{w}^2R_w^2 = P_*, \quad \lim\limits_{z\to 0}g\hat{\theta}\hat{w}R_w^2 = gS_* \end{cases} \tag{5}$$

The system (5) is considered on the unbounded domain $0 < z < \infty$.

## 3. Steady convective, buoyant and spontaneous jets with a pressure force

Consider a steady convective jet from a point source of buoyancy $S_* > 0$ and $P_* = 0$. The analytical solution to the modified model of a steady jet (4) is

$$R_w(z) = \alpha_R z, \quad \hat{w}(z) = \alpha_R^{-\frac{2}{3}}\left(gS_*\right)^{\frac{1}{3}} z^{-\frac{1}{3}}, \quad g\hat{\theta}(z) = \alpha_R^{-\frac{4}{3}}\left(gS_*\right)^{\frac{2}{3}} z^{-\frac{5}{3}} \tag{6}$$

A comparison between solution of the model (1) and the expression (6) was presented in [10]. It shows that the inclusion of the pressure forces in the model decreases the amplitude of vertical velocity and increases the amplitude of buoyancy by about 10%.

The power-law dependences of the vertical velocity and buoyancy on the altitude correspond to the similarity law for the convective jet, presented in [12].

Buoyant jets are of much interest for problems of volcanic outbursts. An analytical solution to the volcanic outburst problem (5) is found as

$$R_w(z) = \alpha_R z, \quad \hat{w}^2(z) = \frac{1}{\alpha_R^2 z^2}\left[P_*^{3/2} + \alpha_R gS_* z^2\right]^{2/3}, \quad g\hat{\theta}(z) = \frac{gS_*}{\alpha_R z}\left[P_*^{3/2} + \alpha_R gS_* z^2\right]^{-1/3} \tag{7}$$

A solution similar to (7) also exists in the model described in [13]. When $P_* = 0$ equalities (7) turn into equalities of convective jet (6).When $g = 0$ equalities (7) turn into equalities of submerge jet [14].

The free-convection sublayer in developed turbulence is the fluid or gas layer adjacent to a heated surface. The surface sublayer has thickness $0 < z/h < 0.1$; where $h$ is the height of the convective layer. It is also described by a constancy of buoyancy flux per unit area $gS_\theta = const > 0$, where $[gS_\theta] = m^2/s^3$.

Let $\Gamma(z) = d\bar{\theta}/dz \le 0$ be a stratification parameter. Then,

$$g\Gamma(z) = g\frac{d\bar{\theta}}{dz} = -\lambda_\theta\left(gS_\theta\right)^{2/3} z^{-4/3} \tag{8}$$

where $0.9 \le \lambda_\theta \le 1.1$, according to the atmospheric field observations (e.g., [15, 16]).

Equality (8) follows from the classical Prandtl turbulence theory [17].

Spontaneous jets arise in unstable atmospheric layers, for more details see [18]. They gain energy from the unstable layer and so do not require any additional heat and momentum sources at the underlying surface.

Extending the modified integral model (5) to the case of unstable stratification, we obtain equations for a steady isolated spontaneous jet that include pressure. Assuming $\lambda_\theta = 1$ in the form (8) we have

$$\begin{cases} \frac{1}{2}\frac{d}{dz}\hat{w}^2 R_w^2 = \frac{2}{3} g\hat{\theta} R_w^2, \quad \frac{d}{dz} g\hat{\theta}\hat{w} R_w^2 = \left(gS_\theta\right)^{2/3} z^{-4/3}\hat{w} R_w^2, \quad R_w = \alpha_R z, \\ \lim_{z\to 0} \hat{w}^2 R_w^2 = 0, \quad \lim_{z\to 0} g\hat{\theta}\hat{w} R_w^2 = 0 \end{cases} \tag{9}$$

The analytical solution of (9) is

$$R_w(z) = \alpha_R z, \quad \hat{w}(z) = \frac{1}{2}\left(gS_\theta\right)^{1/3} z^{1/3}, \quad g\hat{\theta}(z) = \frac{1}{2}\left(gS_\theta\right)^{2/3} z^{-1/3} \tag{10}$$

Equations (10) are consistent with the similarity law for a spontaneous jet obtained in [18].

The exact solution (10) that include a pressure force can be compared with results of a classical integral model [18], that based on the vertical boundary layer approximation. The comparison results show that that the inclusion of the pressure forces in the model conserves the buoyancy and increases the amplitude of vertical velocity by about 15%.

## 4. Ensemble of dynamically identical jets and turbulent diffusivity within a convective boundary layer

It is known that in a turbulent convective layer an ensemble of convective elements forms over a horizontally homogenous heated surface (see [9]). An idea of what an ensemble of thermals looks like can be gained from laboratory experiments at large Rayleigh numbers. The results of laboratory experiments [19], presented in Fig. 3, clearly exhibit the fine structure of the convective layer.

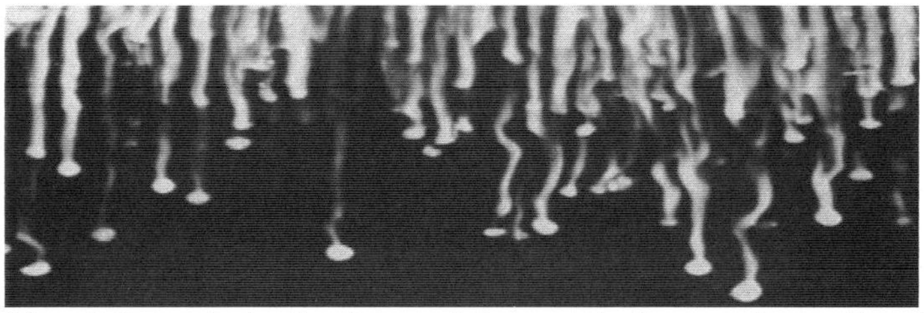

**Figure 3.** Ensemble of thermals in the form of dense salt fingers descending in a water layer, according to [19].

Let $h$ be the height of a convective layer. It should be noted that the ensemble of convective jets existed mainly within the lower part of a convective layer $0 < z/h < 0.5$.Within the upper part of a layer $0.5 < z/h < 1$, the amount of jets is low.

We construct a “generalized” convective jet based on the hypothesis that within the surface layer $0 < z/h < 0.1$ parameters of the “generalized” jet are defined as parameters of the spontaneous jet (10) ; and within the mixing layer $0.1 < z/h < 0.5$ they defined as parameters of the convective jet (6).

Using the results of laboratory experiments [19] and the ensemble model [20], we assume that the convective surface layer is packed with the system of identical “generalized” convective jets.

Suppose that $gS_\theta$ is a buoyancy flux at the underlying surface. According to [21], we introduce Deardorff parameters of velocity and buoyancy

$$w_D = h^{1/3}\left(gS_\theta\right)^{1/3}, \quad g\theta_D = h^{-1/3}\left(gS_\theta\right)^{2/3} \tag{11}$$

Based on the concept of statistical ensemble, we deduce a relationship for the heat eddy diffusivity. In the framework of the semi-empirical theory of turbulence by Prandtl [17], it is assumed that

$$< K_h > = < w > l_P, \quad \frac{l_P}{h} = k_v \frac{z}{h}\left(1 - \frac{z}{h}\right)^2 \tag{12}$$

Where $< w >$ is the average velocity of updrafts; $l_P$ is Prandtl mixing length; $k_v = 0.4$ is von Karman constant.

Substituting the vertical velocity of the “generalized” convective jet into (12) and rewriting deduced relationships with regard to Deardorff parameters (11) yield

$$\frac{<K_h>}{w_D h} \sim \begin{cases} (z/h)^{2/3}\left[1-(z/h)\right]^2 & when \quad 0.1<z/h<0.5. \\ (z/h)^{4/3} & when \quad 0<z/h<0.1 \end{cases} \tag{13}$$

Well-known approximations of the real heat eddy diffusivity $K_h$ are

$$\frac{K_h}{w_D h} \sim \begin{cases} (z/h)\left[1-(z/h)\right]^2 & when \quad 0.1<z/h<0.5. \\ (z/h)^{4/3} & when \quad 0<z/h<0.1 \end{cases} \tag{14}$$

An approximation for $K_h$ in the surface layer $0<z/h<0.1$ was constructed in [22] within the framework of similarity theory. An approximation for $K_h$ in the mixing layer $0.1<z/h<0.5$ was constructed by numerical modelling in [23].

Similarity between approximations (13) and (14) means that the ensemble of "generalized" convective jets is a propulsion of the turbulent diffusivity within the convective boundary layer.

## 5. Ensemble of dynamically identical spontaneous jets and turbulent moments of a convective surface layer

Based on an integral model similar to (10), we deduce the forms of the second, third and fourth statistical turbulent moments of vertical velocity. Then

$$\begin{cases} \dfrac{<\widehat{w}^2>}{w_D^2} = \dfrac{\sigma}{4}\dfrac{(gS_\theta)^{2/3} z^{2/3}}{w_D^2} = \dfrac{\sigma}{4}\dfrac{(gS_\theta)^{2/3} h^{2/3}}{w_D^2}\left(\dfrac{z}{h}\right)^{2/3} = \dfrac{\sigma}{4}\left(\dfrac{z}{h}\right)^{2/3} \\ \dfrac{<\widehat{w}^3>}{w_D^3} = \dfrac{\sigma_{www}}{8}\dfrac{(gS_\theta) z}{w_D^3} = \dfrac{\sigma_{www}}{8}\dfrac{(gS_\theta) h}{w_D^3}\left(\dfrac{z}{h}\right) = \dfrac{\sigma_{www}}{8}\left(\dfrac{z}{h}\right) \\ \dfrac{<\widehat{w}^4>}{w_D^4} = \dfrac{\sigma_{wwww}}{16}\dfrac{(gS_\theta)^{4/3} z^{4/3}}{w_D^4} = \dfrac{\sigma_{wwww}}{16}\dfrac{(gS_\theta)^{4/3} h^{4/3}}{w_D^4}\left(\dfrac{z}{h}\right)^{4/3} = \dfrac{\sigma_{wwww}}{16}\left(\dfrac{z}{h}\right)^{4/3} \end{cases} \tag{15}$$

Where $<\widehat{w}^2>$, $<\widehat{w}^3>$ and $<\widehat{w}^4>$ are statistical turbulent moments, related to an ensemble of spontaneous jets and downdrafts; $\sigma_{ww}/4 \approx 1.1$, $\sigma_{www}/8 \approx 1.1$ and $\sigma_{wwww}/16 \approx 0.9$ are a dimensionless constant, which depends on the relative area of the updraft flows, the density of jets per area unit, and a description method of interference between jets.

We assume that $\overline{w^2}$, $\overline{w^3}$ and $\overline{w^4}$ are the real second, third and fourth order moments of vertical velocity of the atmospheric surface layer. According to results of dimension theory [15, 24] transformed by Deardorff parameters (11) we get

$$\frac{\overline{w^2}}{w_D^2} = \lambda_{ww}\left(\frac{z}{h}\right)^{2/3}, \quad \frac{\overline{w^3}}{w_D^3} = \lambda_{www}\left(\frac{z}{h}\right), \quad \frac{\overline{w^4}}{w_D^4} = \lambda_{wwww}\left(\frac{z}{h}\right)^{4/3} \tag{16}$$

Field data of the second [25], third [25] and fourth [26] moments of vertical velocity and their approximations (16) with $\lambda_{ww} = 1.8$, $\lambda_{www} \approx 1$ and $\lambda_{wwww} \approx 6.5$. are presented in Fig. 4.

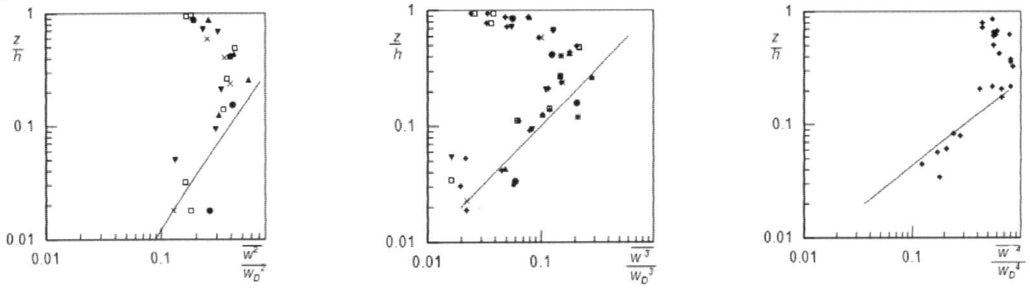


**Figure 4.** Field data of the second [25], third [25] and fourth [26] moments of vertical velocity – the dots. Their approximations (16) – the solid lines.

According to (15), (16) the power-law dependences of statistical and real moments of vertical velocity on the dimensionless altitude $z/h$ are identical. Hence an ensemble of spontaneous jets forms a dependence of momentum and heat diffusivity on the altitude within the convective surface layer.

## 6. Conclusion

Presented results show that the classical model of convective thermals and jets can be modified by including a pressure force.

It is demonstrated that the modified integral model of convective jets retain a power-law dependence of vertical velocity and buoyancy on the altitude, that deduced in the classical models.

For convective jets, within a neutrally stratified atmosphere, an effect of a pressure force increases the amplitude of buoyancy and decreases the amplitude of vertical velocity. The total varying is about 10%. For spontaneous jets within an unstable stratified atmosphere, an effect of a pressure force retain the amplitude of buoyancy and increases the amplitude of vertical velocity by about 15%.

The relatively weak influence of the pressure force on jet's parameters enables the approximation of vertical boundary-layer in the theory of buoyant jets to be regarded as a first approximation.

Ensembles of buoyant jets form vertical profiles of turbulent moments and eddy diffusivity within convective surface layers.

## Acknowledgments

This study was supported by the Russian Foundation for Basic Research, grant no. 15-05-068491-a.

## References

[1] Delichatsios MA 1979 J. *Fluids Mech.* **93(2)** 241–250.
[2] Yu H-Z 1990 *Trans. ASME* **112** 186–191.
[3] Vul'fson A N 2001 *Fluid Dynamics* **36(3)** 418-428.
[4] Vul'fson A N, Borodin O O 2001 *J. of Applied Mech. and Tech. Phys.* **42(2)** 255-261.
[5] Scase M M, Hewitt R E 2012 *J. Fluid Mech.* **697** 455–480.
[6] Craske J, van Reeuwijk M 2016 *J. Fluid Mech.* **792** 1013–1052.
[7] Woodhouse M J, Phillips J C, Hogg A J 2016 *J. Fluid Mech.* **794** 595–638.
[8] Costa A, Suzuki Y J, Cerminara M, … 2016 *J. of Volcanology and Geothermal Res.* **326**, 2-25.
[9] Vulfson A N, Borodin O O 2016 *Physics-Uspekhi* **59(2)**, 109.
[10] Vulfson A N, Nikolaev P V 2017 *Izvestiya, Atmos. and Oceanic Phys.* **53(4)**, 419-427.
[11] Sedov L I 1996 *Mechanics of Continuous Media* (World Scientific).
[12] Zel'dovich B Ya 1937 *Zh. Eksp. Teor. Fiz.* **12** 1463–1465.
[13] Priestly C H B, Ball F K 1955 *Quart. J. Roy. Meteor. Soc.* **81 (348)** 144–157.
[14] Abramovich G N 1963 *Theory of Turbulent Jets* (MIT Press).
[15] Priestly C H B 1959 *Turbulent transfer in the lover atmosphere* (Univer. Chicago Press).
[16] Kader B A, Yaglom A M 1990 *J. Fluid Mech.* **212** 637–662.
[17] Prandtl L 1932 *Beitr. Phys. fr. Atmoshare* **19(3)** 188–202.
[18] Batchelor G K 1954 *Quart. J. Roy. Met Soc* **80 (345)** 339–358.
[19] Huppert H E, Turner J S 1981 *J. Fluid Mech.* **106** 299–329.
[20] Vulfson A N, Borodin O O 2009 *Russian Meteorol. and Hydrol.* **34(8)** 491-500.
[21] Deardorff J W 1970 *J. Atmos. Sci.* **27(8)** 1211–1212.
[22] Obukhov A M 1946 *Trudy Inst. Teor. Geofiz. Akad. Nauk SSSR* **1** 95–115.
[23] Troen I, Mahrt L 1986 *Boundary-Layer Meteorol.* **37** 129–148.
[24] Vulfson A N, Volodin I A, Borodin O O 2004 *Russian Meteorol. and Hydrol.* **10** 1-10.
[25] Lenschow D H, Wyngaard J C, Pennel W T 1980 *J. Atmos. Sci.* **37(6)** 1313–1326.
[26] Gryanik VM, Hartman J 2002 *J. Atmos. Sci.* **59(18)** 2729–2744.

IOP Conf. Series: Journal of Physics: Conf. Series **955** (2017) 012014 doi:10.1088/1742-6596/955/1/012014

# Development of a regional model based on adapted WAVEWATCH III and WRF models for the prediction of surface wind waves on the reservoir and wind

**A Kuznetsova**[1, 2], **G Baydakov**[1], **D Sergeev**[1, 2], **Yu Troitskaya**[1, 2]

[1] Institute of Applied Physics of the Russian Academy of Sciences, 46 Ulyanov st, 603950, Nizhny Novgorod, Russian Federation
[2] Nizhny Novgorod State University, 23 Gagarina av, 603950, Nizhny Novgorod, Russian Federation

alexandra@ipfran.ru

**Abstract**. A regional model for the prediction of surface wind waves on the Gorky Reservoir and of the wind over the water area is proposed. The wind was calculated in the WRF model with and without Large Eddy Simulation block. The WRF model wind calculation results were used as wind forcing of the WAVEWATCH model. A comparison of the mean wave parameters with the measured in the experiment at the Gorky Reservoir was made.

## 1. Introduction

The fundamental need to create a software package for high-quality weather and waves prediction on the inland water body is caused by the fact that today the wave forecast in inland waters is calculated very approximately and rarely corresponds to reality. In this paper, we propose a regional model that uses the adapted to the conditions of middle-sized water body global wave model WAVEWATCH III (WW3) [1], initially tuned to the conditions of the seas and oceans, and then simultaneously uses the adapted atmospheric model WRF [2] as input data for WW3 wind forcing [3]. The simultaneous use of models is undertaken, since the correct setting of the wind provides a more accurate prediction of the parameters of waves in the reservoir, taking into account changes in wind distribution and its heterogeneity over the water area. Thus, the present work demonstrates the first stages of creating a regional model. The next stage in the development of the regional model will be automation of the process of data exchange between the WW3 and WRF models, the implementation is assumed to perform within the OASIS model [4].

The regional model was developed for the water area of the Gorky Reservoir. Gorky Reservoir is an artificial lake in the central part of the Volga River formed by a hydroelectric dam of Gorky Hydroelectric Station between the towns of Gorodets and Zavolzhye. Its lake part extends for ~100 km, ~40 km wide. Our group has a huge amount of data from the field experiments carried out in the south part of the Gorky reservoir from the boat. In the course of the experiment, the profiles of wind speed and surface wave spectra were simultaneously measured. The results of the field experiment is presented in [5].

## 2. Simulation

The third generation numerical models are adapted to the ocean conditions. However, the wave models such as WW3, SWAN [6], WAM [7] are used successfully on large lakes [8, 9, 10]. We have chosen WAVEWATCH III model for the simulation of surface waves on the middle-sized reservoirs, because this model considers the largest number of interactions available in the current model versions. Still for the WAVEWATCH III application to the conditions of the middle-sized reservoir, the tuning of the model is required. Reasons and steps of tuning are the following. It should be carried out in two steps: the adjusting of the wind input source term and the adjusting of the "collision integral". This adjusting is caused by the specific characteristics of the waves at small fetches the middle-sized reservoir: the wind input, which is proportional to the relation of the friction velocity (or 10 m wind speed) to the phase velocity [1] is more intense, as well as a stronger non-linearity is caused by steeper waves. The dissipation due to wave breaking remains unchanged.

In this paper, we use the adapted to the conditions of middle-sized reservoir WAVEWATCH III model [11]. Briefly, there are the following edits: the minimum value of a significant wave height $H_S$ is adjusted in the Fortran code; then the topographic grid of the Gorky Reservoir with dimensions 72 × 108 and increments of $0.00833^0$is used for the reservoir description. The grid is taken from the NOAA data "Global Land One-kilometer Base Elevation (GLOBE)" (figure 2 a). The approximation of deep water is considered. The frequency range is changed to 0.2-4 Hz in accordance with the experimentally observed range, which is splitted in 31 frequencies in the simulation and is modeled by a logarithmic formula for the frequency growth $\sigma_N = (\delta)^{N-1}\sigma_1$, where the growth rate is $\delta = 1.1$ in accordance with the recommendations of [1]; 30 angular directions of the wave field are considered. The waves in the reservoir are simulated for a given Gaussian initial seeding for different wind input parameterizations using the specified topographic data, wind speed and direction, and water-air temperature difference. These parameters can be obtained from the reanalysis data but they are too coarse for our needs. So wind input for the regional model we are working on goes from the WRF simulation. Thus the developing model for regional forecasting takes into account the spatial variability of the wind field using the wind forcing from the atmospheric model WRF.

To apply WRF to the calculation of the wind field over the Gorky reservoir, the following actions are made. In WRF preprocessing system, the preprocessing of the data to prepare input to the real program for real-data simulations is realized. For the geogrid module, the recommended geographical data for the lakes description «modis_lakes» for 4 nested domains in the studied region is used (figure 1).

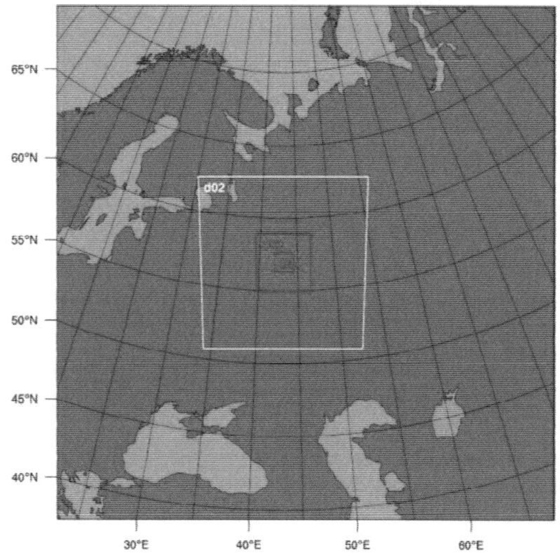


**Figure 1.** The domains used in WRF simulation.

The minimum cell size of the fourth nested domain is 30 arcseconds (it is equivalent to the cell size of the topographical data used in WAVEWATCH III $0.00833^0$) (figure 2 b). These data are used to describe the domains and to interpolate the static geographical information for the given grid. To describe the current weather situation, the meteorological data "NCEP Final Analysis (FNL from GFS) (ds083.2)" with 1 degree resolution is loaded. It is updated every 6 hours and is extracted from the GRIB format using the ungrib program. Metgrid program produces a horizontal interpolation of the extracted meteorological data on the domains grid. The simulations are held on the Yellowstone supercomputer [12].

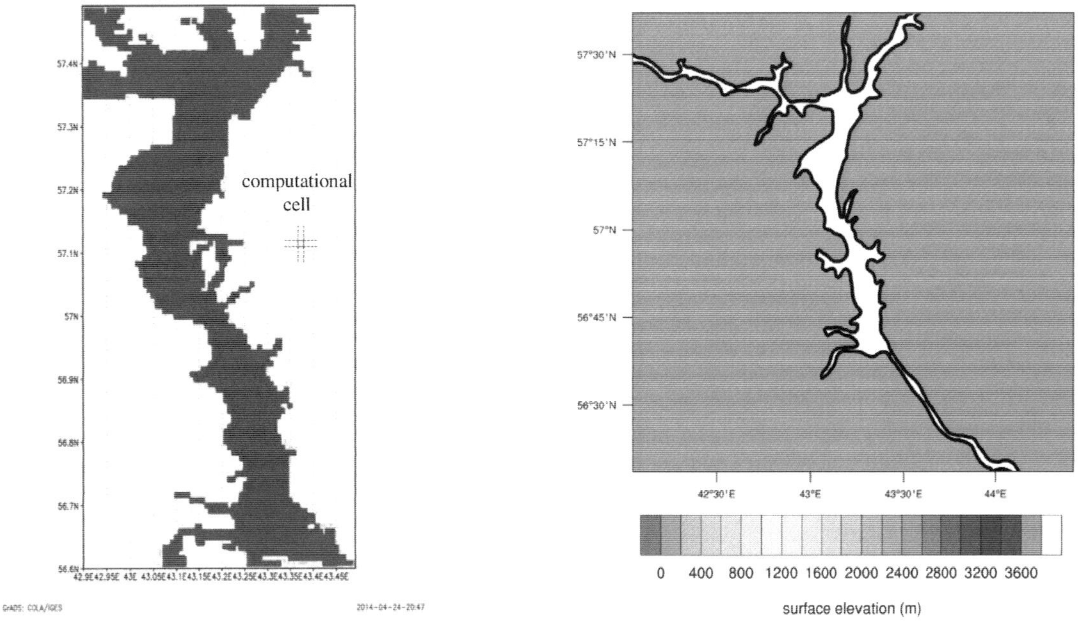


a) b)

**Figure 2.** Topographical grid of Gorky Reservoir as it appears in a) WAVEWATCH III, b) WRF. Computational cell with a size of $0.00833^0$ is shown.

## 3. Results of the simulation and discussions

The specified settings of the WRF model, the simulation of the wind field over the water area were performed with switching on the Large Eddy Simulation (LES) block to the WRF model and switching it off. When the LES unit was switched on, the first three domains were calculated within the Yonsei University scheme (YSU) for the planetary boundary layer (PBL) and within the surface layer scheme based on the Monin-Obukhov similarity theory, taking into account the viscous Carlson-Boland sublayer form. Simulation of the wind speed in the fourth domain were performed with the PBL parameterization turned off, but with taking into account turbulent flows (LES). Without the LES block, all the four domains were calculated using the YSU parameterization. The results obtained in the simulation for the test day 13.06.13 are shown in figure 3. The figure 3 shows the distribution of the wind over the water area of the Gorky reservoir , its value is indicated by color, and its direction is indicated by direction of segment. WRF application to the area containing the Gorky reservoir shows the significant spatial variability of the wind over the water area.(figure 3 a).

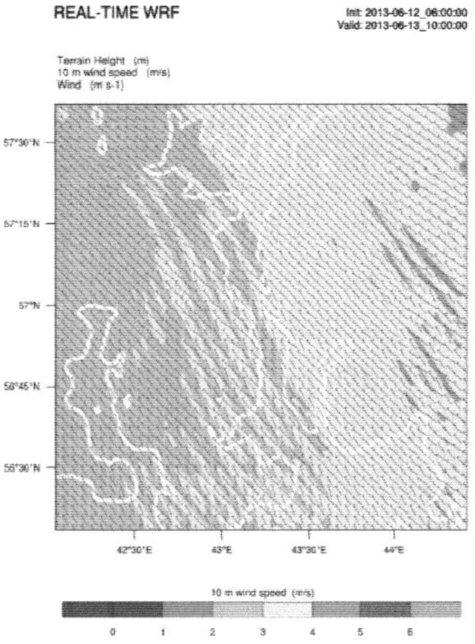


a)

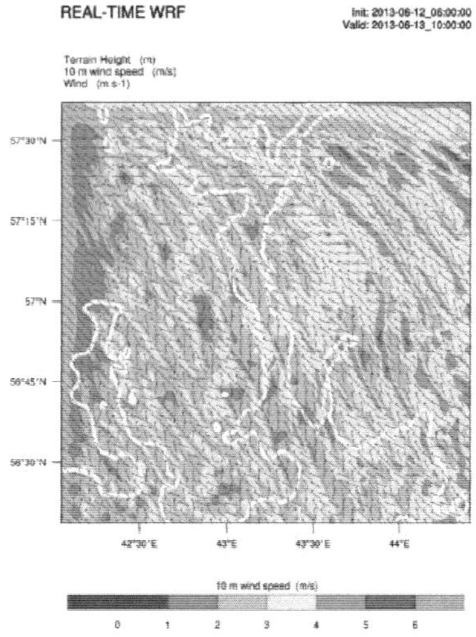


b)

**Figure 3.** The distribution of the wind over the water area of the Gorky reservoir in the test day 13.06.13, WRF simulation. a) WRF, b)WRF+LES.

However, the switching on the LES block to the WRF model with the unchanged size of the topographic grid cell resulted in the formation of numerical error, which are planned to be eliminated in subsequent numerical experiments by reducing the cell size. It is expected that with a smaller cell size, the wind speed simulation with WRF + LES will help to create highly accurate forecasts which takes into account the calculation of turbulent flows.

In addition, a comparison of the calculated wind speed at the point of measurement with the measurement data showed a slight underestimation of the calculated data, as well as insufficient tracking of the variability in time (figure 4).

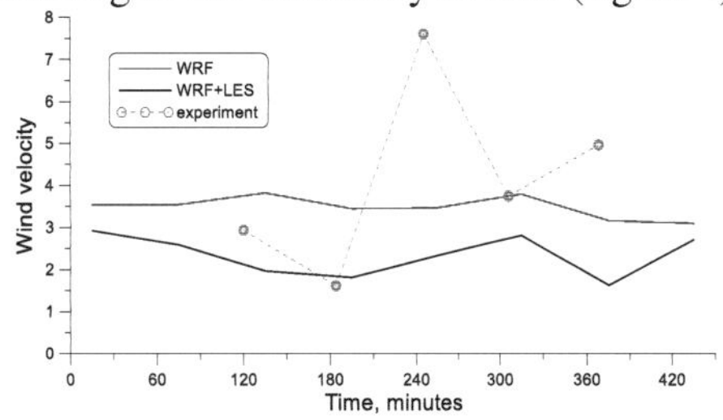


**Figure 4.** Calculated dependence of wind speed U10 on time: blue line - WRF, black line - WRF + LES, points - experiment.

We associate this inaccuracy of wind prediction in WRF model with several factors. First, with a small amount of data assimilated in the test area. In the considered area, there are two weather stations only (Volzhskaya GMO and Yurievets), and they are located on the coast, where the wind speeds are significantly different from those over the waters of the reservoir. Consequently, the data assimilation from the additional sources (eg, private weather stations) along the perimeter of the pond is needed. Installation of the private weather stations and data assimilation are planned as a part of this study.

Then the wind forcing from WRF is used in the adapted WW3. The verification is made on the basis of the experimental data for the following WAVEWATCH III output: significant wave heights and mean wave periods. Both in the model and in the processing of the experimental results, the calculation of $H_S$ is performed by the formula:

$$H_S = 4\sqrt{E} \quad (1)$$

The calculation of mean wave period $T_m$ is performed by the formula:

$$T_m = T_{m0,-1} = \left( \int_{f_{\min}}^{f_r} E(f)df \right)^{-1} \int_{f_{\min}}^{f_r} E(f)f^{-1}df \,. \quad (2)$$

All data is obtained at the point corresponding to the point of observations.

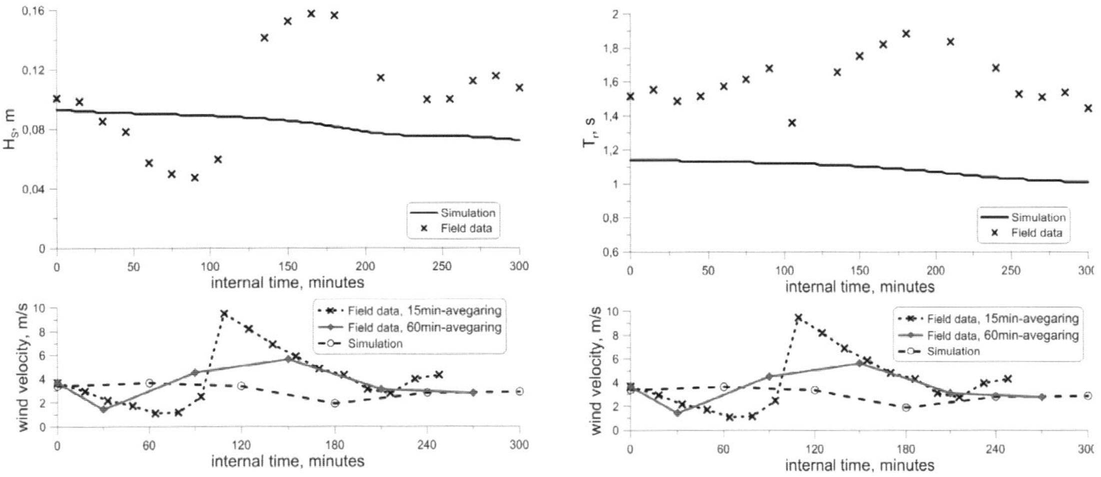


**Figure 5.** Lower plots:wind measured in the experiment and averaged over 15 minutes (dashed black line), wind measured in the experiment and averaged over 60 minutes (black line), wind from WRF, averaged over 60 minutes (grey line). Top plots: a) $H_S$ and b)$T_m$ for the adapted WW3 with WRF wind forcing (solid line), experimental data (crosses).

Figure 5 shows the value of the wind measured in the experiment and averaged over 15 minutes (dashed black line) and over 60 minutes (black line) for the convenience of the comparison with the wind from WRF, which is also averaged over 60 minutes (red line). The significant wave height and mean wave period are compared with those measured in the Gorky Reservoir experiment. It could be seen that the results of simulation adequately describe the mean wave parameters but underestimate it. Then, there is also insufficient tracking of variability in time.

## 4. Conclusion

The regional model for the prediction of surface wind waves on the Gorky Reservoir and of the wind over the water area is proposed. The wind was calculated in the WRF model with and without LES. The use of LES in the WRF model with the unchanged cell size of the topographic grid has led to the formation of numerical errors, which are planned to be eliminated in the subsequent numerical experiments by reducing the cell size. It is expected that with a smaller value of the cell size, the wind speed calculation results with WRF + LES will help to create highly accurate forecasts. The WRF model wind calculation results were used as wind forcing of the WW3 model. A comparison of the mean wave parameters with the measured in the experiment at the Gorky Reservoir was made. In general, the results are satisfactory and meet the needs of operational modeling of wind and waves, but there is an underestimation of the calculated data. Further improvement of calculations is planned due to, first, the elaboration of the used topography. Secondly, among the assumptions made in the calculations, the deep water approximation assumption is made. Taking into account the real bathymetry of the Gorky Reservoir, as well as the inclusion of WW3 source terms associated with the transition to shallow water, can make a positive impact on the results. Thirdly, the data assimilation from the additional sources like private weather stations along the perimeter of the pond will be performed.

**References**

[1] Tolman H and WAVEWATCH III Development Group. User manual and system documentation of WAVEWATCH III version 4.18. //*Environmental Modeling Center, Marine Modeling and Analysis Branch*. 282 pp. + Appendices. 2014

[2] Skamarock W C, Klemp J B, Dudhia J, Gill D O, Barker D M, Duda M G, Huang X-Y, Wang W, and Powers J G, 2008: A Description of the Advanced Research WRF Version 3. *NCAR Tech. Note NCAR/TN-475+STR,* 113 pp

[3] Kuznetsova A M, Baydakov G A, Papko V V, Kandaurov A A, Vdovin M I, Sergeev D A, Troitskaya Yu I, // *Geography, environment, sustainability,* Vol. 9, № 2, 2016, DOI: 10.15356/2071-9388_02v09_2016_02

[4] Craig A, Valcke S, Coquart L, 2017: Development and performance of a new version of the OASIS coupler, OASIS_MCT_3.0, *Geosci. Model Dev.*, https://doi.org/10.5194/gmd-2017-64

[5] Kuznetsova A M, Baydakov G A, Papko V V, Kandaurov A A, Vdovin M I, Sergeev D A, Troitskaya Yu I, (2016) // *Russian Meteorology and Hydrology*, 2016, №2, c.85-97

[6] SWAN team. SWAN – user manual. (2006) //*Delft University of technology, Environmental Fluid Mechanics Section*. 129 pp.

[7] Gunter H, Hasselmann S, Janssen P A E M (1992) The WAM model cycle 4. Technical report No. 4. //*DKRZ WAM Model Documentation*. Hamburg. 101 pp

[8] Alves J H G M, Chawla A, Tolman H L, Schwab D, Lang G, Mann G. //*Wea. Forecasting,* 29, 1473 – 1497. 2014

[9] Lopatoukhin L J, Boukhanovsky A V, Chernyshova E S, Ivanov S V (2004) // *Proc. The $8^{th}$ International Workshop on Waves Hindcasting and Forecasting*, Hawaii

[10] Hesser T J, Cialone M A, Anderson M E (2013) // *The US Army Research and Development Center (ERDC)*, 156 pp.

[11] Kuznetsova A M, Baydakov G A, Papko V V, Kandaurov A A, Vdovin M I, Sergeev D A, Troitskaya Yu I, // *Hindawi Publishing Corporation, Advances in Meteorology*, 2016, vol. 1, article ID 574602, pp. 1-13

[12] Computational and Information Systems Laboratory. (2012) Yellowstone: IBM iDataPlex System (University Community Computing). Boulder, CO: National Center for Atmospheric Research. http://n2t.net/ark:/85065/d7wd3xhc

**Acknowledgments**

This work is supported by RFBR (17-05-41117, 15-45-02580). The numerical simulation is partially supported by Russian Science Foundation (Agreement No. 15-17-20009).

IOP Conf. Series: Journal of Physics: Conf. Series **955** (2017) 012015 doi:10.1088/1742-6596/955/1/012015

# Nonlinear Interaction of Waves in Rotating Spherical Layers

**D Zhilenko[1], O Krivonosova[1] and M Gritsevich[2,3,4]**

[1] Moscow State University, GSP-1, Leninskie Gory, 119991,Moscow, Russia
[2] Department of Physics, University of Helsinki, Gustaf Hällströmin katu 2a, FI-00014 Helsinki, Finland
[3] Dorodnicyn Computing Center, RAS, Vavilova str. 40, 119333, Moscow, Russia
[4] Institute of Physics and Technology, Ural Federal University, Mira str. 19, 620002, Ekaterinburg, Russia

E-mail: olga@imec.msu.ru

**Abstract.** Flows of a viscous incompressible fluid in a spherical layer that are due to rotational oscillations of its inner boundary at two frequencies with respect to the state of rest are numerically studied. It is found that an increase in the amplitude of oscillations of the boundary at the higher frequency can result in a significant enhancement of the low-frequency mode in a flow near the outer boundary. The direction of propagation of the low-frequency wave changes from radial to meridional, whereas the high-frequency wave propagates in the radial direction in a limited inner region of the spherical layer. The role of the meridional circulation in the energy exchange between spaced waves is demonstrated.

## 1. Introduction

Most of the flows in astrophysical objects (stars, planets, and atmospheres) are characterized by the spherical geometry and rotation. These factors can be taken into account in a model spherical Couette flow. Rotation is often nonuniform. The nonuniformity of rotation of the Sun and planets of the solar system is well studied (some data are presented, e.g., in [1]). In the first approximation, the nonuniformity of rotation can be represented as its periodic modulation. The periodic (single-frequency) modulation of the rotation velocity of one of the spherical boundaries in the case of their quite fast unidirectional rotation can result in the formation of inertial waves in a flow between these boundaries [2]. In the case of the absence of average rotation and small amplitudes, the rotational oscillations of the inner sphere at a single frequency produce damping spherical waves [3]. However, the time dependence of the rotation velocity of natural objects can be more complex. In particular, data on the length of days on the Earth [4] indicate a cyclic process in which at least two frequencies or two characteristic time scales can be identified. This fact stimulates our interest in the two-frequency modulation of the rotation velocity. It is known that the interaction between two waves can enhance one of them [5]. Here we study the process of interaction between two waves, which is not masked by other phenomena. The possibility of the generalization of obtained results is discussed in Section 4.

## 2. Numerical methods and field of study

An isothermal flow of a viscous incompressible fluid is described by the Navier–Stokes and continuity equations:

$$\frac{\partial U}{\partial t} = U \times rotU - grad\left(\frac{p}{\rho} + \frac{U^2}{2}\right) - \nu rotrotU, \ divU = 0,$$

where $U$, $p$, ν, and ρ are the velocity, pressure, viscosity, and density of the fluid, respectively. These equations were numerically solved in a spherical coordinate system where the impermeability and no-slip conditions for the azimuthal $u_\varphi$, radial $u_r$, and polar $u_\theta$ components of the velocity have the form $u_\varphi(r = r_{1,2}) = \Omega_{1,2}(t)r_{1,2}\sin\theta$, $u_r(r = r_{1,2}) = 0$, $u_\theta(r = r_{1,2}) = 0$, where subscripts 1 and 2 correspond to the inner and outer spheres, respectively. We used an algorithm of numerical solution [6] based on a conservative finite difference scheme of the discretization of the Navier – Stokes equations in space and a third-order semi-implicit Runge–Kutta scheme for integration in time. The applicability of axisymmetric approximation was confirmed by three-dimensional calculations. Discretization in space was performed on grids uniform in $r$ and $\theta$ with the total number of nodes $4.8\times10^4$. The sensitivity of the results to the parameters of a grid was analyzed, e.g., in [7]. Agreement between experimental results and calculations was achieved with stationary [8] and periodic [9] boundary conditions.

In this work, the structure of a flow formed under the action of oscillations of the inner boundary of the layer is also analyzed by plotting wave surfaces or equal-phase surfaces. To calculate the phase of a flow at each point of the layer as a function of time, we used an approach well known in radio physics [10]. In this approach, the real time series $x(t)$ is considered as a projection of the complex vector $Z(t)$ on the abscissa axis $x(t) = B(t)\cos\Phi(t)$, where $B(t)$ and $\Phi(t)$ are the amplitude and phase of the initial signal $x(t)$. One of the methods for determining the vector $Z(t)$ is as follows: $Z(t) = x(t) + iy(t)$, $y(t) = H[x(t)] = \int_{-\infty}^{\infty} H(t-\tau)x(\tau)d\tau$, and $H(t) = (\pi t)^{-1}$ is the Hilbert transform; in this case $B(t) = \left(x^2(t) + y^2(t)\right)^{1/2}$, $\Phi(t) = arctg(y(t)/x(t))$. When $x(t) = C\sin\omega t$ $B(t) = C$ and $\Phi(t) = \omega t$. Such an approach allows calculating the amplitude and phase of an arbitrary real signal, including a two-frequency signal, as functions of time. In this work, such a method was used to calculate the phase of the azimuthal component $u_\varphi$ of the velocity at each point of the layer $\Psi(t,r,\theta)$ and phase of rotation of the oscillating sphere $\Psi_s(t)$. It was shown in [3] that the described method for analyzing the structure of the flow makes it possible to adequately reproduce inertial waves previously obtained in experiments and numerical calculations.

It was also demonstrated in [3] that harmonic oscillations of the inner sphere with small amplitudes generate radially damping spherical waves (figure 1a). The distance between phase jumps along radii coincides with half of the wavelength λ, which is calculated as the wavelength $\lambda = 2\pi\delta = 2\pi\left(\nu/\pi f\right)^{1/2}$ in the Stokes problem on a flow generated by harmonic oscillations of an infinite plate at the frequency $f$ [11], δ is the thickness of the dynamic boundary layer. When the amplitude of oscillations of the sphere increases, the flow becomes complicated (figure 1b), and wave surfaces ending at the outer boundary appear, as well as nodes at which the azimuthal component of the flow is always zero.

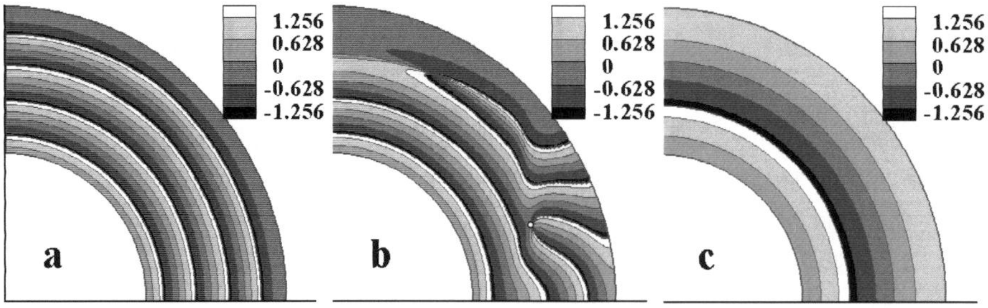


**Figure 1.** Wave surfaces (a) - $A$=1s$^{-1}$, $f$=0.5$s^{-1}$, (b) - $A$=5s$^{-1}$, $f$=0.5$s^{-1}$, (c) - $A$=0.2s$^{-1}$, $f$=0.07$s^{-1}$.

The rotational velocity of the inner sphere $\Omega(t)$ (the subscript 1 is omitted because $\Omega_2(t) = 0$ in this work) varies according to the law $\Omega(t) = A_1\sin(2\pi f_1 t) + A_2\sin(2\pi f_2 t)$, where the subscripts 1 and 2 refer to the high- and low-frequency modes such that $f_2 < f_1$ and $A_2 < A_1$: $f_1 = 0.5\text{s}^{-1}$, $A_1 = 1$ and 5 s¯1 and $f_2 = 0.07\ \text{s}^{-1}$, $A_2 = 0.2\text{s}^{-1}$. Calculations were performed with the dimensional parameters $r_1 =. 0.075\text{m}$, $r_2 = 0.15\text{m}$ ($\sigma = (r_2-r_1) / r_1 = 1$) and $\nu = 5\times10^{-5}\ \text{m}^2\ \text{s}^{-1}$. At these parameters, the flow is unsteady and axisymmetric and there are no other frequencies except for the frequencies $f_1$ and $f_2$ in the azimuthal velocity spectrum.

## 3. Results

At first we consider the structure of the flow at small-amplitude oscillations of the inner boundary. Amplitudes are considered as small when spherical waves propagating in the radial direction appear (figure 1a,c, figure 2a). While $\Omega(t)=A\sin(2\pi ft)$, the phase shift $\psi_0(r,\theta) = \Psi(t,r,\theta) - \Psi_s(t)$ at the point with the coordinates $(r,\theta)$ with respect to the phase of oscillations of the boundary ($\Psi_s(t) = 2\pi ft$) is independent of the time [3]. Figure 2a shows the wave surfaces $\Psi(t_3,r,\theta)$ at a certain time instant $t_3$ (shown in figure 3c) at two-frequency small-amplitude oscillations. These surfaces are also attributed to a spherical wave, but the spherical layer in this case can be divided into three sublayers: a sublayer adjacent to the inner sphere where $\Psi(t_3,r,\theta) > 41$, a sublayer adjacent to the outer sphere where $\Psi(t_3,r,\theta) < 9,5$, and a narrow transient layer (shown in figure 2a in white) with a large phase gradient $9.5 < \Psi(t_3,r,\theta) < 41$, which increases in time (figure 2c shows the dependence of the phase of oscillations on the radius in the equatorial plane for three time instants, where the transient layer is shown in gray). The effective frequency $\tilde{f}$ introduced as $2\pi\tilde{f}(t_2 - t_1) = \Psi(t_2,r,\theta) - \Psi(t_1,r,\theta)$ is $\tilde{f} \cong f_1$ in the inner sublayer, $\tilde{f} \cong f_2$ in the outer sublayer, and $f_2 < \tilde{f} < f_1$ in the transient sublayer. Figures 3a–c show the time dependence of the dimensionless azimuthal component of the velocity $u_\varphi /|u_{\varphi\ max}|$, phase $\Psi(t,r,\theta)$ and its linear approximation at three points lying in each sublayer. It is seen that the rate of phase increase, i.e., the effective frequency $\tilde{f}$, varies depending on the character of variation of $u_\varphi$. The closer the time dependence of the velocity to harmonic oscillations, the closer the dependence $\Psi(t,r,\theta)$ to a linear time function and the closer the frequency $\tilde{f}$ to the frequency of harmonic oscillations.

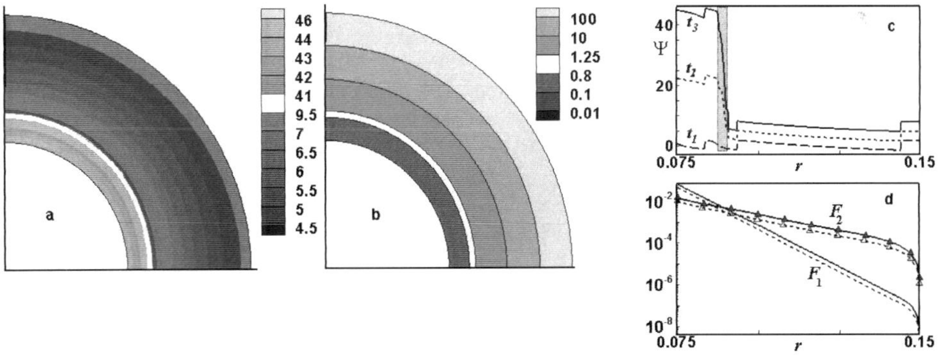


**Figure 2.** Two-frequency oscillations: $f_1 = 0.5\text{s}^{-1}$, $A_1 = 1\text{s}^{-1}$ and $f_2 = 0.07\ \text{s}^{-1}$, $A_2 = 0.2\text{s}^{-1}$. (a) - $\Psi(t_3,r,\theta)$, (b) - $F_2/ F_1$, (c) - $\Psi(t,r,\theta=\pi/2)$, (d) - $F_1(r)$ and $F_2(r)$ at $\theta=\pi/2$- solid lines and $\theta=\pi/4$- dashed lines.

The ratio of the amplitudes $F_2$ and $F_1$ of the low- and high-frequency oscillation modes, respectively, at each point of the layer calculated using the Fourier transform in the transient region is in the interval $0.8.< F_2/ F_1 <. 1.25$ (Figure 2b). Figure 2d shows the dependence of the amplitudes $F_1$ and $F_2$ on the radius at two values of the meridional angle: at the equator and at an angle of 45° to the rotation axis. The amplitudes $F_1$ and $F_2$ decrease exponentially as $\exp(-kr)$ in most of the thickness of the layer and the damping ratio $k$ is independent of the meridional angle. The damping ratio $k$ increases with the frequency ($k_1 = 184.5\text{m}^{-1}$, $k_2 = 75.24\text{m}^{-1}$); i.e., the high-frequency mode damps more rapidly. As a result, a larger contribution to the flow near the inner sphere comes from the high-frequency mode, but it damps more rapidly, whereas the low-frequency mode makes a larger contribution to the flow behind the transient layer, where the amplitudes of the modes become closer to each other. The ratios of the amplitudes and the dependences of the amplitudes on the radius (see Figure 2b, d) coincide for the considered two-frequency oscillations and for two harmonic oscillations of the inner sphere (see Figure 1a, c); i.e., the modes of two-frequency small-amplitude oscillations do not interact with each other.

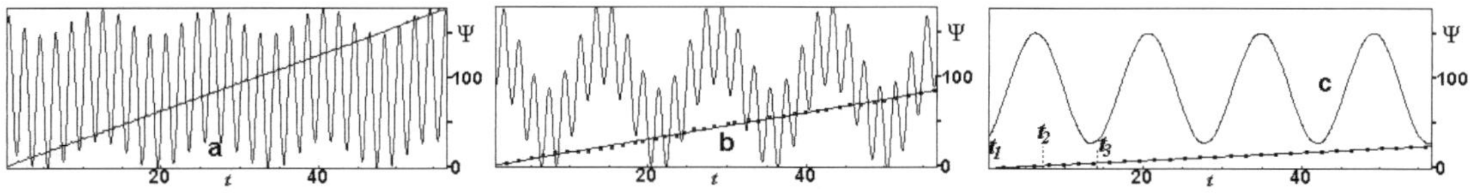


**Figure 3.** $u_\varphi/|u_{\varphi\,max}|$ (sinusoids), Ψ (squares) and its linear approximations (straight lines) (a) - in the inner $r$ = 0.0752m, (b) - in the transient $r$ = 0.09m, and (c) - in the outer $r$ = 0.132m sublayers.

We now consider how the structure of the flow changes when the amplitude of only the high-frequency mode increases $A_1$ =5s$^{-1}$ (figure 4). The spherical layer can also be divided into three regions (figure 4a). The first region is adjacent to the inner sphere, but its boundary has now a convexity near the equatorial plane. In this region, oscillations in time occur at $\tilde{f}$ close to $f_1$. The region adjacent to the outer sphere is characterized not only by oscillations at $\tilde{f} \cong f_2$ but also by a change in the direction of propagation of waves from the radial direction to the meridional direction. Wave surfaces in this region are almost perpendicular to the outer sphere. Figure 4b shows zero $u_\varphi$ contours at time instants, corresponding to maximal deviation of oscillating sphere. It is seen that these contours in the inner region are close to circles, whereas contours in the outer region are arranged in the radial direction. The comparison of images in Figure 4a, b indicates that the form of wave structures completely corresponds to the velocity field. The third transient region where $f_2 < \tilde{f} < f_1$ is between the inner and outer regions (white colour in figure 4a).

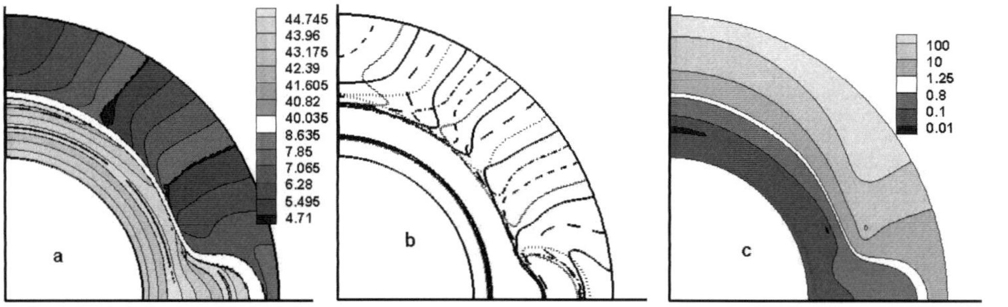


**Figure 4.** Two-frequency oscillations: $f_1$ = 0.5s$^{-1}$, $A_1$ =5s$^{-1}$, $f_2$= 0.07 s$^{-1}$, $A_2$ = 0.2s$^{-1}$. (a) - $\Psi(t_3,r,\theta)$, (b) - contours $u_\varphi$ = 0, (c) - $F_2/F_1$.

The amplitudes $F_1$ and $F_2$ of the high- and low-frequency modes calculated using the Fourier transform in the transient region are close to each other (figure 4c). The dependences of the amplitudes on the radius in this case are shown in figure 5a. Figure 5b shows the amplitudes $F_1$ and $F_2$ calculated for harmonic oscillations of the sphere with the parameters of the high-frequency mode ($A_1$ = 5s$^{-1}$, $f_1$ = 0.5s$^{-1}$) and low-frequency mode ($A_1$ = 0.02s$^{-1}$, $f_1$ = 0.07s$^{-1}$). At $A_1$ =5s$^{-1}$ , high-frequency oscillations cannot be treated as small-amplitude and the structure of wave surfaces changes (cf. figure 1a, b).

The damping ratio determined from the dependence $F_1(r)$ (figure 5b) depends on the meridional angle. In the equatorial plane, $k_1$ (π/2) =63.6m$^{-1}$ is almost one third of the value $k_1$ =. 184.5m$^{-1}$ at small-amplitude oscillations (figure 2d), whereas the latter value is significantly smaller than the damping ratio $k_1$ (π/4) =233.9m$^{-1}$. The dependences $F_1(r)$ in figure 5a, b almost coincide with each other, whereas the dependences $F_2(r)$ in these figures are different. The damping ratio near the inner sphere in Figure 5a depends on the meridional angle and it is close to zero in the middle radial part of the layer. Thus, at the considered two-frequency oscillations of the inner sphere, not only the low-frequency mode is enhanced but also its radial propagation direction from the inner sphere to the outer

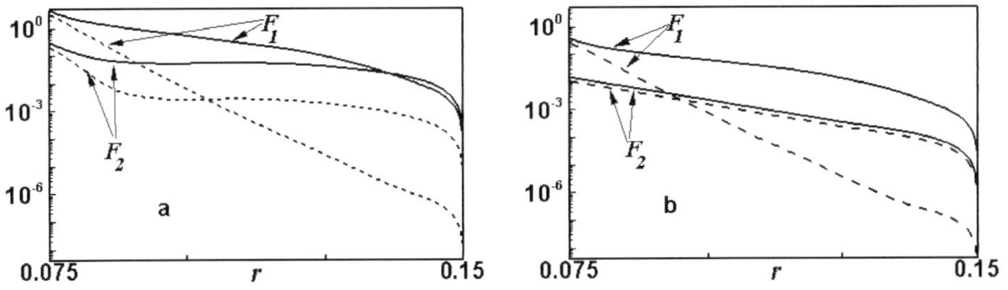


**Figure 5.** Radius dependence of the amplitudes at $\theta=\pi/2$ solid lines and at $\theta=\pi/4$ dashed lines. (a) - two-frequency oscillations: $f_1 = 0.5s^{-1}$, $A_1 = 5s^{-1}$ and $f_2 = 0.07\ s^{-1}$, $A_2 = 0.2s^{-1}$; (b) - independent harmonic oscillations with the parameters of the high- and low-frequency modes.

one changes to the meridional propagation direction from the equator to a pole along the outer sphere. The modes are again spatially separated: the low- and high-frequency waves propagate in the regions adjacent to the outer and inner spheres, respectively.

## 4. Discussion of the results

The enhancement of the low-frequency mode in the flow at an increase in the amplitude of the high-frequency mode in the rotational velocity signal can be considered as manifestation of energy transfer from smaller scales to larger scales in the process of nonlinear interaction between the modes. Energy transfer from smaller scales to larger scales is similar to processes in quasi-two-dimensional turbulence. An inverse energy transfer cascade characteristic of quasi-two-dimensional turbulence was previously observed in a spherical layer at the periodic modulation of the rotation velocity of the inner sphere [9] in the frequency range limited from below by the modulation frequency and from above by the sphere rotation frequency. A significant difference of the flow formed at two-frequency oscillations of the inner boundary with respect to the rest state from the flow considered in [9] is the absence of the mean flow in the azimuthal direction and any instabilities. A common property of these flows is the presence of averaged large-scale meridional circulations. In our case, it is single-vortex circulation directed from the poles to the equator of the inner sphere.

The intensity of meridional circulation at small-amplitude oscillations is low. The behavior of both modes is similar to that at the corresponding harmonic oscillations (see figure 2d); i.e., the interaction between modes is absent. Because of the difference between the damping ratios, the high-frequency mode prevails near the inner sphere as a source of oscillations, whereas the low-frequency mode prevails far from the source of oscillations near the outer sphere; the interface between these two regions coincides with the contours of equal amplitudes of the modes (figure 2b). When the amplitude of the high-frequency mode in the signal of sphere velocity increases, the intensity of circulation increases by an order of magnitude, the radial velocity near the equator increases by two orders of magnitude, and the equatorial jet turns near the equator of the outer sphere, as is indicated by wave surfaces, e.g., in figure 1b and figure 4a. Such an intense flow in the meridional plane significantly changes the damping ratios of the azimuthal component of the velocity. The damping ratios near the equatorial plane decrease significantly (see figure 5a) because fluid acquiring a high kinetic energy near the inner sphere is rapidly transferred to the region adjacent to the outer sphere. However, the low-frequency mode prevails near the outer sphere and this mode is enhanced.

Thus, circulation in the meridional plane plays a decisive role in energy transfer from smaller scales to larger scales. In particular, the enhancement of the low-frequency mode is due to intense meridional circulation, which is in turn enhanced at an increase in the amplitude of any mode in the rotation signal. A feature of the meridional circulation is its presence in any flow limited by rotating spherical boundaries: in wide and very thin layers, at the rotation of only the inner boundary, and at the unidirectional rotation of both spherical boundaries. Large-scale meridional circulation exists in the Earth's atmosphere [12]. Since the enhancement of the low-frequency mode is, as mentioned above, due to intense circulation, this is the reason to expect that the enhancement of the low-frequency mode considered in this work can be manifested in large-scale geo flows.

## 5. Conclusions

The results of the numerical studies have shown that a flow in a spherical layer at two-frequency rotational oscillations of the inner sphere with respect to the rest state is separated into two regions. The high-and low-frequency modes dominate near the inner and outer spheres, respectively. The position of the boundary between the regions is independent of time and coincides with amplitude contours of oscillations at each of the modes. The waves with small amplitudes are spherical as in the case of periodic oscillations with wave surfaces perpendicular to the radius. With an increase in the amplitude of the high-frequency mode in the signal of the velocity of the sphere, the division of the flow into two region holds, but the wave structure of the flow, as well as the ratio of the amplitudes of the low- and high-frequency modes, changes. The direction of propagation of the wave in the region adjacent to the outer sphere changes from radial to meridional and the low-to-high-frequency mode amplitude ratio increases. The considered phenomenon is the energy transfer from high-frequency velocity oscillations with the larger amplitude to the low-frequency mode with the smaller amplitude. The energy transfer between the modes involves a meridional circulation and the unsteady flow remains stable in this case. The excess of the amplitude of the low-frequency mode over the amplitude of the high-frequency mode increases with a decrease in the low frequency and depends significantly on the meridional angle: it increases by two orders of magnitude from the equator to the pole.

## Acknowledgments

This work was supported by the Russian Foundation for Basic Research project no. 160500004. MG acknowledges support from the ERC Advanced Grant No. 320773, and the Russian Foundation for Basic Research, project no. 16-07-01072. Research at the Ural Federal University is supported by the Act 211 of the Government of the Russian Federation, agreement No 02.A03.21.0006.

## References

[1] Noir J, Hemmerlin E, Wicht J, Baca S M, and Aurnou J M 2009 *Phys. Earth Planet Inter.* **173** 141
Sturrock P A, Buncher J B, Fischbach E, Gruenwald J T, Javorsek D, Jenkins J H, Lee R H, Mattes J J, and Newport J R 2010 *Solar Phys.* **267** 251
[2] Bryan G 1889 *Phil. Trans. London* **180** 187
Staquet C and Sommeria 2002 *J. Ann. Rev. Fluid Mech.* **34** 559
[3] Zhilenko D Yu and Krivonosova O E 2015 *Dokl. Phys.* **60**
[4] Sidorenkov N S 2004 *Herald Russ. Acad. Sci.* **74** 402
[5] Phillips O M 1981 *J. Fluid Mech.* **106** 215
Zakharov V E and Shamin R V 2010 *JETP Lett.* **91** 62
Talipova T G, Pelinovsky E N and Kharif K 2011 *JETP Lett.* **94** 182
[6] Nikitin N 2006 *J. Comp. Phys.* **217** 759
[7] Zhilenko D Yu and Krivonosova O E 2015 *Tech. Phys. Lett.* **41**, 5
[8] Zhilenko D Yu and Krivonosova O E and Nikitin N V 2007 *Fluid Dyn.* **42**, 886
[9] Zhilenko D Yu and Krivonosova O E 2015 *JETP Lett.* **101**, 527
[10] Zinov'ev A L and Filippov L I 1975 *Introduction to the Theory of Signals and Circuits* (Moscow: Vyssh. Shkola) p 319
[11] Landau L D and Lifshitz E M 1986 *Fluid Mechanics* (Moscow: Nauka) p 534
[12] Gill A E 1982 *Atmosphere-Ocean Dynamics* (Cambridge: Academic) p 208

# Oscillations of oblate drop between heterogeneous plates under uniform electric field

**M A Kashina**[1] **and A A Alabuzhev**[1,2]
[1]Perm State University, Perm 614990, Russia
[2]Institute of Continuous Media Mechanics UB RAS, Perm 614013, Russia

E-mail: alabuzhev@mail.ru

**Abstract.** The forced oscillations of the incompressible fluid drop under the action of the uniform electric field are considered. In equilibrium, the drop has the form of a circular cylinder bounded axially by the parallel solid planes; the contact angle is right. An incompressible fluid of different density surrounds the drop. The external electric field acts as an external force that causes motion of the contact line. In order to describe this contact line motion, the modified Hocking boundary condition is applied: the velocity of the contact line is proportional to the deviation of the contact angle and the speed of the fast relaxation processes, whose frequency is proportional to twice the frequency of the electric field. The case of heterogeneous plates is investigated. We assume that the Hocking parameter depends on the polar angle in this case. The function describing the change in the coefficient of the interaction between the plate and the fluid (the contact line) is expanded in a series of the Laplace operator eigenfunctions.

## 1. Introduction

In most articles on electrowetting-on-dielectric (EWOD) (see fig. 1) [1-4], the fundamental equation in the EWOD study is the Young–Lippmann equation for the contact angle $\vartheta$ written as:

$$\cos\vartheta = \cos\vartheta_0 + E_w, \ \cos\vartheta_0 = \sigma_{ic}^{-1}\left(\sigma_{ip} - \sigma_{cp}\right), \ E_w = 0.5CV^2\sigma_{ic}^{-1}, \ C = \varepsilon\varepsilon_0 d^{-1}, \quad (1)$$

where $\vartheta_0$ is the contact angle without applied voltage – the equilibrium contact angle, which is defined by the well-known Young's equation, $\sigma$ is the interfacial tension between the drop of conducting fluid ($c$), the isolating (surrounding) fluid ($i$) and the dielectric plate ($p$), $E_w$ is the EW number (it represents the ratio of the electrostatic energy to the liquid – surrounding fluid interfacial energy), $C$ is capacitance per unit area, $V$ is the value of the applied DC voltage, $d$ is the thickness of the dielectric film, $\varepsilon_0$ and $\varepsilon$ are the vacuum and dielectric layer permittivity, respectively. For the alternating current (AC) instead of $V^2$ in equation (1) the square value of the voltage, or effective voltage, $U^2$ is used [1]. Although the Young–Lippmann equation (1) is now universally recognized as a basic equation of EWOD treatment, different approaches have been proposed to derive it [1,4]. Now EWOD has found wide application in various fields, such as electronic display technology [5,6], variable-focus liquid lenses [7,8], digital (droplet) microfluidic devices for bioanalysis (lab-on-a-chip) [9,10], etc.

However, the available experimental results proved to be very much different from the theoretical predictions of the Young–Lippmann equation (1). It is anticipated that a zero contact angle will be

[1] To whom any correspondence should be addressed.

obtained after some critical voltage angle (complete wetting and the contact angle remains unchanged), but in fact the value of the contact angle obtained in experiments is always finite [1,4].The mechanism of the contact angle saturation is still not clearly understood and is the subject for discussion [1].

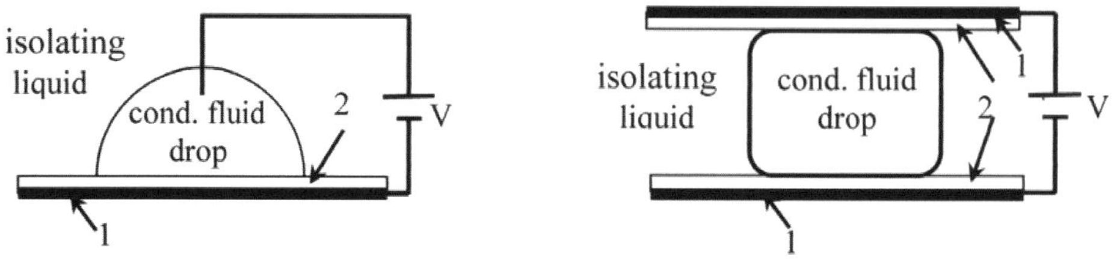


**Figure 1.** Typical schematic configurations of EWOD-devices.
1 – electrode, 2 – dielectric layer.

In [11], for a cylindrical drop we proposed another effective boundary condition on the basis of Hocking's equation [12]:

$$\frac{\partial \zeta^*}{\partial t^*} = \pm \Lambda^* \left( \frac{\partial \zeta^*}{\partial z^*} + A^* \cos\left(2\omega^* t^*\right) \right), \tag{2}$$

where $\zeta^*$ is the deviation of the drop interface from the equilibrium position, $z^*$ is the axial coordinate, $\Lambda^*$ is a phenomenological constant (the so-called wetting parameter or Hocking parameter), having the dimension of velocity, $A^*$ is the effective amplitude, $\omega^*$ is the AC frequency. The second term in boundary condition (2) is the external action, which is written in the same fashion as in the Young–Lippmann equation (1): $\cos\vartheta = E_w = \tilde{E}_w V^2$ as $\cos\vartheta_0 = 0$ in our problem, so that $\zeta_z \sim \operatorname{ctg}\vartheta = E_w \big/ \sqrt{1 - E_w^2} = \tilde{E}_w V^2 \big/ \sqrt{1 - \tilde{E}_w^2 V^4} \approx \tilde{E}_w V^2 + O\left(V^6\right) \approx \tilde{E}_w V^2 \sim V^2$. In article [11] we also investigated the effect of inhomogeneous electric field. Also a general discussion of the effective boundary conditions can be found in articles [13,14].

Another important problem is the non-uniform wetting of the surface, along which the contact line moves [15,16]. Heterogeneous substrates can cause different effects [17], for example, hysteresis of the contact angle [18,19]. However, most these articles consider the effect of mechanical (acoustical) vibrations [16].

This study is intended as an extension to work [11]. We consider the behavior of a droplet sandwiched between two heterogeneous flat plates under the applied uniform ac voltage. In order to describe the motion of the contact line the modified boundary condition (2) is used: the velocity of the contact line is proportional to the deviation of the contact angle and the speed of the fast relaxation processes, whose frequency is proportional to twice the electric field frequency. In the future, we plan to compare the obtained results with the experimental data [20].

## 2. Problem formulation

The problem formulation largely coincides with that presented in article [13]. We consider the dynamic behavior of the incompressible liquid drop of density $\rho_i^*$ surrounded by another liquid of density $\rho_e^*$ (here and in the following, the quantities with subscript $i$ refer to the drop, and those with subscript $e$ – to the surrounding liquid). The system is bounded by two parallel solid surfaces (see figure 2) with the inter-plate spacing $h^*$. The equilibrium contact angle $\vartheta_0$ between the lateral surface of the drop and the solid surface is equal to $0.5\pi$. The external uniform alternating electric field acts as an external force that causes the contact line motion [11].

Because of the problem symmetry it is convenient to introduce the cylindrical coordinates $r^*$, $\alpha$, $z^*$. The azimuthal angle $\alpha$ is reckoned from the $x$-axis. Let the surface of the droplet be described by the equation $r^* = R_0^* + \zeta^*\left(\alpha, z^*, t^*\right)$. Assuming a potential liquid motion, we introduce the velocity potential $\vec{v}^* = \vec{\nabla}\varphi^*$. Taking length $R_0^*$, height $h^*$, density $\rho_e^* + \rho_i^*$, time $\sigma^{-1/2}\sqrt{\left(\rho_e^* + \rho_i^*\right)R_0^{*3}}$, velocity potential $A^*\sqrt{\sigma}\left(\left(\rho_e^* + \rho_i^*\right)R_0^{*3}\right)^{-1/2}$, pressure $A^*\sigma\left(R_0^*\right)^{-2}$, and deviation of the surface $A^*$ as characteristic quantities, we pass to dimensionless variables and obtain the following linear problem

$$p_j = -\rho_j\varphi_{jt}, \ \Delta\varphi_j = 0, \ j = i, e, \tag{3}$$

$$\Delta = \frac{1}{r}\frac{\partial}{\partial r}\left(r\frac{\partial}{\partial r}\right) + \frac{1}{r^2}\frac{\partial^2}{\partial \alpha^2} + b^2\frac{\partial^2}{\partial z^2},$$

$$r = 1: \ \left[\varphi_r\right] = 0, \ \zeta_t = \varphi_r, \ \left[p\right] = \zeta + \zeta_{\alpha\alpha} + b^2\zeta_{zz}, \tag{4}$$

$$z = \pm\frac{1}{2}: \ \varphi_z = 0, \tag{5}$$

$$r = 1, \ z = \pm\frac{1}{2}: \ \zeta_t = \mp\lambda\left(\alpha\right)\left(\zeta_z + a\cos\left(2\omega t\right)\right), \tag{6}$$

where $p$ is the fluid pressure, $\lambda\left(\alpha\right)$ describes the heterogeneous condition of plates, $a$ is forced oscillation amplitude, the square brackets denote a jump in the quantity at the interface between the external liquid and the drop. The boundary-value problem (3)–(6) involves six parameters: the aspect ratio, the dimensionless density, the wetting parameter, the AC frequency and amplitude

$$b = R_0 h^{-1}, \ \rho_i = \rho_i^*\left(\rho_e^* + \rho_i^*\right)^{-1}, \ \rho_e = \rho_e^*\left(\rho_e^* + \rho_i^*\right)^{-1}, \ \lambda = \Lambda^*\sigma^{-1/2}b\sqrt{\left(\rho_e^* + \rho_i^*\right)R_0^*},$$

$$\omega = \omega^*\sigma^{-1/2}\sqrt{\left(\rho_e^* + \rho_i^*\right)R_0^{*3}}, \ a = A^*C\sigma^{-3/2}\sqrt{\left(\rho_e^* + \rho_i^*\right)R_0^{*3}}\Big/2.$$

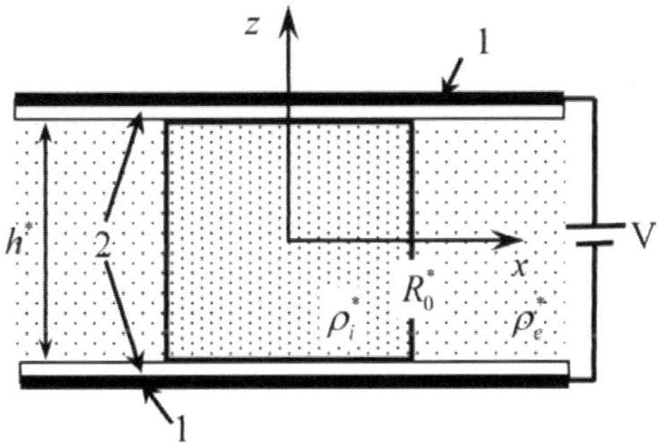


**Figure 2.** Problem geometry (1 – electrode, 2 – dielectric layer).

## 3. Forced oscillations

The function $\lambda\left(\alpha\right)$ is expanded into the Fourier series in eigen functions of the Laplace operator. Let us consider a particular case of the heterogeneous plates: $\lambda\left(\alpha\right) = \lambda_0\left|\sin\left(k_\lambda\cos\left(\alpha\right)\right)\right|$, where $k_\lambda$ is the wavenumbers. Note, that we used the above-mentioned function in [11] for non-uniform electric filed. The solutions for the velocity potential and the surface deviation are written as

$$\varphi_i\left(r, z, t\right) = \mathrm{Re}\left(i2\omega\sum_{m=0}^{\infty}\sum_{k=0}^{\infty}a_{mk}\,\mathrm{I}_m\left(\left(2k+1\right)\pi br\right)\sin\left(\left(2k+1\right)\pi z\right)\cos\left(2m\alpha\right)e^{i2\omega t}\right), \tag{7}$$

$$\varphi_e\left(r, z, t\right) = \mathrm{Re}\left(i2\omega\sum_{m=0}^{\infty}\sum_{k=0}^{\infty}b_{mk}\,\mathrm{K}_m\left(\left(2k+1\right)\pi br\right)\sin\left(\left(2k+1\right)\pi z\right)\cos\left(2m\alpha\right)e^{i2\omega t}\right), \tag{8}$$

$$\zeta(z,t)=\mathrm{Re}\left(\left(\sum_{m=0}^{\infty}\sum_{k=0}^{\infty}c_{mk}\sin\left((2k+1)\pi z\right)\cos(2m\alpha)+d_0\sin\left(\frac{z}{b}\right)+\right.\right.$$
$$\left.\left.+\sum_{m=1}^{\infty}d_m\,\mathrm{sh}\left(\frac{\sqrt{4m^2-1}}{b}z\right)\cos(2m\alpha)\right)e^{i2\omega t}\right), \tag{9}$$

where $\mathrm{I}_m$ and $\mathrm{K}_m$ are the $m$-th order modified Bessel functions. Substituting solutions (10)–(12) into (3)–(6), we obtain the linear system of non-homogeneous equations for the unknown amplitudes $d_0$ and $d_m$. Other coefficients $a_{mk}$, $b_{mk}$, $c_{mk}$ are calculated through these amplitudes.

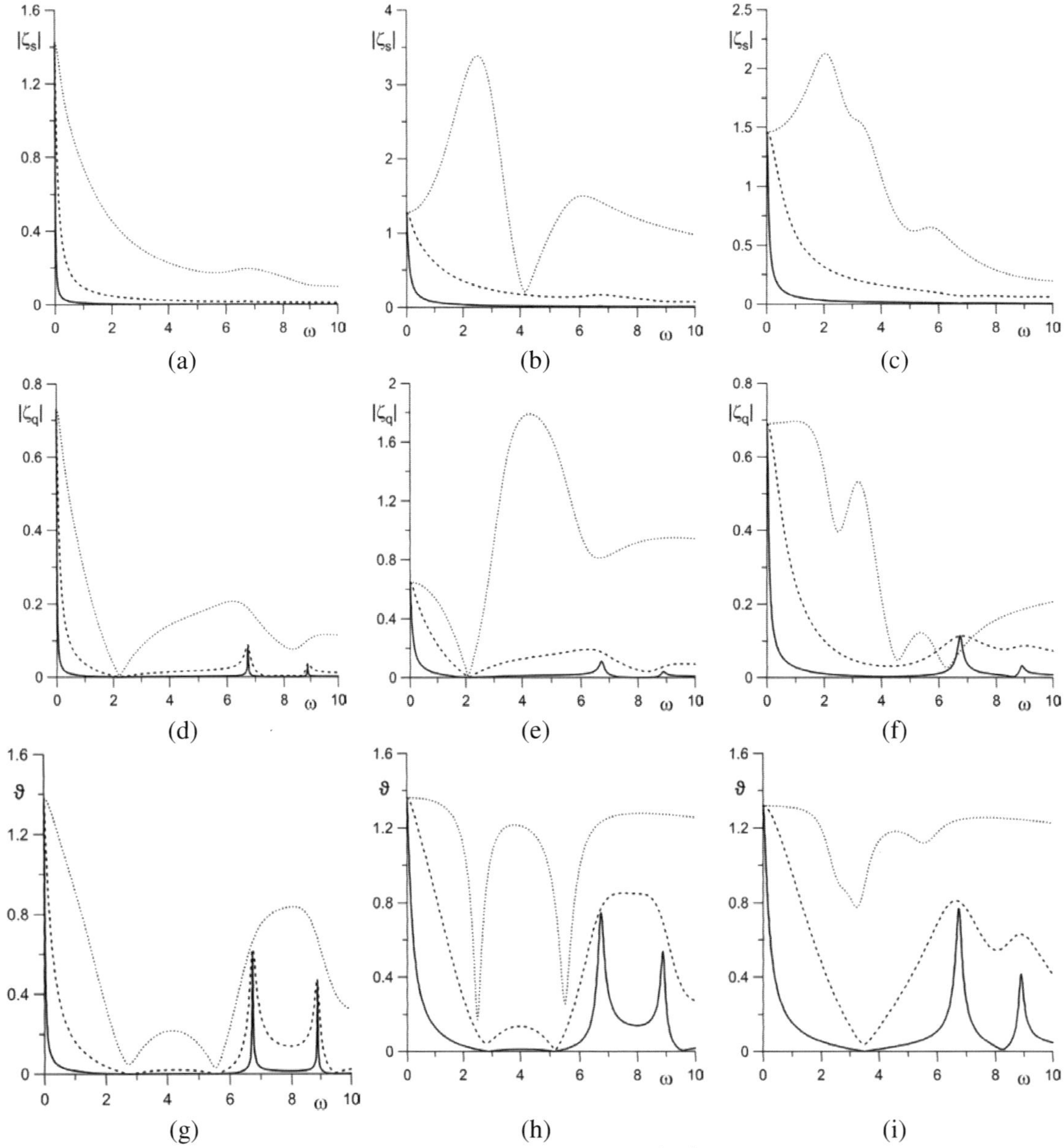

**Figure 3.** Deviation of contact line $|\zeta_s|$ (a-c), drop surface $|\zeta_q|$ (d-f) and contact angle $\vartheta$ at upper plate (g-i) vs frequency $\omega$
($a=5$, $b=1$, $\rho_i=0.7$), $\lambda_0=0.1$ – solid line, $\lambda_0=1$ – dotted, $\lambda_0=10$ – dashed.

(a,d,g) – $k_\lambda = 0.1$, (b,e,h) – $k_\lambda = 1$, (c,f,i) – $k_\lambda = 10$.

The dependence of the amplitude of the contact line and the surface oscillations at $z = 0.25$ and the contact angle at the upper plate on the frequency of the driving force is given in figure 3 for different values of the Hocking parameter and wavenumber. As it is seen from the plots in the fig.3a-c, there are no "antiresonances", i.e. the amplitude of the contact line goes to zero. These resonances were observed for a homogeneous substrate both in the case of mechanical vibrations [21,22] and in the case of electric field [11]: at certain values of $\omega$ the drop motion is independent of the wetting parameter and the contact line remains motionless at any $\lambda$. Thus, the eigenfrequencies of different oscillation modes change because of the heterogeneous surface and this effect of "antiresonances" drops out.

The wavenumber $k_\lambda$ plays the role of the effective wetting coefficient $\lambda_0$. The smaller the Hocking parameter $\lambda$, the greater the interaction between the contact line and the substrate and the smaller the amplitude of oscillations (see fig. 3a,c). Consequently, the long-wave heterogeneity of the substrate leads to a decrease in the oscillation amplitude. The deviation of the contact angle is less than $0.5\pi$ (see fig. 3g,i), i.e. complete wetting does not occur.

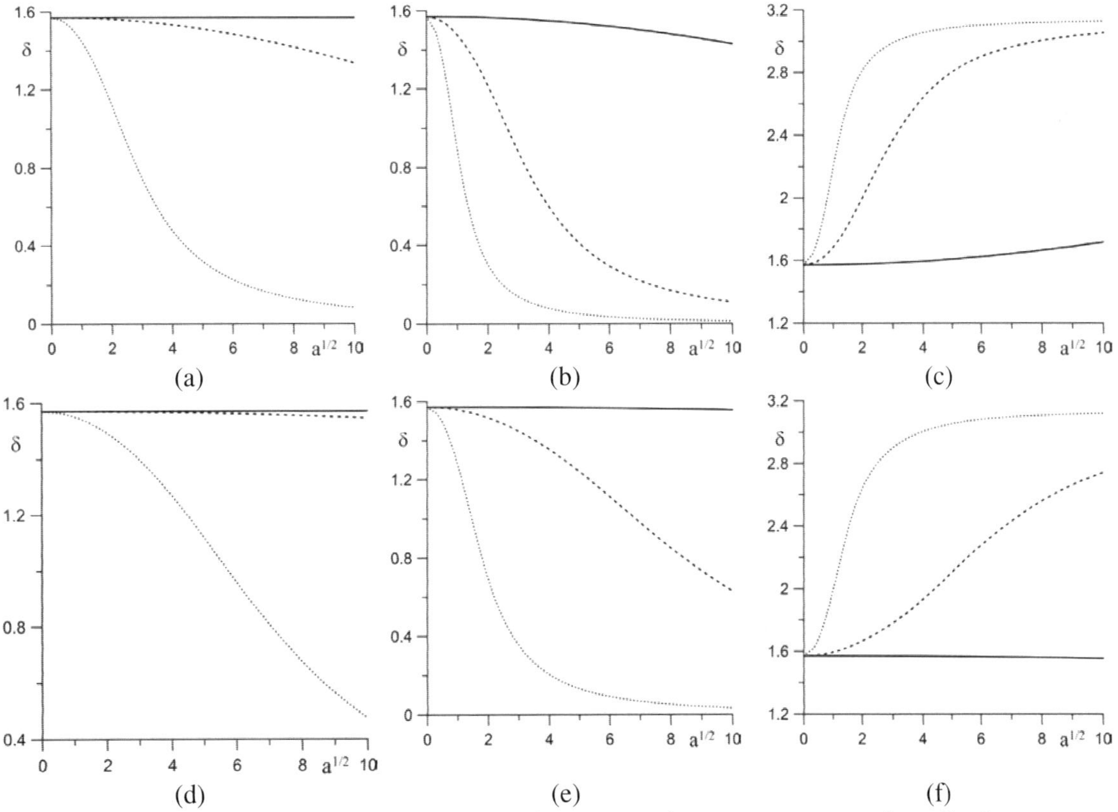


**Figure 4.** Deviation of contact angle at upper plate vs square root of amplitude ($b = 1$, $\rho_i = 0.7$), $\lambda_0 = 0.1$ – solid line, $\lambda_0 = 1$ – dotted, $\lambda_0 = 10$ – dashed.
(a-c) $\omega = 1$, (d-f) $\omega = 2$.
(a,d) – $k_\lambda = 0.1$, (b,e) – $k_\lambda = 1$, (c,f) – $k_\lambda = 10$.

In figure 4, the deviation of the contact angle is plotted against the square root of the amplitude $a$ (i.e. proportional to AC potential $V$ ) for different values of the Hocking parameter $\lambda$ and AC frequency $\omega$ . The responses obtained qualitatively agree with the experimental data. However, the maximum deviation of the contact angle tends to $\pi/2$, i.e. $\vartheta \to 0$ or $\vartheta \to \pi$, whereas the contact angle found in the experiments is finite.

## 4. Conclusions

The behavior of the cylindrical drop between two solid plates has been considered taking into account the dynamics of the contact angle under the action of uniform electric field. The solid plates have heterogeneous surfaces described by the function $\lambda(\alpha)=\lambda_0\left|\sin\left(k_\lambda \cos(\alpha)\right)\right|$. The main purpose of this paper is to develop a method for studying the forced oscillations of the drop on heterogeneous substrates and determining the contact angle. The investigation of forced oscillations has shown that the wavenumber $k_\lambda$ plays the role of the effective wetting parameter.

For small values of the parameter $\lambda_0$, at which the energy of interaction between the contact line and plate is strong, the oscillations amplitude is small. In the opposite case, the amplitude of the surface forced oscillations is large and tends to infinity in the limit $\lambda_0 \to \infty$. There are no "antiresonant" frequencies, i.e. such external frequencies, at which the contact line does not move and the contact angle does not change.

## Acknowledgments

This work was supported by the Russian Science Foundation (project 14-21-00090).

## 5. References

[1] Mugele F and Baret J-C 2005 *J. Phys.: Condens. Matter.* **17**
[2] Berge B 1993 *Comptes Rendus Acad. Sci. II* **317**
[3] Chen L and Bonaccurso E 2014 *Adv. Colloid Interface Sci.* **210**
[4] Chevalliot S, Kuiper S and Heikenfeld J 2012 *J. of Adhesion Sci. Tech.* **26** 1909–30
[5] Hayes R A and Feenstra B J 2003 *Nature* **425** 383–5
[6] Roques-Carmes T, Hayes R A, Feenstra B J and Schlangen L J M 2004 *J. Appl. Phys* **95** 4389–96
[7] Kuiper S and Hendriks B H W 2004 *Appl. Phys. Lett.* **85**, 1128–30
[8] Li C and Jiang H 2014 *Micromachines* **5** 432-41
[9] Hua Z, Rouse J L, Eckhardt A E, Srinivasan V, Pamula V K, Schell W A, Benton J L, Mitchell T G and Pollack M G 2010 *Anal Chem* **82** 2310-16
[10] Li J, Wang Y,Chen H and Wan J 2014 *Lab Chip* **14** 4334-37
[11] Alabuzhev A A and Kashina M A 2016 *J. Phys.: Conf. Ser.* **681** 012042
[12] Hocking L M 1987 *J. Fluid Mech.* **179** 253-66
[13] Shklyaev S and Straube A V 2008 *Phys. Fluids* **20** 052102
[14] Bostwick J B and Steen P H 2014 *J. Fluid Mech.* **760** 5-38
[15] Savva N and Kalliadasis S 2013 *J. Fluid Mech.* **725** 462–91
[16] Savva N and Kalliadasis S 2014 *J. Fluid Mech.* **754** 515-49
[17] Brunet P, Eggers J and Deegan R D 2007 *Phys. Rev. Lett.* **99** 144501
[18] Fayzrakhmanova I S and Straube A V 2009 *Phys. Fluids* **21** 072104
[19] Fayzrakhmanova I S, Straube A V and Shklyaev S 2011 *Phys. Fluids* **23** 102105
[20] Mampallil D, Eral H B, Staicu A, Mugele F and van den Ende D 2013 *Phys. Rev. E* **88** 053015
[21] Lyubimov D V, Lyubimova T P and Shklyaev S V 2006 *Phys. Fluids* **18** 012101
[22] Alabuzhev A A 2016 *Appl. Mech. Tech. Phys.* **53** 9–19

# Computer modelling of technogenic thermal pollution zones in large water bodies

**Ya N Parshakova[1], T P Lyubimova[1, 2]**
[1] Institute of Continuous Media Mechanics UB RAS, Acade. Koroleva 1, Perm 614013, Russia
[2] Perm State University, Bukireva 15, Perm 614990, Russia

E-mail: parshakova@icmm.ru

**Abstract.** In the present work, the thermal pollution zones created due to discharge of heated water from thermal power plants are investigated using the example of the Permskaya Thermal Power Plant (Permskaya TPP or Permskaya GRES), which is one of the largest thermal power plants in Europe. The study is performed for different technological and hydrometeorological conditions. Since the vertical temperature distribution in such wastewater reservoirs is highly inhomogeneous, the computations are performed in the framework of 3D model.

## 1. Introduction

The density stratification effects caused by inhomogeneities of temperature and mineralization fields can play an important role in the formation of both hydrological and hydrochemical regimes of surface water bodies. However, they can not be described in the framework of hydrodynamical models based on the shallow water approximation. Such effects can be accurately described only using the hydrodynamic models in three-dimensional formulation.

The significance of the problem under consideration is confirmed by the fact that it came into the view of researchers as early as in 30-ies of the XX century which gave an impetus to the development of the first applied models of "plane" hydrodynamics for its solution [1,2]. There are a number of factors why finding solutions to these problems is a challenging task. Among these are the fractality of morphometry of natural water bodies, considerable difference in the scales of natural and technological parameters, essential variability of hydrometereological factors.

Earlier, the restricted efficiency of the computational means was one of the biggest limiting factors. That is why, starting from the pioneering works of N.M Bernadskii [1,2] and until recently, the most widely used approach to these problems was 2D modelling based on the shallow water approximation (see, for example [3-7]). In [3], the authors outline the methods of computational hydrodynamics which are based on 1- and 2-D approaches. The compact finite difference methods are described in в [4]. In [5], a model of sea ice motion taking into account the atmospheric and water flows due to temperature inhomogeneities is constructed based on the data of Arctic and Antarctic regions. A 2D-numerical simulation of the spread of thermal pollution in the coastal area of the Red Sea is performed in [6]. An effect of thermal pollution produced by Iran thermal power plant is numerically investigated in [7] with the use of 2D approach. The solution is constructed for three different scales - 25 km x 25 km, 12 km x 12 km, 7 km x 7 km. Consideration is given to different scenarios of meteorological and season conditions. However, numerous data of field observations strongly suggest the need for revising the formulation which is based on the two-dimensional representation of the examined fields and homogeneity of the depth distribution of water temperature. Therefore, to obtain more adequate results, it is necessary to change to 3D models.

In works [8-10], a three dimensional numerical simulation of turbulent mixing of water masses of different temperatures was carried out based on the solution of the Navier-Stokes equations and LES model of turbulence. Computations were done with the use of a three – stage scheme with parallelization. A dependence of the computing time on the number of grid nodes and processors was obtained. The computational aspects of the LES model of turbulence were discussed in [11-12].

The authors of the present paper developed hydrodynamical models of surface water bodies, which are based on 3D numerical simulations [13-15]. In the present paper, with reference to the Permskaya thermal power plant (TPP) (Permskaya GRES), we investigated the temperature fields generated due to discharge of heated waters depending on technological and hydrometeorological parameters. It is constructed 3D model for the domain of 10 km which includes the intake and discharge channels of the Permskaya TPP.

## 2. Computational technique.

The 3D hydrodynamical model was built for the reservoir part of linear dimensions of 1000 m adjacent to the Perm TPP and including the points of water intake and water discharge. Software package ANSYS Fluent was used for the 3D simulations at the computer cluster URAN of the IMM UB RAS. The problem was solved in the framework of a non-stationary non-isothermal approach on the basis of the $k-\varepsilon$ model describing turbulent pulsations. We implement the Reynolds-averaged Navier-Stokes equations:

$$\frac{\partial \rho}{\partial t}+\frac{\partial}{\partial x_i}(\rho u_i)=0 \tag{1}$$

$$\begin{aligned}\frac{\partial}{\partial t}(\rho u_i)+\frac{\partial}{\partial x_j}(\rho u_i u_j)=-\frac{\partial p}{\partial x_i}+\frac{\partial}{\partial x_j}\left[\mu\left(\frac{\partial u_i}{\partial x_j}+\frac{\partial u_j}{\partial x_i}-\frac{2}{3}\delta_{ij}\frac{\partial u_l}{\partial x_l}\right)\right]+\\ +\frac{\partial}{\partial x_j}\left[\mu_t\left(\frac{\partial u_i}{\partial x_j}+\frac{\partial u_j}{\partial x_i}\right)-\frac{2}{3}\left(\rho k+\mu_t\frac{\partial u_l}{\partial x_l}\right)\delta_{ij}\right]+\rho g_i\end{aligned} \tag{2}$$

Here, $\rho$ is the density, $x_i$ are coordinates (we use Cartesian coordinate system), $u_i$ are the velocity components, $\mu$ is the kinematic viscosity, $\mu_t$ is the turbulent viscosity.

The turbulence kinetic energy $k$ and rate of its dissipation $\varepsilon$ are obtained from the following transport equations:

$$\frac{\partial}{\partial t}(\rho k)+\frac{\partial}{\partial x_i}(\rho k u_i)=\frac{\partial}{\partial x_j}\left[\left(\mu+\frac{\mu_t}{\sigma_k}\right)\frac{\partial k}{\partial x_j}\right]+G_k+G_b-\rho\varepsilon \tag{3}$$

$$\frac{\partial}{\partial t}(\rho \varepsilon)+\frac{\partial}{\partial x_i}(\rho \varepsilon u_i)=\frac{\partial}{\partial x_j}\left[\left(\mu+\frac{\mu_t}{\sigma_\varepsilon}\right)\frac{\partial \varepsilon}{\partial x_j}\right]+C_{1\varepsilon}\frac{\varepsilon}{k}(G_k+C_{3\varepsilon}G_b)-C_{2\varepsilon}\rho\frac{\varepsilon^2}{k} \tag{4}$$

In equations (3)-(4), $G_k$ represents the generation of turbulence kinetic energy due to the mean velocity gradients $G_k=\mu_t S^2$ where $S$ is the modulus of the mean strain rate tensor, defined as $S=\sqrt{2S_{ij}S_{ij}}$, $S_{ij}=0.5\left(\partial u_j/\partial x_i+\partial u_i/\partial u_i\right)$, $G_b$ is the turbulence kinetic energy due to buoyancy, which is calculated as

$$G_b=g_i\left(\beta\frac{\mu_t}{\Pr_t}\frac{\partial T}{\partial x_i}\right) \tag{5}$$

where $\mu_t$ is the turbulent viscosity determined as: $\mu_t=\rho C_\mu k^2/\varepsilon$, where $C_\mu$ is a constant.

Simulation of turbulent heat transfer is performed using the Reynolds model similarly to that of turbulent momentum transfer. Hence, the equation of energy is expressed as

$$\frac{\partial}{\partial t}(\rho E)+\frac{\partial}{\partial x_i}\left[u_i(\rho E+p)\right]=\frac{\partial}{\partial x_j}\left(k_{eff}\frac{\partial T}{\partial x_j}+u_i(\tau_{ij})_{eff}\right) \quad (6)$$

where $E = ch + \frac{p}{\rho}$ denotes total energy, $h = C_p T$ denotes system enthalpy, $k_{eff}$ denotes effective thermal conductivity, and $(\tau_{ij})_{eff}$ is the stress tensor deviator defined as

$$(\tau_{ij})_{eff}=\mu_{eff}\left(\frac{\partial u_j}{\partial x_i}+\frac{\partial u_i}{\partial x_j}\right)-\frac{2}{3}\mu_{eff}\frac{\partial u_k}{\partial x_k}\delta_{ij} \quad (7)$$

The model constants $\Pr_t$, $G_{1\varepsilon}$, $C_{2\varepsilon}, C_\mu$, $\sigma_k$ and $\sigma_\varepsilon$ were taken to have the following values [16]: $\Pr_t = 0.85$, $C_{1\varepsilon} = 1.44$, $C_{2\varepsilon} = 1.92$, $C_\mu = 0.09$, $\sigma_k = 1.0$, $\sigma_\varepsilon = 1.3$.

The spatial discretization scheme of second-order accuracy was applied. Simulation of temporal evolution was carried out using an explicit second-order scheme.

Boundary conditions imposed on the edges of the computation domain were as follows. At the bottom and at the banks the no-slip conditions and fixed temperature were imposed: $u_1 = u_2 = u_3 = 0, \quad T = T_0$,

At the inlet to the computational domain the main flow velocity was assumed to have one nonzero component which was taken to be constant over the inlet cross-section, and the temperature was assumed to be equal to the background temperature of the reservoir water: $u_i = V_i$, $T = T_0$,

At water intake and water discharge points, the water velocity and temperature were taken to be constant: $u_i = V_{\text{intake}_i}$, $T = T_0$ at the input channel entrance and $u_i = V_{\text{entrance}_i}$, $T = T_{\text{entrance}}$ at its outlet. The upper boundary of the fluid was assumed to be free and non-deformable, and the effect of wind is taken into account by specifying the tangential stresses according to Ekman formula [17], the linear thermal transfer law was applied accounting for the surface heating from the surrounding air, the heat transfer coefficient was chosen from the analysis of the obtained in-situ measurement data.

A computational grid was generated with Gambit 2.4 package of ANSYS Fluent. The number of nodes through the depth of the computational domain was taken to be 21. The non-uniform mesh was constructed using the bottom morphometric data obtained from in-situ measurements in 2014. In a horizontal direction, the computational mesh consisted of tetragonal elements distributed uniformly along the entire length, with the characteristic linear size of 20m. The mesh included 400 hundred thousands of nodes.

To adapt the morphological data available in a coordinate-depth format to the capabilities of the mesh generator Gambit, the reservoir bottom morphology was represented as a set of simple geometrical objects of some specified resolution, which were then introduced into the file.

A code has been written to produce a batch file for the Gambit grid generator of the ANSYS Fluent package from the data array describing the reservoir bottom morphology. Thus, the complex geometry of the computational domain is realized. The proposed code is of general character and is applicable to the construction of similar geometries and in other tasks.

Some preliminary studies have been conducted to get a numerical solution to the problem of interest using ANSYS Fluent. First, we obtained a stationary solution to the examined problem ignoring wind effects. This solution takes several hundreds of iterations to rapidly converge. Then, the problem was solved using a non-stationary approach with the time step of 2 seconds in the presence of wind effects.

## 3. Results of calculations

3D numerical modeling was conducted for different scenario conditions of the impact of the Perm TPP on the Kamskii reservoir. The calculation results for two of these critical scenarios which are of most interest from the ecological and technological points of view are given below. According to the current regulatory documents determining the permissible loads on fishery facilities and the Kama Reservoir, it is not allowed to increase the water temperature to more than 28°C in summer.

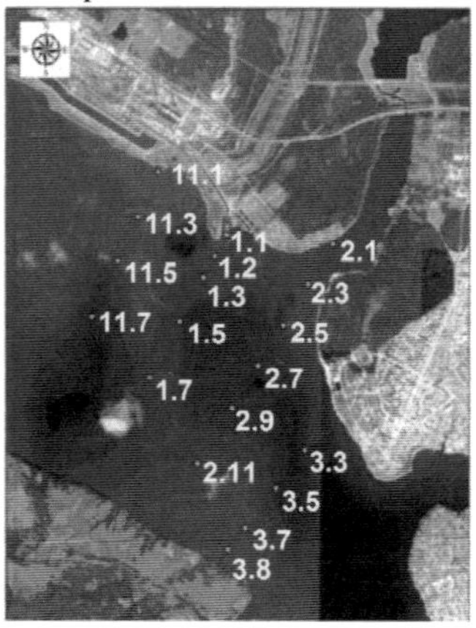


**Figure 1.** Map with the control points location. Each point is denoted by two numbers, of which the first corresponds to the number of a control line, and the second to the point number along the line.

### 3.1 First scenario

The goal of this study is to assess the zone of thermal pollution in the "normal" mode of operation of Permskaya GRES and for the most probable meteorological and hydrological conditions.

Technological parameters: 3 operating power units (2 stream-power units and 1 combined-cycle unit under construction), flow rate of discharged water - 42.5 $m^3$/sec, and discharge water temperature – 32.4 °C.

Under these conditions, the formation of the flow structure in the surface layer is governed by the wind action and the discharge of heated waters from the Permskaya GRES. In general, the flow acquires a unidirectional structure rather quickly. When approaching the mixing zone, the surface flow velocities are ~0.1-0.15 m/s. The flows formed by the discharge of heated waters have a velocity of about 0.2-0.4 m/sec (Fig. 2a). The wind practically does not shif the stream to the bank of the river, intensive propagation of the heated waters all over the reservoir takes place, due to which there is intensive mixing of waters up to the depth of 3-4 meters is observed (Figure 3). Figures 3a-d show the depth-distribution of the water temperature in the control points 1, 2, 3 and 11 for the first scenario 1. As one can see from Figures 11a, b, c, in the control sections 1, 2, 3, the flow of warm water is directed downstream, there is a significant heterogeneity of the depth-distribution of temperature. At the entrance to the working channel (Fig. 11 g, control point 11), temperature inhomogeneities in depth and width of the reservoir are not observed. However, along the left bank, were the outflow channel is located, there is a significant increase in temperature downstream. The area at which the temperature increases with respect to background values by 3 degrees or more is about 1.2 km2.

### 3.2 Second scenario

The goal of calculations for this water discharge variant is to determine the zone of thermal impact in the presence of southern winds at most probable meteorological and technological parameters. Technological parameters are the same as in variant 1. Hydrological and meteorological parameters are the same as in variant 1 except for the wind direction. Sourth-east wind of velocity of 3 m/s and 8 m/s.

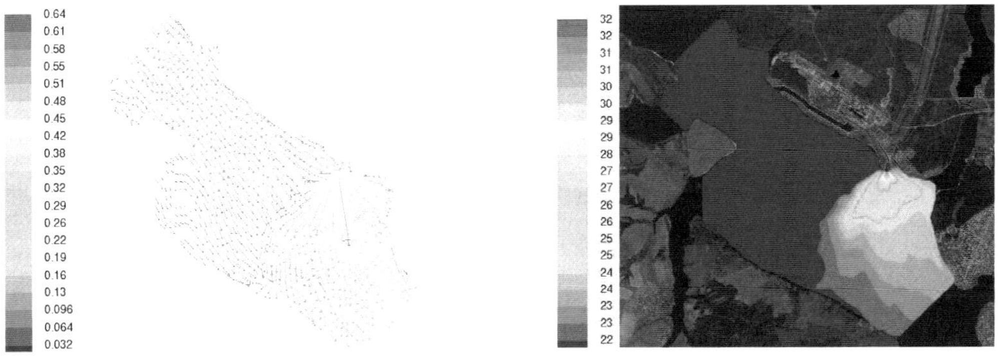

**(a)** **(b)**

**Figure 2.** Numerical data for the first scenario: (a) velocity vector field in the surface layer of the Kama reservoir, (b) temperature field (° C) in the surface layer of the Kama reservoir (the lines denote the boundaries of temperature rise by 3 and by 5 ° C relative to the background temperature)

Under these conditions, a rather complex flow structure is formed, both near the surface and at the depth. This is due to the fact that the runoff, wind and the discharge flows (formed due to the discharges of the Permskaya GRES) have nearly the same velocities, but they are multidirectional. In addition, there are the density effects due to the different temperatures of the streams. Therefore, there is no any distinguished flow direction at depths of up to 5-6 meters.

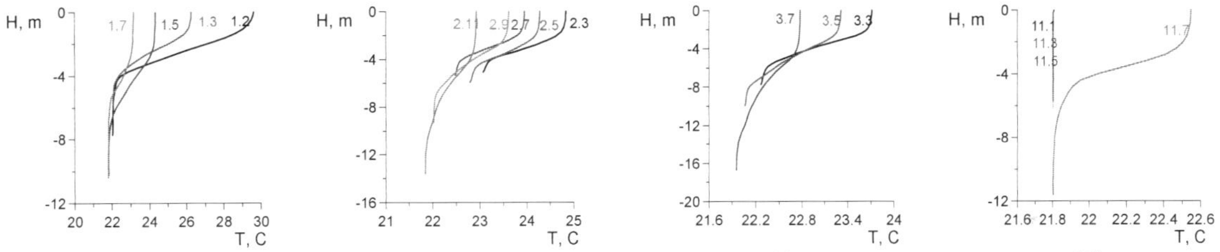


**(a)** **(b)** **(c)** **(d)**

**Figure 3.** Depth-distributions of water temperature in the control points for the first scenario: a) the first control point series, b) the second, c) the third, d) the eleventh

In the case of southeastern winds, the most unfavorable situation with the arrival of warm water in the supply channel of Permskaya GRES is observed (Figure 4). In this case, the increase in wind velocity reduces the time-period needed for warm water to reach the supply channel and increases the intensity of the arrival of warm water and the increase in the flow rate exerts the opposite effect on these processes. Decrease in the flow rate increases the probability of warm water arrival to the water intake facilities. As follows from the results of calculations, the reverse flows from the outlet channel to the supply channel are observed only in the upper layer of thickness, as a rule, no more than 2-3 m.

In general, when the second scenario is implemented, the maximum thermal pollution effect is observed: the increase in temperature with respect to the background values by 3 and 5 ° C occurs, respectively, in the water area of 2.8 and 1.5 km$^2$. In the case of significant wind in the direction opposite to the direction of the flow in the river, a three-dimensional vortex is formed within a few hours, the horizontal dimension of which is equal to the distance between the interfaces of the supply and return channels with the reservoir, and the vertical dimension is equal to the depth of the river. The presence of this vortex leads to the motion of warm water against the flow in the river. In this case, less than in a day, warm water reaches the site of the intake of the cooling channel, which is extremely undesirable from a technological point of view. There is also a significant temperature inhomogeneity in depth, and the temperature gradient is greatest near the bottom of the river.

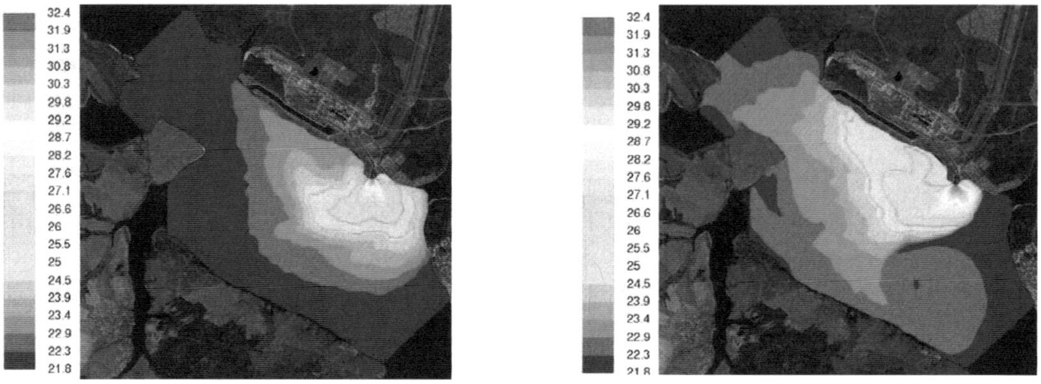


**(a)** **(b)**

**Figure 4.** Temperature field (° C) in the surface layer of the Kama reservoir (the lines indicate the boundaries of temperature increase by 3 and by 5 ° C relative to the background temperature), (a) SE wind 3 m/s, (b) SE wind 8 m/s

## 4. Conclusion

The study of hydrodynamic characteristics and the calculation of the propagation of thermal pollution zone in the river basin has been carried out using the example of Permskaya GRES. Numerical modelling based on the three-dimensional model yielded hydrodynamic characteristics of directions and velocities of flow. Calculation of the propagation of thermal pollution zone due to the discharge of heated water from Permskaya GRES were performed for different wind directions. It was found that the arrival of warm water into the supply channel during the ice-free period can be observed when the following conditions are realized: there is fairly strong south-eastern wind (W> 3 m / sec) during the long enough time (t> 12-15 hours); the flow rate of the water discharge in the Kama HPP section is low, Q <1000 m3 / sec; the temperature of the discharged water from the Permskaya GRES water is at least 30 ° C; the discharge rate is greater than 40 m3/s.

**Acknowledgments**
The work was supported by Russian Scientific Foundation (grant No. 17-77-20093).

## References

[1] Bernadksii N M 1933 *Mater. on hydrology, hydrography and water resource. of the USSR*, **20**
[2] Bernadksii N, Proskuryakov B 1933 *Theory and Practice of Calc.s of the Cooling Pond.* **1**
[3] Fletcher C A 1988 *Special Techniques for Differential Flow Categories* **2** 493
[4] Tolstykh A I 1990 *Compact Difference Scheme and Their Applications to Fluid D. Problems*
[5] Parkinson C L 1979 *J. Geophys. Res.* **84** 311
[6] Chieh S H 1987 *J. Hydraulic Engineer.* **113** (8) 1032
[7] Abbaspour M, Javid A H, Moghimi P, Kayhan K 2005 *J. of Envir. Science and Techn.* **2** (1) 13
[8] Issakhov A 2011 *J. of Physics* **318** (4) 42051
[9] Issakhov A 2013 *Power, Control and Optimization* **239** 165
[10] Issakhov A 2014 *Scientific World J.* doi 10.1155/2014/678095
[11] Tennekes H, Lumley J L 1972 *The MIT Press*
[12] Lesieur M, Metais O, Comte P 2005 *Large Eddy Simulation of Turbulence*
[13] Lepikhin A P, Lyubimova T P, Parshakova Y N, Tiunov A A 2012 *J. of Min. Science* **48** (2) 390
[14] Lyubimova T, Lepikhin A, Konovalov V, Parshakova Ya, Tiunov A 2014 *J. Hydrology* **508** 328
[15] Lyubimova T, Lepikhin A, Parshakova Ya, Tiunov A 2016 *J. of Hydrology*. **534** 579
[16] Launder B E, Spalding D B 1972 Lectures in Mathematical Models of Turbulence
[17] Welander P 1968 *Tellus XX* **1** 1

# Numerical simulation of solitary waves on deep water with constant vorticity

**A S Dosaev[1], M I Shishina[2] and Yu I Troitskaya[1]**

[1] Institute of Applied Physics of the Russian Academy of Sciences, 46 Ul'yanov Street, 603950, Nizhny Novgorod, Russia
[2] Nizhny Novgorod Planetarium named after G. M. Grechko, 20 Revolutsionnaya Street, 603002, Nizhny Novgorod, Russia

dosaev@appl.sci-nnov.ru

**Abstract**. Characteristics of solitary deep water waves on a flow with constant vorticity are investigated by numerical simulation within the framework of fully nonlinear equations of motion (Euler equations) using the method of surface-tracking conformal coordinates. To ensure that solutions observed are stable, soliton formation as a result of disintegration of an initial pulse-like disturbance is modeled. Evidence is obtained that solitary waves with height above a certain threshold are unstable.

## 1. Introduction

In theoretical studies of water waves it is common to consider the fluid motion as purely irrotational, yet wave dynamics can be strongly affected by interaction with shear currents. In nature shear is generated near the water surface due to the action of surface wind stress, or near the bed of a river or of the sea. An important special case is that of a linear shear flow, i.e. a flow with a uniform vorticity distribution, because in such a system only potential part of velocity field is time-dependent, and usual techniques developed for potential waves can be applied to study its dynamics.

Periodic stationary waves with constant vorticity were studied within a deep water setting in, for example, [1], periodic waves on water of finite depth in [2], and solitary waves on finite depth in [3]. These works describe a spectacular phenomenon of steep bulb-like waves that can propagate along the current. In the present paper we confine our attention to deep water waves propagating against a linear shear current. It is known that long waves propagating against the current are weakly dispersive, and according to weakly nonlinear theory on deep water their evolution is described by Benjamin-Ono equation [4], which has Lorentzian-shaped soliton solutions. The main aim of our work is to ensure stability of obtained solitary waves. To this end, we model soliton formation as a result of disintegration of an initial pulse-like disturbance.

To simulate a two-dimensional flow with constant vorticity one can employ the same efficient technique based upon use of conformal coordinates that was developed for potential waves. Exact equations of motion in conformal variables were derived in [5]. Here we use a slightly different formulation, similar to [6], which was obtained in [7] and applied there by the authors to modelling of dynamics and modulational instability of oscillatory nonlinear waves.

IOP Conf. Series: Journal of Physics: Conf. Series **955** (2017) 012018 doi:10.1088/1742-6596/955/1/012018

## 2. Basic equations

Let a two-dimensional ideal fluid flow occur on a current of depth $H_0$ which has a free surface and a constant vorticity. We consider the waves on the water surface which have the lengths much less than $H_0$, thus, for this waves the deep water condition is fulfilled and the fluid layer can be considered to be infinitely deep. Let the ideal incompressible fluid occupy the domain which is bounded by the free surface $F(x,y,t)=0$ (where $t>0$ is time) in the $(x,y)$ plane. The fluid flow velocity vector is $\mathbf{v}=\{v_x,v_y,0\}$. Cartesian velocity components are represented through effective potential $\varphi=\varphi(x,y,t)$ and its conjugate harmonic function $\Phi=\Phi(x,y,t)$ as

$$v_x=\frac{\partial\varphi}{\partial x}-\omega y=\frac{\partial\Phi}{\partial y}-\omega y, \qquad v_y=\frac{\partial\varphi}{\partial y}=-\frac{\partial\Phi}{\partial x} \tag{1}$$

where $\omega$ is a constant vorticity. Taking into account relations (1) from the condition of incompressibility of the fluid ($\operatorname{div}\vec{v}=0$) it follows that $\varphi$ satisfies the Laplace equation:

$$\Delta\varphi=0. \tag{2}$$

It is necessary to find the solution of the Laplace equation (2) with the boundary conditions

$$\left.\left(\frac{\partial\varphi}{\partial t}+\frac{1}{2}\left|\nabla\varphi-\omega y\cdot\mathbf{x_0}\right|^2+\omega\left(\Phi-\frac{\omega y^2}{2}\right)+\frac{p_a}{\rho}+gy\right)\right|_{F(x,y,t)=0}=0, \tag{3}$$

$$\left.\left(\frac{\partial F}{\partial t}+\frac{\partial\varphi}{\partial x}\frac{\partial F}{\partial x}+\frac{\partial\varphi}{\partial y}\frac{\partial F}{\partial y}-\omega y\frac{\partial F}{\partial x}\right)\right|_{F(x,y,t)=0}=0, \tag{4}$$

$$\left.\varphi_y\right|_{y\to-\infty}=0, \qquad \left.p\right|_{F(x,y,t)=0}=p_a. \tag{5}$$

Let an incompressible fluid occupy a domain $G$: $G=\{-\infty<x<+\infty,\ -\infty<y\le f(x,t)\}$ in the $(x,y)$ plane. We consider a complex function $z=z(\xi,\eta,t)=x(\xi,\eta,t)+iy(\xi,\eta,t)$ which is analytic in the domain $G$. We carry out a conformal mapping of the domain $G$ onto the half-plane $-\infty<\xi<+\infty$, $-\infty<\eta\le 0$ in the plane $\zeta=\xi+i\eta$ by means of the function $z(\xi,\eta,t)=\xi+i\eta+\tilde{z}(\xi,\eta,t)\equiv\xi+i\eta+\tilde{x}(\xi,\eta,t)+i\tilde{y}(\xi,\eta,t)$. The functions $\tilde{x}(\xi,\eta,t)$, $\tilde{y}(\xi,\eta,t)$ are harmonic in the half-plane $\operatorname{Im}\zeta\le 0$ and satisfy the Cauchy-Riemann conditions in this half-plane. The functions $y$ and $\tilde{x}$ are connected by the relation $y=\mathrm{H}(\tilde{x})$, where $\mathrm{H}$ is the Hilbert transform. The Jacobian $J=x_\xi^2+y_\xi^2\ne 0$. The functions $\varphi(\xi,\eta,t)$, $\Phi(\xi,\eta,t)$ are harmonic in the half-plane $\operatorname{Im}\zeta\le 0$ and satisfy the Cauchy-Riemann conditions in this half-plane. The "complex potential" $W=\varphi+i\Phi$ is an analytic function in the half-plane $\operatorname{Im}\zeta\le 0$, and $W=\varphi+i\mathrm{H}(\varphi)$ on the real axis. As a result of the conformal mapping, equations (3) and (4) take the form (at $\eta=0$ and $p_a=0$):

$$\left.z_t\right|_{\eta=0}=-iz_\xi(1+i\mathrm{H})\left.\left(\frac{\Phi_\xi-\omega yy_\xi}{J}\right)\right|_{\eta=0}, \tag{6}$$

$$\left.\left[\varphi_t+gy+\omega\Phi=\varphi_\xi\,\mathrm{H}\left(\frac{\Phi_\xi-\omega yy_\xi}{J}\right)+\frac{\omega yx_\xi}{J}\varphi_\xi-\frac{1}{2J}\left((\varphi_\xi)^2-(\Phi_\xi)^2\right)\right]\right|_{\eta=0}. \tag{7}$$

We introduce the new functions $z(\theta,t)=x(\theta,t)+iy(\theta,t)$, $\theta=\xi+i\eta$, $R(\theta,t)$, $V(\theta,t)$, using the formulas ([6]): $R=\frac{1}{z_\theta}$, $V=i\frac{W_\theta}{z_\theta}$. The functions $R(\theta,t)$, $V(\theta,t)$ are analytic in the lower half-plane and satisfy the boundary conditions: $V(\theta,t)\to 0$, $|\theta|\to\infty$, $\operatorname{Im}\theta\le 0$; $R(\theta,t)\to 1$, $|\theta|\to\infty$, $\operatorname{Im}\theta\le 0$.

The functions $R(\theta,t)$, $V(\theta,t)$ satisfy the following system of integro-differential equations (see [7]):

$$\begin{cases} R_t = i\left(R_\theta U - RU_\theta\right) \\ V_t = i\left(UV_\theta - R\,\mathrm{P}'\left(|V|^2 - 2\omega\,\mathrm{Im}\,z\,\mathrm{Im}V\right)\right) + g(R-1) + i\omega V \end{cases} \tag{8}$$

where $U = \mathrm{P}\left(VR^* + V^*R\right) - 2\omega\mathrm{P}\left(\mathrm{Im}z\cdot\mathrm{Im}R\right)$, P is the projection operator: $\mathrm{P}(f) = 1/2\left(1+i\mathrm{H}\right)f$ and prime denotes the derivative with respect to the complex variable $\theta$.

Equations in conformal variables (6)-(7) or equations in the Dyachenko variables (8) only contain integro-differential operators that are local in Fourier representation, which allows for very efficient computation using Fast Fourier Transform. In all numerical simulations described here periodic boundary conditions are implied.

## 3. Stationary waves

Let's consider a solution to (6)-(7) in a form of a stationary wave, that is, a wave propagating with constant phase velocity $c$ without changing its form:

$$x = \xi + \tilde{x}(\xi - ct),\; y = y(\xi - ct),\; \varphi = \tilde{\varphi}(\xi - ct) - gb_0 t. \tag{9}$$

Substituting (9) into (6)-(7) one can obtain an equation for stationary wave profile [9]:

$$gy + \frac{\left(c + \omega\left(\mathrm{H}\left(yy_\xi\right) + yx_\xi\right)\right)^2}{2J} - \frac{1}{2}c^2 = gb_0. \tag{10}$$

Equations of motion (2)-(5) contain only two dimensional parameters $g$ and $\omega$, and we can eliminate them by changing to appropriate time and length scales

$$X = x/\lambda_g, Y = y/\lambda_g, T = |\omega|t, \text{ where } \lambda_g = \frac{g}{\omega^2},$$

to obtain a dimensionless system. It means that dimensionless stationary wave profiles are universal and can only depend on dimensionless phase velocity $C = c/c_0$ and period $\Lambda = \lambda/\lambda_g$. Profiles of solitary waves, which can be considered as waves with infinite period, constitute therefore a one-parametric family of solutions, and all their dimensionless characteristics (say, wave height $a/\lambda_g$ or velocity $c/c_0$) can be found as a function of single parameter.

To solve (10) numerically we have generalized an iterative method from [8] to the case of nonzero vorticity. Let's rewrite (10) as $gy + M/2J = q$, $M = \left(c + \omega\left(\mathrm{H}\left(yy_\xi\right) + yx_\xi\right)\right)^2$. The iterative method is based on the observation that not only Jacobian $J = x_\xi^2 + y_\xi^2$ can be computed for a given mapping $z(\theta)$, but also mapping $z(\theta)$ can be "recovered" from its Jacobian by the following manipulations: $h = \left(1 + i\mathrm{H}\right)\frac{\ln J}{2}$, $z = \partial_\xi^{-1}\exp h$. To obtain a numerical solution of (10) for chosen $\lambda$ and $c$, we initialize the procedure with some seed $J_{[0]} = \varepsilon\cos\left(2\pi\xi/\lambda\right)$, and then perform in a loop:

1. $h_{[n]} = \left(1 + i\mathrm{H}\right)\left(\frac{1}{2}\ln J_{[n]}\right)$
2. $z_{[n]}' := \exp h_{[n]}$ and zeroth harmonic of $z_{[n]}'$ is set to 1, then $z_{[n]} := \partial_\xi^{-1} z_{[n]}'$ and zeroth harmonic of $z_{[n]}$ is adjusted so that $\int yx_\xi d\xi = 0$ holds,
3. $q_{[n]} := g\frac{s_{[n]}y_{[n]}(\lambda/2) - y_{[n]}(0)}{s_{[n]} - 1}$, where $s = \frac{M(0)J(\lambda/2)}{M(\lambda/2)J(0)}$,

4. $J_{[n+1]} := -\ln \frac{2(q_{[n]} - gy_{[n]})}{M_{[n]}}$.

If the sequence $z_{[n]}$ converges it allows us to obtain the required wave profile with a good precision.

Figure 1 shows example profiles for various wavelengths, with wave height chosen to be close to the limiting value for each given wavelength. It is clear that wave crest becomes more localized with increase of wave period. However, as figure 2 shows, the domain in the parameter space where we were able to get a solution is rather limited. It therefore appears impossible to study much longer ( $\lambda > 10^2 \lambda_g$ ) and steeper waves with that method.

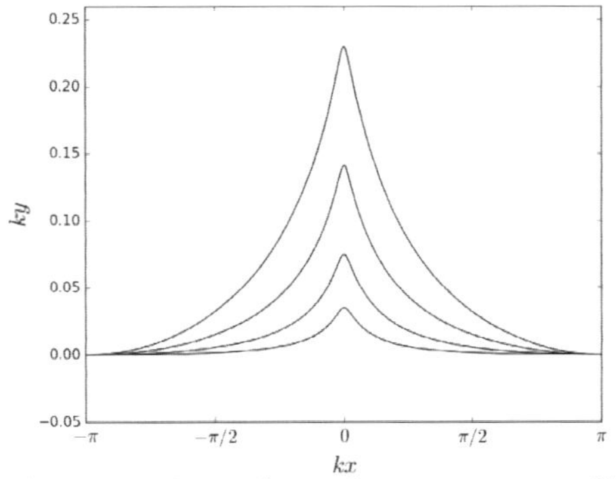


**Figure 1.** Example stationary wave profiles for various wavelengths (from the top down, lowest points of all profiles are aligned): $10\lambda_g$, $20\lambda_g$, $40\lambda_g$, $80\lambda_g$.

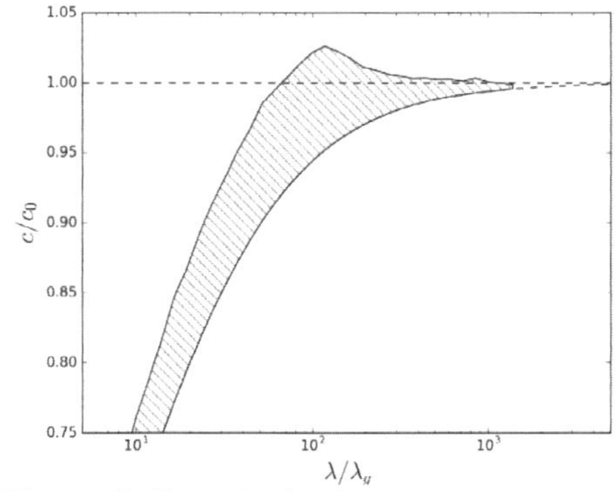


**Figure 2.** Domain in the parameter space where the iterative method for solving stationary wave equation converges. The lower boundary of the domain corresponds to the linear dispersion relation.

## 4. Initial disturbance disintegration

To observe formation of solitons we modeled evolution of initial pulse-like disturbances

$$y(x) = \frac{a}{1 + x^2 / d^2} w(x) \quad (11)$$

within the framework of equations (8). Window function $w(x) = 1/2 + \cos(2\pi x / L)/2$ is used here to smooth the jump of derivatives on the ends of the global period. Global period $L$ is chosen to be at least $10^3$ times greater than $d$. A pulse with width $d$ and height $a$ satisfying $ad = 4\lambda_g^2$ would correspond to a soliton of Benjamin-Ono equation for our system, but we have also run a few simulations with $d$ and $a$ not satisfying that constraint.

Figure 3 shows characteristics of solitons as functions of their height. Figure 4 compares exact solitons with solitons of Benjamin-Ono equation having same height and asymptotic. A noteworthy feature of these results is the fact that despite a wide range of initial parameters tested in our simulations, no waves higher than $\approx 0.42\lambda_g$ has been formed. Together with the evidence obtained below that there may exist a solution to (10) with greater height, it may indicate that solitary waves of height greater than the $\approx 0.42\lambda_g$ threshold are unstable. Other possible explanations include effects of the periodic boundary conditions and insufficient precision of the numerical model.

As we found out by conducting the simulations, even for the initial conditions that are close to the soliton of the Benjamin-Ono equation, it can be difficult to predict the characteristics of the resulting wave. To improve this, we can prescribe the initial conditions as a sum

$$y = f(x) = \sum_i \frac{a_i}{1 + x^2/d_i^2} w(x) \tag{12}$$

which allows us to fit both a sharp crest of an exact soliton and its $y(x) \sim 1/x^2$ tails. Then for a given vector of parameters $\{a_i, d_i\}$, having found a conformal parameterization $z(\xi)$ of the profile (12), we can compute the residual of stationary wave equation (10):

$$r(a_1, \ldots, d_1, \ldots) = gy + \frac{\left(c + \omega\left(x_\xi y + \mathrm{H}\left(yy_\xi\right)\right)\right)^2}{2J} - \frac{c^2}{2}$$

and use it as a criterion $\int |r|^2 d\xi \to \min$ for finding optimal $\{a_i, d_i\}$. Characteristics of the obtained profiles are in good agreement with the results of numerical simulations as is shown at figure 3.We can formally carry out this procedure to obtain profiles even higher than $0.42\lambda_g$. As a numerical experiment with an approximate profile of initial height $a = 0.47\lambda_g$ has shown, height and width of such initial disturbance remain almost unchanged for quite a long time ( $|\omega| t \sim 2 \cdot 10^4$ ), but then the wave quickly collapses.

## 5. Conclusion

In the present work we have studied the formation of solitons from an initial disturbance and compared their characteristics to the values predicted by weakly nonlinear theory. Results of the simulations indicate that solitary waves on deep water with constant vorticity may have an amplitude threshold above which they become unstable.

**Acknowledgements**
This research was supported by RFBR (grant No. 16-05-00839), and by the President of Russian Federation (grant No. MK-2041.2017.5). Numerical experiments were supported by RSF grant No. 14-17-00667, data processing was supported by RSF grant No. 15-17-20009.

## References

[1] Simmen J A and Saffman P G 1985 Steady deep-water waves on a linear shear current *Stud. Appl. Maths.* **73** 35-57
[2] Teles da Silva A F and Peregrine D H 1988 Steep, steady surface waves on water of finite depth with constant vorticity *J. Fluid Mech.* **195** 281-302
[3] Vanden-Broeck J M 1994 Steep solitary waves in water of finite depth with constant vorticity *J. Fluid Mech.* **274** 339-348
[4] Shrira V 1986 On the nonlinear waves on the surface of constant vorticity fluid layer *Soviet Physics, Doklady* **31(2)** 107-109
[5] Ruban V P 2008 Explicit equations for two-dimensional water waves with constant vorticity *Phys. Rev. E* **77** 037302
[6] Zakharov V E, Dyachenko A I and Vasilyev O A 2002 New method for numerical simulation of a nonstationary potential flow of incompressible fluid with a free surface *Eur. J. Mech. B-Fluids* **21(3)** 283-291
[7] Dosaev A S, Troitskaya Y I and Shishina M I 2017 Simulation of surface gravity waves in the Dyachenko variables on the free boundary of flow with constant vorticity *Fluid Dynamics* **52(1)** 58-70
[8] Chalikov D V and Bulgakov K Y 2014 Stokes waves at a finite depth *Fundamentalnaya i Prikladnaya Gidrofizika* **7(4)** 3-15 (in Russian).
[9] Shishina M I 2016 Steady surface gravity waves on a constant vorticity flow free surface *Processes in Geomedia* **8** 71-77

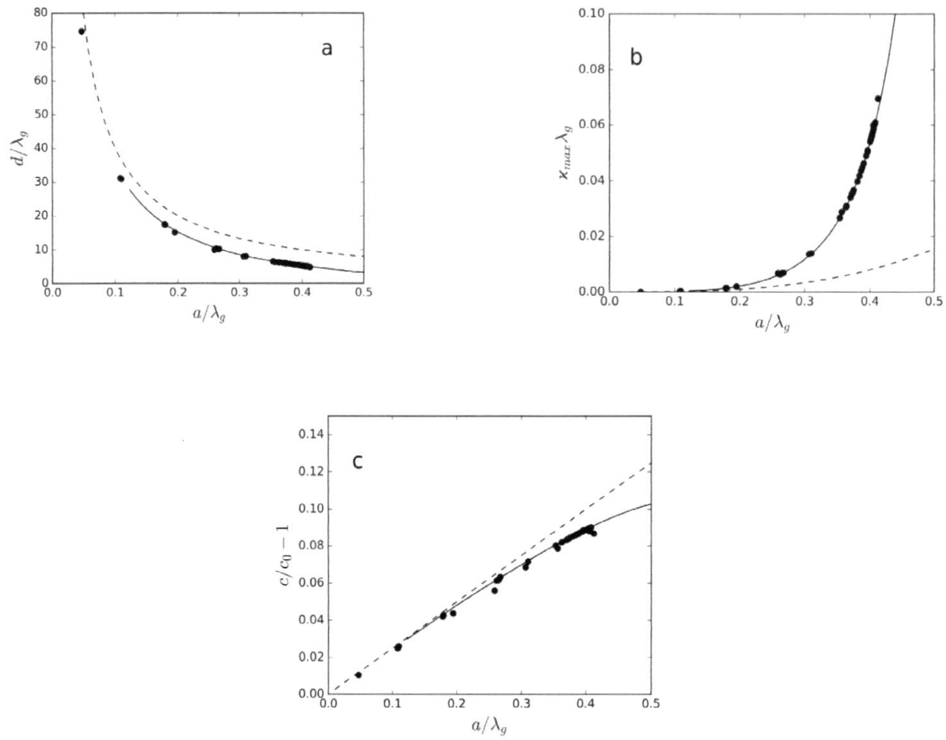


**Figure 3.** Characteristics of solitons as functions of height, a) width, b) crest curvature, c) velocity. Dashed line – solitons of Benjamin-Ono equation, black markers –solitons obtained from the numerical simulations, solid line – approximate solutions of the stationary wave equation (10).

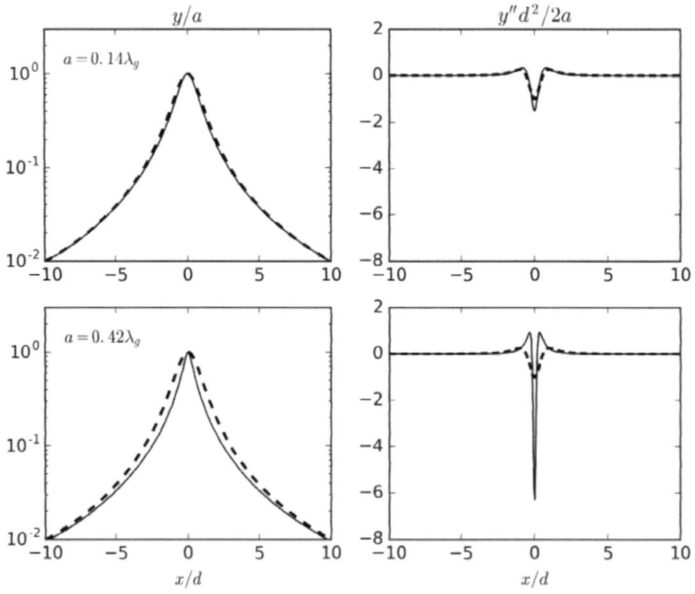


**Figure 4.** Comparison between exact solitons (solid lines) and solitons of the Benjamin-Ono equation (dashed lines)

# Numerical modeling of thermal regime in inland water bodies with field measurement data

**D. Gladskikh[1, 2], D. Sergeev[1], G. Baydakov[1], I. Soustova[1], Yu. Troitskaya[1]**

[1] Institute of Applied Physics of the Russian Academy of Sciences, Division of Geophysical Research
[2] Nizhny Novgorod State Technical University n.a. R.E. Alekseev, Institute of Radio Electronics and Information Technologies, Department of Information Systems and Technology

daria.gladskikh@gmail.com

**Abstract**. Modification of the program complex LAKE, which is intended to compute the thermal regimes of inland water bodies, and the results of its validation in accordance with the parameters of lake part of Gorky water reservoir are reviewed in the research. The modification caused changing the procedure of input temperature profile assignment and parameterization of surface stress on air-water boundary in accordance with the consideration of wind influence on mixing process. Also the innovation consists in combined methods of gathering meteorological parameters from files of global meteorological reanalysis and data of hydrometeorological station. Temperature profiles carried out with CTD-probe during expeditions in the period 2014-2017 were used for validation of the model. The comparison between the real data and the numerical results and its assessment based on time and temperature dependences in control points, correspondence of the forms of the profiles and standard deviation for all performed realizations are provided. It is demonstrated that the model reproduces the results of field measurement data for all observed conditions and seasons. The numerical results for the regimes with strong mixing are in the best quantitative and qualitative agreement with the real profiles. The accuracy of the forecast for the ones with strong stratification near the surface is lower but all specificities of the forms are correctly reproduced.

## 1. Introduction

Vertical temperature distribution is one of the main characteristics of natural geophysical objects. It has a great impact on biology and ecology of inland water bodies such as lakes and water reservoirs, it significantly affects eutrophication (enrichment of a water body with nutrients). In [1-3] the influence of temperature distribution on proliferation of microorganisms is established, particularly proliferation of cyanophyta and diatoms. Due to their small size, lakes and water reservoirs are sensitive to mixing and wind-wave coupling, so the thermal regime is characterized by strong seasonal and short-term variability, which was researched in [4-6]. It should also be noted, that there is a growing number of requests for forecasts of thermal regimes in inland water bodies [7] which cannot be obtained from global meteorological forecasts.

There are different models, one- and multi-dimensional ones, which are used for computation of the hydrological parameters, such as thermal regimes of inland water bodies. Models based on one-dimensional heat equation have fewer requirements for computing resources than multi-dimensional ones, and demonstrate satisfactory agreement with experimental evidence. On the other hand, these models require validation in accordance with the parameters of the given water body, and field measurement data are used for this purpose. One-dimensional model is submitted in the program complex LAKE, developed in [8] and used in this research. Modification of the program complex and its validation in accordance with the parameters of mid-sized lowland water bodies with an example of lake part of Gorky water reservoir (56°42′N 43°19′E) are provided. The data from the hydrometeorological stations closest to the water object and the files of meteorological reanalysis NCEP/NCAR were used for validation.

## 2. Model description and numerical implementation

The model is based on the one-dimensional heat equation:

$$c\rho\frac{\partial T}{\partial t} = \frac{1}{h^2}\frac{\partial}{\partial \xi}\left(\lambda\frac{\partial T}{\partial \xi}\right) + c\rho\frac{dh}{dt}\frac{\xi}{h}\frac{\partial T}{\partial \xi} - c\rho\frac{1}{h}B_w\frac{\partial T}{\partial \xi} - \frac{1}{h}\frac{\partial S}{\partial \xi}. \qquad (1)$$

The symbols are: c is heat capacity of the water, $\rho$ – its density, $\lambda$ - eddy diffusivity, T – temperature, $B_w = dh_0/dt$ – water balance at free surface of lake, S - solar radiation flux penetrating the water body. The beginning of the vertical downward z-coordinate is supposed to be coupled with water surface, so we can turn from z to the new independent variable $\xi = {}^{z}/_{h}$, where h=h(t) is a depth of water body, and t is a time.

The following generally accepted exponential relation is used for the calculation of the solar radiation flux penetrating the water body:

$$S(\xi) = S(0)\exp(-\alpha_e h\xi),$$

where $S(0)$ is solar radiation flux on the surface of water, $\alpha_e$- extinction coefficient. Temperature gradient (heat flux) is a boundary condition for the equation (1) on the air-water boundary and is calculated from the equation of heat balance:

$$S(1-\alpha) + E_a - E_s - H_s - LE_s = -\frac{\lambda}{h}\frac{\partial T}{\partial \xi}, \qquad (2)$$

where S is total solar radiation flux, $E_a$ – longwave radiation flux, $E_s$ – radiation of surface, $H_s$ and $LE_s$ are overt and covert heat fluxes respectively.

The finite-difference scheme of the equation (1) is as follows:

$$c\rho\frac{T_i^{j+1} - T_i^j}{\Delta t} = \frac{1}{(h^j)^2}\frac{\lambda_{i+\frac{1}{2}}T_{i+1}^{j+1} - (\lambda_{i+1/2} + \lambda_{i-1/2})T_i^{j+1} + \lambda_{i-1/2}T_{i-1}^{j+1}}{\Delta\xi^2}$$
$$+ c\rho\frac{\Delta h}{\Delta t}\frac{\xi}{h^j}\frac{T_{i+1}^{j+1} - T_{i-1}^{j+1}}{2\Delta\xi} - c\rho B_w^j\frac{1}{h^j}\frac{T_{i+1}^{j+1} - T_{i-1}^{j+1}}{2\Delta\xi} - \frac{1}{h^j}\frac{S_{i+1/2}^j - S_{i-1/2}^j}{\Delta\xi}.$$

Indexation of nodes in difference grid is used here: superscript represents the number of timestep, and subscript – the number of node on the vertical coordinate. The system of linear equations (4) has a tridiagonal matrix and can be solved with Thomas algorithm or cyclic reduction.

Equation of heat balance (2) is used to calculate the temperature on the air-water boundary and has the following submission in difference form:

$$E_a^* + S^*(1-A) - \delta\sigma\left(T_s^{j+1}\right)^4 - H_s\left(T_s^{j+1}, T_a^*\right) - LE_s\left(q_s^{j+1}\left(T_s^{j+1}\right), q_a^*\right) = -\lambda\frac{T_{s+1}^{j+1} - T_s^{j+1}}{h^j\Delta\xi}.$$

The symbols are the following: index s is the value on the surface, $(s+1)$ – the value of the node next to the surface, a is the value of the node on the atmosphere level where meteorological parameters are defined, asterisk * is values, which are given from field measurement data.

Eddy diffusivity $\lambda$ is the main mechanism to describe vertical heat and mass exchange in water objects. There are different ways of its parameterization, we use parameterization based on equations of kinetic energy of turbulence and its dissipation rate, or "$E-\lambda$"-parameterization. $\lambda$ is defined as follows:

$$\lambda = c\rho k,$$

where k is the coefficient of turbulence evaluated as

$$k = C_e \frac{E^2}{\varepsilon}.$$

Here $E = 1/2\left[\overline{(u')^2 + (v')^2 + (w')^2}\right]$ is kinetic energy of turbulence, the upper line is symbol of averaging, the values with primes are deviations from mean values, ε is dissipation rate of kinetic energy of turbulence, $C_e$ is the dimensionless coefficient. This equation is used to estimate E:

$$\frac{\partial E}{\partial t} = \frac{\alpha_E}{h^2}\frac{\partial}{\partial \xi}k\frac{\partial E}{\partial \xi} + \frac{\xi}{h}\frac{dh}{dt}\frac{\partial E}{\partial \xi} + P - \varepsilon, \quad (3)$$

where $\alpha_E$ is the dimensionless constant, and P is total generation of turbulence energy through the shear of velocity and effect of density stratification.

Dissipation rate of turbulence energy is defined with the following equation:

$$\frac{\partial \varepsilon}{\partial t} = \frac{\alpha_\varepsilon}{h^2}\frac{\partial}{\partial \xi}k\frac{\partial \varepsilon}{\partial \xi} + \frac{\xi}{h}\frac{dh}{dt}\frac{\partial \varepsilon}{\partial \xi} + C_1\frac{\varepsilon}{E}(P - \varepsilon), \quad (4)$$

where $\alpha_\varepsilon$ is the dimensionless constant, and $C_1$ is the function of Reynolds number, which is defined as follows:

$$Re = \frac{(2E/3)^2}{\nu\varepsilon}.$$

To approximate the kinetic energy of turbulence equation and its dissipation rate equation, Crank-Nicolson numerical scheme is used. Equations (3-4) are replaced with the finite-difference analogues:

$$\delta_{\bar{t}}E_{i+1/2}^{j+1} = \frac{\alpha_E}{h_j}\delta_\xi\left(k_{i+1/2}^{j+1}\delta_{\bar{\xi}}E_{i+1/2}^{j+1/2}\right) + \frac{\xi_{i+1/2}}{h^j}\delta_{\bar{t}}h^{j+1}\delta_{\xi}E_{i+1/2}^{j+1/2} + P_{i+1/2}^{j+1/2} - \varepsilon_{i+1/2}^{j+1/2},$$

$$\delta_{\bar{t}}\varepsilon_{i+1/2}^{j+1} = \frac{\alpha_\varepsilon}{h_j}\delta_\xi\left(k_{i+1/2}^{j+1}\delta_{\bar{\xi}}E_{i+1/2}^{j+1/2}\right) + \frac{\xi_{i+1/2}}{h^j}\delta_{\bar{t}}h^{j+1}\delta_{\xi}E_{i+1/2}^{j+1/2} + C_{1\,i+1/2}^{\,j+1}\frac{\varepsilon_{i+1/2}^{j+1}}{E_{i+1/2}^{j+1}}\left(P_{i+1/2}^{j+1/2} - \varepsilon_{i+1/2}^{j+1/2}\right).$$

Boundary condition for equation of kinetic energy of turbulence on the air-water boundary is as follows:

$$-\frac{k}{h}\frac{\partial E}{\partial \xi} = k_{we}\left(\frac{\tau}{\rho}\right)^{\frac{3}{2}}, \quad (5)$$

where $k_{we}$ is coefficient of turbulence enhancement by wave breaking and τ is surface stress. Assumption about continuity of impulse flow is used for zonal and meridional components of current velocity on the air-water boundary, and these components are computed with the theory of similarity by Monin-Obukhov, whereas these components on the boundary water-soil are defined with Chezy formula [9].

## 3. The data and parameters used for computation

The input parameters and data, which are recorded during all computation time, are set at the start of the program. Also meteorological data are used but they are read on the every step of computation.

The input temperature profile was defined with temperature difference between the surface and the bottom and the thickness of mixed layer in previous version. Three points were accounted: temperature on the surface, on the lower boundary of mixed layer and on the water-soil boundary. This limited the possibility of using the results from field measurements. In the modified version the input profile is arbitrary defined as a two-dimensional array of points from the file with values of temperature on every step of the depth. This modification allows to reproduce thermocline stratification typical for mid-sized lowland water bodies. The values of main physical parameters used in the model are either set through field measurement data or adjusted on the basis of correspondence between numerical results and real data. Therefore, the coefficient of solar radiation extinction in water body is set through measurements with Secchi disk drawing on the theory developed in [10]. However, the coefficient of turbulence enhancement by wave breaking is taken equal to 10 for the best match between the forecast and the experimental evidence. Parameterization of this parameter is foreseen, and the dependence between its value and the value of wind velocity has to be determined for this purpose.

The parameters of meteorological conditions are as follows: air temperature, atmospheric pressure, specific air humidity, precipitation, zonal and meridional wind components, downward solar (shortwave) radiation and downward longwave radiation. These parameters are collected with combined method: information is taken from the data of hydrometeorological station in Gorodets and the files of global meteorological reanalysis NCEP/NCAR [11] with spatial resolution 1°. The timestep for meteodata is 6 hours. Information about wind velocity is used to compute surface stress on air-water boundary included in boundary condition for kinetic energy (5). In previous version empirical parameterization proposed in [12] was applied, but it ignored features of underlying surface:

$$\tau = 1.273 * 10^{-3} * w_r$$

where $w_r = \sqrt{u^2 + v^2}$, $u$ and $v$ – wind components. We use updated formula based on multi-year measurements of wind conditions on the experimental sites of Gorky water reservoir (see [13]):

$$\tau = 1.274 * 10^{-3} * w_r + 3.4 * 10^{-4} * w_r^2 + 4.9 * 10^{-5} * w_r^3.$$

## 4. Validation and results

Field measurement data were used for validation of modified model. These data were obtained from the observations during expeditions on the experimental sites of Gorky water reservoir in the period 2014-2017. Measurement of the vertical temperature profiles were carried out with CTD-probe. Several measurements were made during every expedition, the results were averaged. 2014 and the current year were chosen from a large volume of data, because the seasons had a great difference in the thermal regimes due to the difference in seasonal meteorological conditions (see fig.1). 2014 was characterized by intensive solar radiation, a high temperature and a slightly weaker wind. So the temperature profiles obtained in 2014 were characterized by significant temperature difference and strong stratification, whereas the profiles obtained in 2017 are homogeneous in depth owing to weak heat and strong mixing.

Time and temperature dependences in control points (surface, 1 meter and 8 meters) were reviewed to analyze numerical results. More of points are taken closer to the surface than to the bottom, because this layer is more exposed to variability and has a significant impact on eutrophication.

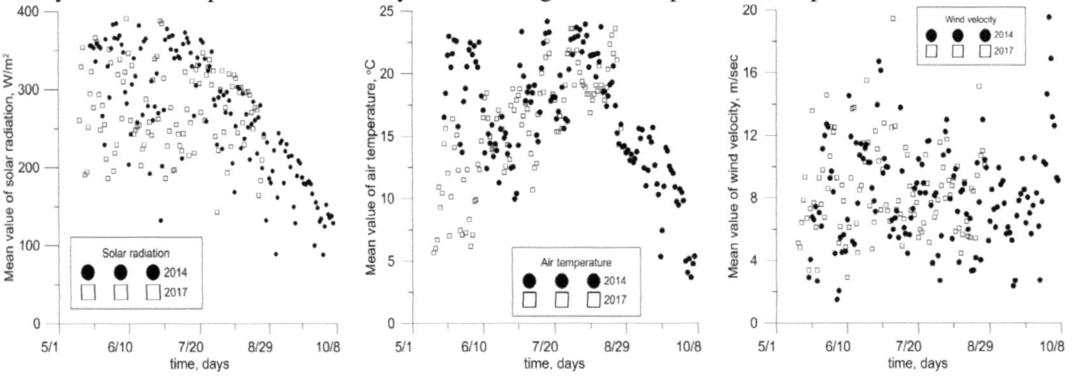


a b c

**Figure 1.** Average daily values of a) solar radiation, b) air temperature, c) wind velocity in 2014 and 2017.

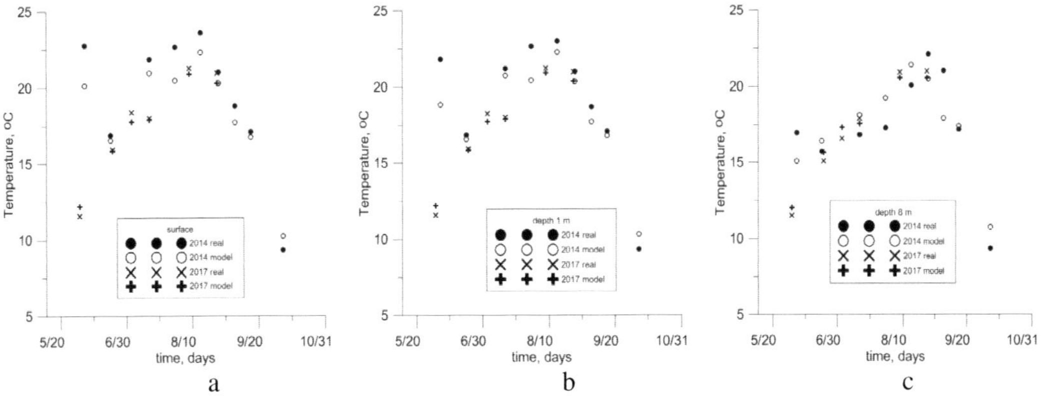


**Figure 2.** Values of temperature in control points: a) surface b) depth 1 meter c) depth 8 meters in the real data and in the model results.

It is evident that the greater the depth, the less the difference is. It confirms that variability and complexity of processes are greater near the surface due to wave-wind coupling. Also the forms of profiles were compared (see fig. 3). Numerical simulation reproduces general forms of distribution for all realizations. However, the best agreement is obtained for the regimes with strong mixing, when stratification is practically nonexistent.

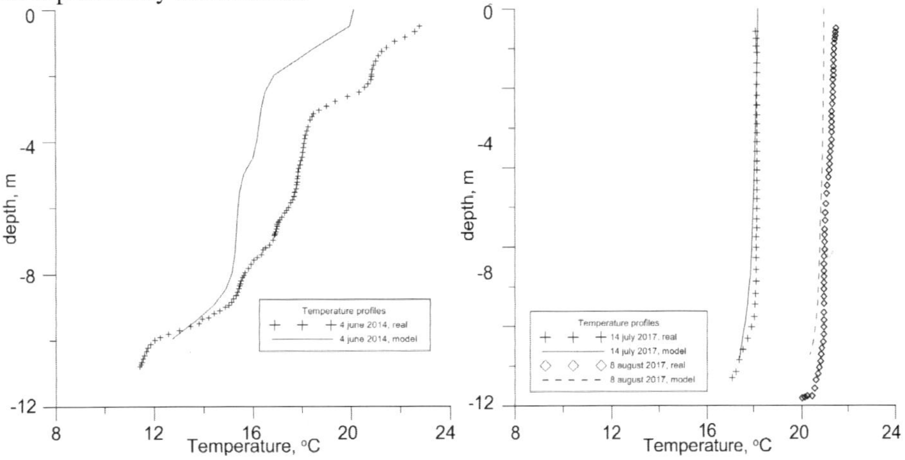


**Figure 3.** Real and model temperature profiles a) with strong stratification, 2014, b) with strong mixing, 2017.

As for the integral characteristics of comparison between the real data and the model results, standard deviation does not exceed 0,8°C for all performed realizations for 2017 year, and 1,5°C for 2014 year, which is more accurate than the results in the previous version [8].

## 5. Conclusion

Modification of the program complex LAKE, which is proposed in [8] and intended to compute the thermal regimes, and the validation of the complex in accordance with the parameters of mid-sized lowland water bodies with an example of Gorky water reservoir were realized. The modification caused changing the procedure of input data assignment and led to greater consideration of wind influence on mixing. The combined method was used for validation: field measurement data, files of global meteorological reanalysis and data of hydrometeorological station were used. Numerical simulation with the modified model gives results which are in good agreement with the real seasonal variations of thermal regime in lake part of Gorky water reservoir.

However, regimes with strong mixing and weak heat are reproduced more accurately, than regimes with strong heat and stratification, so the model results for 2017 match the field measurement data better, than the model results for 2014. The results were obtained with the same physical parameters. It can be explained by not quite adequate simulation of processes with weak wind and intensive solar radiation or by incorrect meteorological data, particularly data from reanalysis by reason of coarse grid.

## Acknowledgements

The work was partly supported by Russian Science Foundation project 15-17-20009 (numerical modeling) and RFBR projects 17-05-41117, 15-45-02580 (carrying out field measurements).

## References

[1] Gorbunov M Yu 2011 Vertical distribution of bacteriochlorophylls in humic lakes of the Volga-Kama National biosphere reserve (Tatarstan Republic) *Povolzhskiy Journal of Ecology* vol 3 pp 280-293

[2] Pezheva M H, Halilova F A, Zhantegolov D V, Kazanchev S C 2014 Major factors affecting the development of microorganisms in fishponds of Kabardino-Balkarian republic *Modern problems of science and education* vol 5 part 5 pp 1025-1029

[3] Sutorikhin I A, Bukaty V I, Akulova O B 2013 Seasonal Dynamics of Spectral Light Attenuation in Lakes of the Altai Krai *Izvestiya of Altai State University Journal* vol 1-2 (77) pp 184-188

[4] Stepanenko, V M, Martynov A, Jöhnk K D, Subin Z M, Perroud M, Fang X, Beyrich F, Mironov D and Goyette S 2013 A one-dimensional model intercomparison study of thermal regime of a shallow, turbid midlatitude lake *Geosci. Model Dev.* vol 6 pp 1337–1352 doi:10.5194/gmd-6-1337-2013

[5] Mironov D V, Golosov S D, Zilitinkevich S S, Kreiman K D and Terzhevik A Yu 1991 Seasonal changes of temperature and mixing conditions in a lake *Modelling air-lake interaction. Physical Background* ed. S S Zilitinkevich (Springer-Verlag, Berlin) pp 74-90

[6] Abbasi A, Annor F O and Giesen N V 2016 Investigation of temperature dynamics in small and shallow reservoirs, case study: Lake Binaba, Upper East Region of Ghana Water 8(3) p 84 doi:10.3390/w8030084.

[7] http://www.flake.igb-berlin.de/Lake17/

[8] Stepanenko V M 2005 Numerical modeling of thermal regime in shallow lakes Computational technologies vol 10 (Special issue. Proceedings of the International Conference and the School of Young Scientists “Computational and informational technologies for environmental sciences” (CITES 2005) Tomsk, 13-23 March 2005) part 1 pp 100-106

[9] Chebotarev A E 1975 Obshchaya gidrologiya (Leningrad)

[10] Ivanov A V, Troitskaya Yu I, Papko V V, Sergeev D A, Baydakov G A, Vdovin M I, Kazakov V I, Kandaurov A A, Afanasieva I M, Donskova O A, Shuvalova N M 2015 Stratification as a factor of influence on water quality of a plane reservoir *Privolzhsky Nauchnyy Journal* vol 2 pp 149–156

[11] https://www.esrl.noaa.gov/psd/data/gridded/data.ncep.reanalysis.html

[12] Beljaars A C M and Holtslag A A M 1991 Flux parameterization over land surfaces for atmospheric models *J. Appl. Meteor.* vol 30 pp 327–341.

[13] Kuznetsova A , Baydakov G A, Papko V V, Kandaurov A A, Vdovin M I, Sergeev D A, Troitskaya Yu I 2016 Adjusting of wind input source term in WAVEWATCH III model for the middle-sized water body on the basis of the field experiment *Hindawi Publishing Corporation, Advances in Meteorology* vol 2016 pp 13 article ID 574602 doi:10.1155/2016/8539127

IOP Conf. Series: Journal of Physics: Conf. Series **955** (2017) 012020 doi:10.1088/1742-6596/955/1/012020

# Fractal tomography and its application in 3D vision

**N Trubochkina**[1]

[1] National Research University Higher School of Economics, 121354, Russia

ntrubochkina@hse.ru

**Abstract**. A three-dimensional artistic fractal tomography method that implements a non-glasses 3D visualization of fractal worlds in layered media is proposed. It is designed for the glasses-free 3D vision of digital art objects and films containing fractal content. Prospects for the development of this method in art galleries and the film industry are considered.

## 1. Introduction

The method of fractal tomography - the method of artistic static and dynamic glasses-free 13D vision, based on the projection onto transparent screens of 3D multifractal sections, is offered. Unlike holography, the proposed method provides not a glasses-free 3D vision of an object "around", as in holography, but a vision of 3D spaces and objects through, and the total image changes when the observer's position changes. The proposed method uses the "cutting" of a 3D fractal space or object. The calculated images on the slices are projected onto several transparent screens displayed in the depth of the 3D fractal space or object.

## 2. Prehistory

3D vision in artificial systems, as a rule, is associated with stereokinetomatograph [1], [2], Autostereoscopic systems [3], holographic [4] and systems simulating a holographic image [5]. Biologically, the 3D vision is based on the fact that the eyes, as the input devices of a biological computer, get slightly displaced relative to each other two images of the visible world. The brain, analyzing and taking into account differences in graphic images, "calculates" the resulting picture, which already contains three-dimensional objects and creates a sense of depth. Stereoscopic vision was first described in 1838 by Charles Whetstone [6], but it should be noted that artists and scientists have been studying the three-dimensional vision long before this. Based on the analysis of the prehistory of the problem, it is clear that for the artificial (technical) realization of the glasses-free 3D vision, the brain needs to be provided with information: 1. about the presence of a 3D object (function F0); 2. and give the brain to understand that the visible space (object) has depth (functions F1-F8).

### 2.1. Signs of depth in 3D vision

In order to transform the sense of depth into functions (F1-F8) and algorithms, it is necessary to consider a combination of 3D vision features, "signals" for the brain, to which the following set of properties can be attributed: 1. F1 - the variable shape of various objects with a slight lateral displacement of the observer; 2. F2 - the presence of perspective. Remote objects appear to be smaller than closely located objects. "Parallel" lines intersect at infinity; 3. F3 - overlapping of objects. If one object closes part of the second object, then the brain understands that the first object is located closer. Overlap allows you to evaluate the relative positioning of objects in the image along the observation axis; 4. F4 - the shadows. Presence of areas of different illumination. Shadows, allow to add information to the brain about the relative location and height of objects; 5. F5 - the presence of parallax is determined by the change in the visible part and position of the object relative to the distant background, depending on the position of the observer. 6. F6 - the clarity of close textures on objects. Depending on the clarity or fuzziness of textures on 3D objects, the brain can estimate the distance to the object. The vague texture on the object, as a rule, speaks of its remoteness; 7. F7 - allowance for the thickness of the air. The greater the depth, the through the thicker layer of air filled with impurities (fog, suspended matter) are visible objects. Remote objects, as a rule, are covered with haze, and closely located objects are seen more clearly; 8. F8 - accommodation (focusing). The closer the 3D object, the more intense the muscles of the eyes, to ensure the sharpness of the image. Tension of the eyes gives the brain information about different distances to objects. At distant objects, eyes are relaxed. Artists and directors are familiar with the listed properties of 3D vision, and use them in their pictures, photographs and films.

## 3. Technical and information needs of artificial 3D vision without glasses

To implement artificial glasses-free 3D vision, which is the essence of the proposed method of mathematical (fractal) calculated tomography, consider its technical (devices) and information (functions, algorithms) needs: 1. technical needs include on what you can represent, reproduce calculated graphic information; 2. and information needs include the functions of 3D properties glasses-free 3D vision, as well as technologies and algorithms for preparing graphic information containing the signs of 3D vision. In addition to the raster and holographic method of the glasses-free 3D vision, there is another option that combines the solution of 2 problems: 1. the technical problem. Creating a multi-layer system for displaying specific graphic information containing all the features of 3D vision; 2. Information task. Synthesis of specific graphic information for each layer of the technical solution.

### *3.1. Overview and analysis of multi-layer visualization solutions*

An example of a layered presentation of graphic information can serve as multimedia exhibitions, where a multi-screen moving image is projected onto all possible planes of the room [7,8]. Another example of multi-layer visualization and attempts to create a 3D vision effect are rocking technologies. For the first time, the scenes, emphasizing, among other things, the depth of the stage, were applied in 1619 at the Farnese Theater in Parma, Italy [9]. In modern theater, the stage is equipped with only soft backstage, which can be projected, as a rule, independent images [10]. The peculiarity of these projects is that flat images of independent objects are applied in layers.

### *3.2. The closest analogue of the technical solution*

The closest analogue of the proposed 3D system is a multiscreen display of this type is Patent No. 2429513 - Three-dimensional display [11], where "a display with a three-dimensional screen consisting of a package of light-scattering liquid crystal modulators and a microprojector-based video projector forming images of sections of a three-dimensional object in the planes of location of light-scattering liquid crystal modulators". Features of this solution is a small area of 3D visualization, a large hardware cost and complexity with dynamic visualization. It is proposed to form a three-dimensional object using a series of screens, on each of which a corresponding section of the object is reproduced.

## 4. Technical implementation of the fractal tomography method

When creating a demo version of the installation for glasses-free 3D visualization from the prototype, the principle of projection on transparent screens and layered visualization of sections of 3D objects was taken. The **main difference** from the prototype is the way to create **graphic information in layers**. A schematic view of the installation is shown in Figure 1. This figure shows five transparent slightly diffusing screens, or planes, onto which different planes (sections) of a 3D object synthesized

mathematically or by a special method will be projected. When creating the model of the technical solution, the following assumptions were made: The screen does not absorb light, passes 90% of light, and everything else - 10% is reflected with a uniform indicatrix. The last image has less brightness than the first, since each screen absorbs a part of the light coming from the previous screens. It follows that, the number of screens should be limited, and the brightness of each subsequent should be increased, based on the absorption of light by the material of the screens. But this is an ideal case. In reality, we must also take into account the scattering of the transmitted light and its reflection from the next layers towards the viewer. For the experiment, three projection planes of the images were selected, based on the available number of projectors. The installation scheme with three projectors is shown in Figure 2. Since the projector is tilted at a large angle to the screen (Figure 3), the distance from the projector to the screen plane L1, L2 is different for the bottom and top parts of the screen, and therefore the width of the image h1, h2 on them will be different.
So there are trapezoidal distortions.

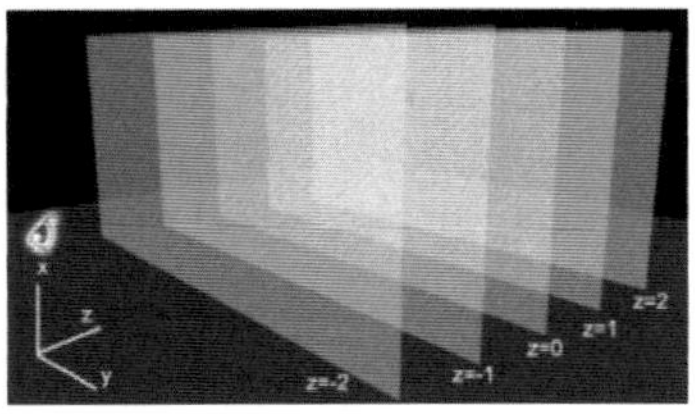

**Fig. 1.** Type of installation containing 5 transparent screens

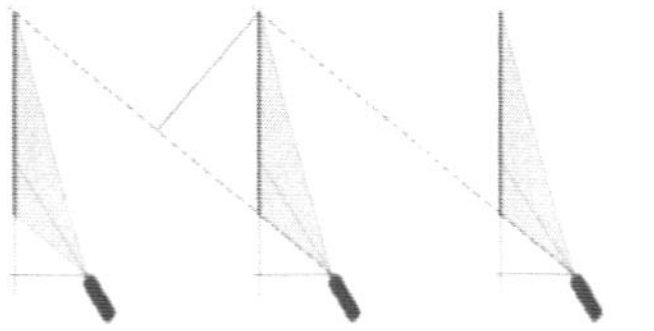

**Fig. 2.** Installation scheme with 3 transparent low-diffusing screens and 3 projectors

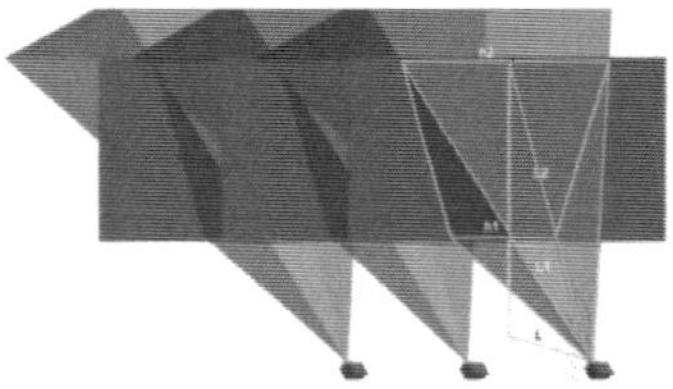

**Fig. 3.** The appearance of distortions of the trapezium type

Solve this problem by either introducing pre-distortion into the projected picture, or using the built-in projector to correct such distortions, or by using an ultra-wide-angle projector specially designed for this type of display, located at a distance L from the screen.

### *4.1. Technology and algorithms for creating and converting graphic information for layers of a multi-layer installation of fractal tomography*

The content or information component of the method of fractal tomography is 3D multifractals, mathematical 3D objects that have the ability to computer "cut" and "endow" sections (slices) of this object with all the properties of artificial 3D vision (F1-F8).

### *4.2. Create 3D object - 3D fractal world*

For example, consider the creation of the main 3D object, the sections of which will be subjected to various mathematical transformations, for the information realization of all the signs of an artificial 3D vision. For the experiment, a 3D multifractal consisting of 4 fractal functions (FF) was selected, which visualizes a 3D fractal world with objects similar to trees (1). The result of the calculations is shown in Figure 4. Consider the various possibilities of this 3D fractal world for use in the fractal tomography method, and the technology that provides its 3D-free 3D visualization.

{ FF1=_Rotate(X=0.75; Y=0; Z=0.375); FF2=_RotateC(X=22.875; Y=-48.375; Z==190.125);
FF3=_FlipXYc(c=1); FF4=AmazingSurf(Scale=1.2203125; MinR=0.125; Fold=1.15;
X=5.421875; Y=1.9296875; Z=-14.4296875; ScaleVary=0; Sphere_or_Cylinder=0) ) (1)

### *4.3. The ability to "cutting» 3D World*

To illustrate the calculation of the cross-section (slice) of a 3D fractal world, consider the "cutting" model depicted in Figure 5. The results of the cutting - sections or sections of the 3D multifractal are shown in Figure 6. The cross sections (slices) of the 3D multifractal are presented in 5 layers (for the transparent screens of the installation in Figure 5).

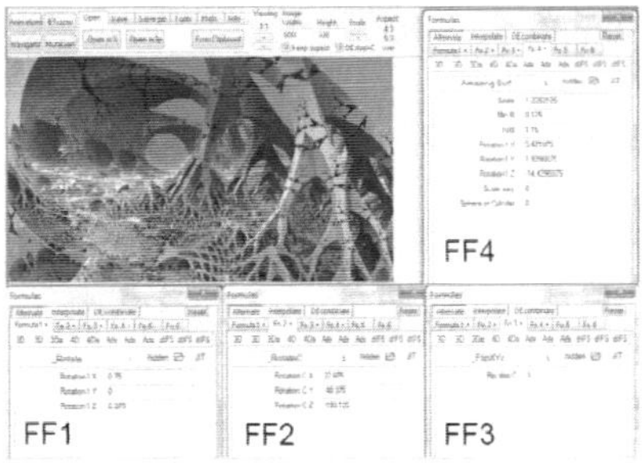


**Fig. 4**. 3D mathematical object - fractal world - 3D multifractal and fractal functions, describing it

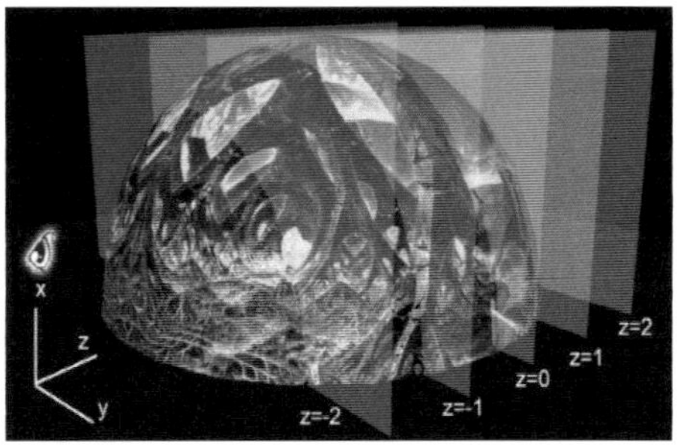


**Fig. 5.** Calculation of sections (slices) of 3D multifractal

The "cutting" of the 3D fractal object was carried out in the Mandelbulb3D program for different sections in different ways: 1. the base section z = 0 was calculated at the maximum value of the depth parameter (Depth); 2. the front sections (z = -1, z = -2) were obtained by decreasing the Depth calculation parameter of the multifractal; 3. the rear sections were obtained in the calculation using the Cutting function with the parameter for the Z axis of the coordinate of the cutoff of the rear section of the fractal 3D object.

*4.4. Algorithms for calculating the cross sections of a 3D multifractal with the implementation of 3D vision techniques with the signs F1-F8*

In addition to the main feature of the multifractal - 3D model of the object, after establishing the possibility of its mathematical and computer "cutting", let's consider the techniques (algorithms) of providing in the sections of 3D fractal world eight depth signs for 3D vision: 1. F1: the shape and size of various objects that are changed in layers; 2. F2: Presence; 3. F3: Overlapping objects in layers; 4. F4: Areas of different illumination. Shadows; 5. F5: The presence of parallax; 6. F6: Clarity of close textures on objects; 7. F7: Accounting for the content and thickness of the air to the object in the section of the 3D object; 8. F8: Accommodation.

*4.4.1. F0: the volume of objects - 3D fractal world or 3D computer model.*

The sign of dimensionality is embedded in the mathematical (for 3D multifractal) and computer (3D computer model) description of the object itself. Changing the spatial parameters gives a change to the 3D view of the object.

*4.4.2. F1: the shape and size of various objects that are changed in layers;*

This feature (F1) is provided by calculating the cross sections (cut) of the 3D multifractal corresponding to the depth of space. For computer implementation, the function of the cutoff from the back Depth (Depth) and the cutoff function from the front Cutting are used.

*4.4.3. Complex realization of signs of volumetricity and shadows: F0: 3D objects or models of objects (model + shadows) and F4: Areas of different illumination. Shadows*

As already noted, the bulk (feature F0) of the multifractal cross section is a consequence of its 3D mathematical model, and various shadows of the multifractal that enhance the bulkiness (sign F4) are obtained by including the Hard Shadow and Ambient Shadow modes.

## 5. Experiments on combining layers of 3D fractal world

To find the optimal variant of combining in 3 layers of 5 sections of the 3D fractal world, 3 algorithms were considered.

### 5.1. 1st algorithm of 3D visualization of the fractal world

The first algorithm for 3D visualization of the fractal world and the result of its use are shown in Figure 6. The algorithm is as follows: 1. on the transparent screen close to the viewer, the 3D cross section of the 3D multifractal is projected (Depth = 10000, Cutting (Z) = 0; Z is the coordinate by which the world is cut off. Fractal world is visible behind the plane of the section; 2. on the middle transparent screen, the cut from the front of the 3D multifractal slice is shown, Depth = 10000, Cutting (Z) = 0,14; 3. on the rear transparent screen, a more cross-cut 3D multifractal slice is shown, Depth = 10000, Cutting (Z) = 0.3.

### 5.2. 2nd algorithm of 3D visualization of the fractal world

The 2nd algorithm for 3D visualization of the fractal world and the result of its use are shown in Figure 7. It differs from the first by the sequence of the arrangement of sections (sections) on transparent screens. The 2nd algorithm for 3D visualization of a fractal world is as follows: 1. on a transparent screen close to the viewer, the cross-section of the 3D multifractal maximized at the rear (Depth = d1) is projected; 2.on the middle transparent screen, the 3D section of the 3D multifractal, Depth = d2, d2 <d1; 3. on the rear transparent screen, the base 3D multifractal Depth = 10000 is not cut off from behind or from the front. In Figure 7, clearer foregrounds of the 3D fractal world are visible, which creates a stronger effect of no glasses 3D visualization by fractal tomography. Hence it can be concluded that on the front transparent screens you need to have slices (sections) having darkened noninformative areas through which the images on the rear screens will be better seen.

### 5.3. 3rd algorithm of 3D visualization of the fractal world

The result of the 3rd algorithm of the fractal world 3D visualization use are shown in Figure 8.

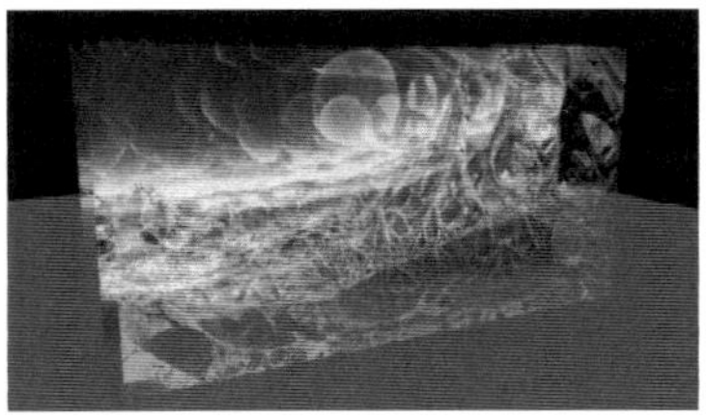

**Fig. 6.** A side view of the layered image of 3D multifractal according to the first algorithm

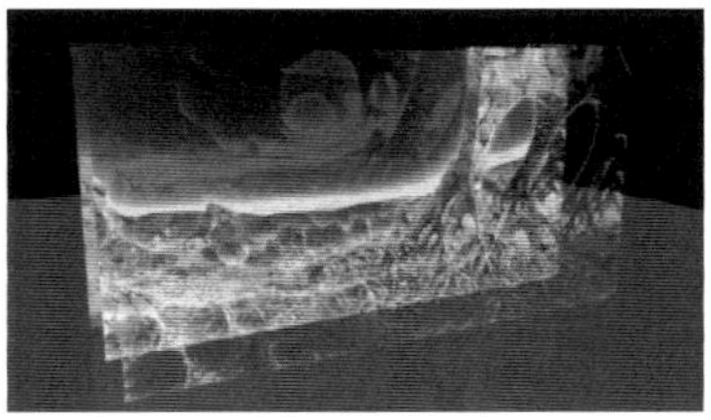

**Fig. 7**. A side view of the layered image of 3D multifractal according to the second algorithm

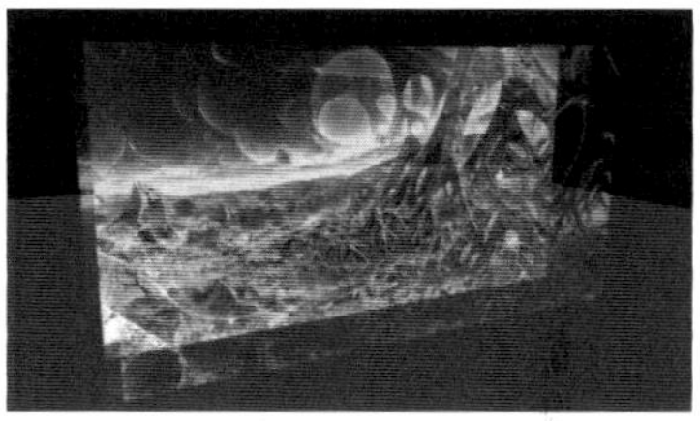

**Fig. 8**. A side view of the layered image of 3D multifractal according to the 3rd algorithm

It differs from the first and second algorithms in the sequence of arrangement of sections (sections) of 3D multifractal on transparent screens. The sequence of images on the screens in depth is as follows: the front section - the base 3D fractal world - the back cut. The third algorithm of 3D visualization of a fractal world is as follows: 1. on the transparent screen closest to the viewer, the maximum cross-section of the 3D multifractal; 2. on the middle transparent screen, the base 3D multifractal, Depth = 10000; 3. on the rear transparent screen, a more cross-cut 3D multifractal slice is shown, Depth = 10000, Cutting (Z) = 0.3. This algorithm is used when not only the front, but also the 3D multifractal backgrounds are important. The check of the developed method of fractal tomography was carried out on the mobile demo installation assembled by the author.

## 6. Mobile demo installation of the method of fractal tomography and Dynamic 3D visualization

In a real mobile installation, projectors with an image located on the axis of the lens were used. Figure 9b shows that the trapezoidal distortions are not eliminated. Since the usual projectors were used in the experiment, correction of trapezoidal distortions was carried out during the installation of the film by introducing distortions opposite to those that would be introduced when the projector was displayed at a large angle to the screen. The result is shown in Figure 9a. As a result of the correction, rectangular stratified cross sections were obtained, creating a total 3D non-zero visualization of the regular rectangular shape (Figure 9c). Using the developed method and a multi-screen installation, not only static but also dynamic 3D defect-free visualization of the fractal world in layered media was demonstrated, which can create the prerequisites for the development of a new cinema, created according to the "layered" principle of graphical information sections, using fractal mathematics as described in this article, as well as the additional use in the layers of 3D computer objects and real shootings in the technology of chromakey, and this is the project of the 3D free-glasses vision a new full-fledged artistic cinema.

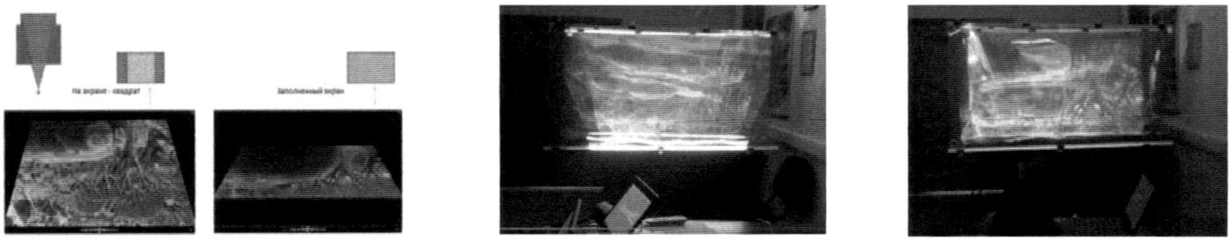

a b c

**Fig. 9.** View of a real mobile demo installation, with dynamic 3D visualization containing three screens from the perspective of the viewer

## 7. Conclusion

The method of artistic fractal tomography for glasses-free 3D visualization of art objects in light galleries and films in 3D cinemas is proposed. A layered model of free-glasses 3D visualization of fractal worlds is proposed, where each layer corresponds to a cut (cross-section) along the depth of the multifractal. An algorithm for calculating the cross sections of 3D fractal worlds has been developed, which makes it possible to obtain its cross sections in depth. Algorithms for layered free-glasses 3D visualization for various display requirements are developed. The basics of technical implementation for free-glasses 3D visualization of fractal worlds in layered media are developed. A demo version of the mobile installation for the free-glasses 3D visualization of fractal worlds was created, which can be used both in art galleries of fine art and show films with fractal content on it.

## References

[1] S. M. Provornov. Glava VIII. Stereoskopicheskoe kino // Kinoproekcionnaya tekhnika. — 2-e. — M.,: «Iskusstvo», 2004. — T. 1. — S. 123—130. — 458 s.
[2] S. Rozhkov Sistemy stereokinematografa, primenyavshiesya v SSSR (rus.) // «Mir tekhniki kino»: zhurnal. — 2006. —№ 1. — S. 37. — ISSN 1991-3400.
[3] Gorelov A.M., SHirankov A.F. Avtostereoskopicheskaya proekcionnaya sistema dlya formirovaniya polnocennogo 3D-obraza. Inzhenernyj zhurnal: nauka i innovacii/(12)2012.
[4] ZHerdev A.YU., Odinokov S.B., Lushnikov D.S., Solomashenko A.B. Funkciya rasseyaniya gologrammnoj stereogrammy, Inzhenernyj zhurnal: nauka i innovacii/(12)/2012.
[5] http://www.hologram3ddisplay.com/
[6] Kak smotret' 3D: so stereoskopom i bez nego // Mir 3D/3D World. № 1. 2012.
[7] http://grushenka.livejournal.com/33178.html
[8] http://carrieres-lumieres.com/fr/home
[9] K.A. Smolina «Sto velikih teatrov mira» Moskva. «Veche». 2002.
[10] http://bse.sci-lib.com/article067268.html
[11] Patent RU 2 429 513 C1. Trekhmernyj displej. Andreev Aleksandr L'vovich (RU), Sobolev Aleksandr Georgievich (RU), Kompanec Igor' Nikolaevich (RU).
[12] http://www.fractalforums.com/index.php?action=downloads;sa=downfile&id=64

IOP Conf. Series: Journal of Physics: Conf. Series **955** (2017) 012021 doi:10.1088/1742-6596/955/1/012021

# Identification of current-carrying part of a random resistor network: electrical approaches vs. graph theory algorithms

**Yu Yu Tarasevich, A S Burmistrov, V A Goltseva, I I Gordeev, V I Serbin, A A Sizova, I V Vodolazskaya, and D A Zholobov**

Astrakhan State University, 20A Tatishchev Street, Astrakhan, 414056, Russia

E-mail: tarasevich@asu.edu.ru

**Abstract.** A set of current-carrying bonds of a random resistor network (RRN) is called the (effective) backbone. The (geometrical) backbone can be defined as a union of all self-avoiding walks between two given points on a network or between its opposite borders. These two definitions provide two different approaches for identification of backbones. On the one hand, one can treat an arbitrary network as RRN and calculate potentials and currents in this RRN. On the other hand, one can apply to the network some search algorithms on graphs. Each of these approaches are known to have both advantages and drawbacks. We have implemented several different algorithms for backbone identification. The algorithms were applied to backbone identification for different system sizes and concentrations of conducting bonds. Our analysis suggests that a universal algorithm suitable for any problem is hardly possible to offer. Most likely, each particular task needs a specific algorithm.

## 1. Introduction

For many decades, the physical properties of inhomogeneous media have attracted a lot of attention in the scientific community [1], particularly, due to numerous applications such as the production and use of nanocomposites [2, 3]. Theoretical prediction of the effective properties of such composites is very important for the analysis of material performance and for the design of new materials. Particular interest is paid to the electrical properties of binary materials. The theories and models relating to the electrical conductivity of mixtures of conducting and insulating species continue to attract great interest from researchers. One of the possible ways to theoretical study of binary mixtures is consideration them as random resistor networks (RRN). When a potential difference is applied between two points of such the network or between two bus bars attached to the opposite borders of the RRN, the set of current-carrying bonds is called the backbone. The remaining bonds are called dangling ends or dead ends. The backbone can be also defined as the union of all the self-avoiding walks. These two definitions provide two different approaches for identification of backbones. On the one hand, one can use Ohm's law or Kirchhoff's rules to calculate potentials and currents in the RRN [4, 5]. On the other hand, one can apply search algorithms on graphs [6, 7, 8, 9, 10, 11, 12]. In fact, the last algorithms of backbone identification belong to maze solving algorithms which, particularly, are applied to wire routing on chips [13]. Each of these two approaches have both advantages and drawbacks. To the best of our knowledge, a systematic comparison and analysis of the algorithms devoted to identification of current-carrying part of the RRN is absent up to now. Particularly, the physical community is not familiar with search algorithms on graphs. In the present conference paper, we try to partially eliminate this gap. Our consideration is restricted only to two-dimensional (2D) RRN (planar graphs). The rest of the paper is constructed as follows. All necessary definitions are presented in section 2. In section 3, different method of backbone identification are briefly described and compared. Section 4 summarizes the main results.

## 2. Basic definitions

Consider an arbitrary 2D network of sites (nodes) connected by bonds (links). Each bond is treated a resistor with a specified electrical conductivity, $\sigma$. In the simplest case, conductivities of all resistors are equal. Suppose that this network is a subject to a potential difference. There are two natural possibilities [14], viz.,

- the "bus-bar geometry", when two parallel conducting bars (buses) are attached to the opposite borders of the network and potentials (say, $V$ and 0, $V > 0$) are applied to these buses [9, 10, 15] (figure 1(a)),
- the "two-point geometry", when a potential difference is applied to two distinct sites, so that an electrical current, $I$, injected at a one site (source) and the same current withdrawn from the other site (sink) [8] (figure 1(b)).

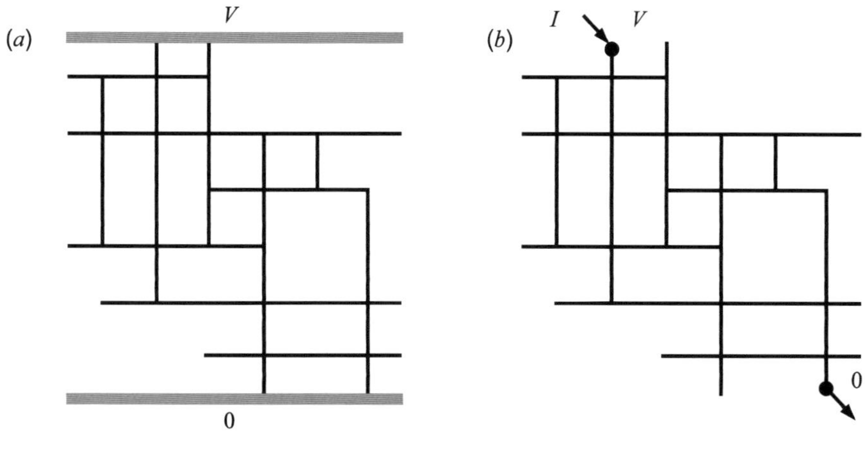


**Figure 1.** Example of a random network. The network was produced from a square lattice by removing of randomly chosen bonds. Each presented bond is associated with a given conductance. ($a$) bus-bar geometry; the (super)conducting buses are shown in gray. ($b$) two-point geometry; source and sink sites are shown as closed circles.

Basically, only a certain fraction of the bonds of a particular RRN is current-carrying. This current-carrying part of the network is called the *backbone* or *effective backbone*; the remaining bonds are called *dead ends* or *dangling ends*. Some of the bonds belonging to the backbone may carry the total current. These bonds are called *red bonds* or *singly connected bonds*; when they are cut the current flow stops [16].

Alternatively, the *geometrical backbone* can be defined as a union of all self-avoiding walks (SAW) between two given points (case of two-point geometry) or between two opposite borders (case of bus-bar geometry) [9]. Notice, that this definition is not quite equivalent to the definition of the backbone as a set of current-carrying bonds due to presence of so-called *perfectly balanced bonds*. The electrical current through a perfectly balanced bond is absent because potential

IOP Conf. Series: Journal of Physics: Conf. Series **955** (2017) 012021 doi:10.1088/1742-6596/955/1/012021

difference between its ends is equal to zero [5]. Thus, the effective backbone is defined as the set of bonds that carry a current, while the geometrical backbone is the set of bonds that either carry a current, or are perfectly balanced [17] (figure 2). A particular part of the backbone, so-called *elastic backbone*, is a union of all shortest paths between two given sites.

**Figure 2.** Example of the random network with bus-bar geometry. Effective backbone is shown in blue, dead ends are shown in gray, difference between geometric and effective backbones is shown in magenta. The effective backbone has been extracted using a program based on the direct electrifying algorithm [5]. The geometrical backbone has been extracted using a program based on the algorithm [15].

## 3. Algorithms for backbone identification

### *3.1. Effective backbone*

The straightforward idea how to find the effective backbone is application of Ohm's law to each bond belonging to the network [4, 18]. The potential difference between two neighboring sites $i$ and $j$ produces electrical current, $I_{ij}$,

$$I_{ij} = \sigma_{ij}(V_i - V_j). \tag{1}$$

Here for generality conductivities of all bonds $\sigma_{ij}$ are assumed to be different. Since for any site $i$,

$$\sum_j I_{ij} = 0, \tag{2}$$

the potential of the site $i$ is

$$V_i = \sum_j \sigma_{ij} V_j / \sum_j \sigma_{ij}. \tag{3}$$

When conductivities of all bonds are equal (say, $\sigma_{ij} = 1$), denominator of (3) is simply the degree (or valency) of the site $i$

$$V_i = \deg^{-1} i \sum_j V_j. \tag{4}$$

Thus, there is a set of linear algebraic equations (SLAE), i.e., (4) and (1) for each site. This SLAE may be huge because the number of equation is of order of square of the network linear size. There may be different techniques how to build up such the SLAE [5]. When the particular RRN has several connected components, it is better to build up the SLAE for each connected component separately to avoid singular matrix [5]. In fact, it means that a kind of cluster labeling, e.g., [19], should be performed before the backbone identification. Since potentials of some sites are known (the sites at which the potentials are applied), (4) may be solved, e.g., iteratively. All bonds with non-zero currents belong to the backbone, whereas all bonds with zero currents form the dead ends. The question is what current should be treated as zero? Any solver of SLAE involves floating-point arithmetics with inevitably round-off errors. A small current in a bond may be either ghostly (round-off error) or real. It is hardly possible to propose a strict rule on how to separate "sheep from goats". Some parts of a geometrical backbone may carry negligible current of order of round-off errors and, hence, could be omitted. Despite a bond is not perfectly balanced, it may be almost perfectly balanced, i.e. balanced of order of computational precision. Nevertheless, when not a geometrical properties of the backbone but the effective conductivity of RRN is of primary interest, precise determination of current distribution is not so important. Hence, the methods of electrifying are suitable for calculations of effective electrical properties rather than extracting a subset of percolation cluster, i.e. its backbone.

*3.2. Geometric backbone*

"Burning" algorithm [8] is intended for backbone identification in two-point geometry. Its name means that the algorithm mimics a forest fire propagation. It can be divided into three steps. Let us suppose that the potential difference is applied to the sites $P$ and $Q$. At the first step, starting from $P$ and using breadth-first search (BFS) the distances between all sites and the site $P$ are determined, each site obtain a label corresponding its distance from $P$ (figure 3($a$)). At the second step, all shortest paths between $P$ and $Q$ are determined. Starting from the site $Q$, each neighboring site which has a label less then the label of the current site is marked as belonging to the shortest paths. After these two steps, the elastic backbone is determined (figure 3($b$)). In fact, first two steps correspond to the well-known Lee algorithm [20]. In this algorithm, the first step is called "wave expansion", the second step is called "backtrace". At the third step, the elastic backbone should be expanded to the geometrical backbone. This last step is most tricky. During the third step, the sites are handled starting from the sites in which the loops in the first step are closed. A site can only marked as potentially belonging to the backbone if its label is less than the labels of the sites that are lit up at time $t_i$, but in addition, the growing backbone can not burn. The question arises: if we start from a given site, an elastic backbone is reached only on one site or on several sites? In the case of multiple sites, all the sites that were handled are added to the growing backbone. This is repeated over and over until no additional part can be added to the growing backbone (figure 3($c$)). Different interpretations and implementations of the third step lead to the appearance of various modifications of the algorithm [8, 21, 22, 23, 24, 25].

A recursive depth-first search (DFS) algorithm [9] which simultaneously generates a percolation cluster and finds its backbone is based on the best-known rule for traversing mazes, i.e., the wall follower or "left-hand rule". When a pedestrian walks through a singly-connected labyrinth touching its wall by his/her left hand, an exit will be undoubtedly found if any. When a path is passed twice in opposite directions, this path is a dead end. In fact, this is a kind of well-known "Ariadne's clew algorithm" (see, e.g., [26]). This walk through a maze should be repeated many times to find all possible paths. Each time, when a walker reaches an opposite border of a maze, a hand should be changed, i.e. when a walk from south to north corresponds to left-hand rule, then a walk from north to south corresponds right-hand rule (figure 4). Recursive implementation of the algorithm is simple but has an essential drawback connected with limited depth of recursive calls. The algorithm can be implemented without recursive calls but, in this case, a stack should be operated by hand.

A matching algorithm has been proposed [7] and applied to backbone identification for site percolation on hypercubic lattices up to 5 dimensions. The algorithm deals with a directed graph. Unfortunately, a brief sketch of the main ideas about the matching algorithm [7] is not enough for implementation of the algorithm, because several important issues have not been discussed in that paper.

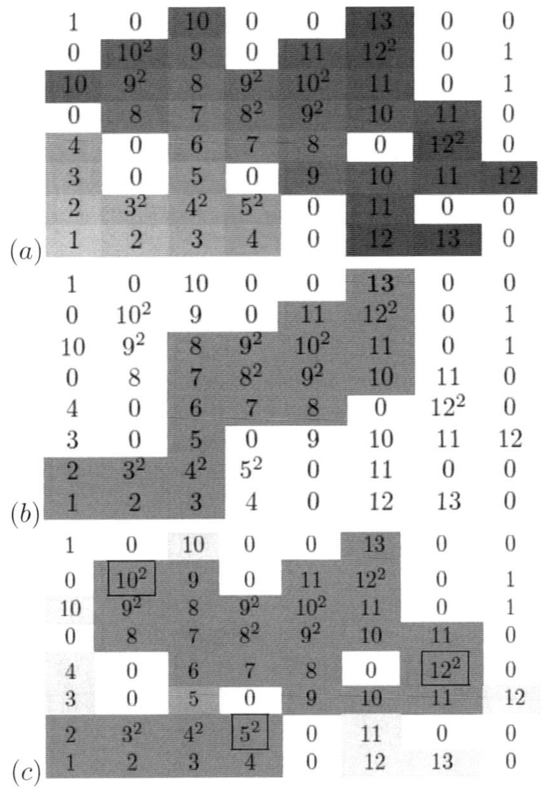


**Figure 3.** Example of application of "burning" algorithm [8] to the particular random network with two-point geometry. (*a*) After the first step, each site has the label corresponding to the distance from lower left corner to the site. Powers correspond to number of shortest paths from the starting site to the current site. (*b*) After the second step, the elastic backbone, i.e., union of all shortest paths between two given sites, is found. (*c*) After the third step, the geometric backbone, dead ends (yellow), and red bonds (red) are marked.

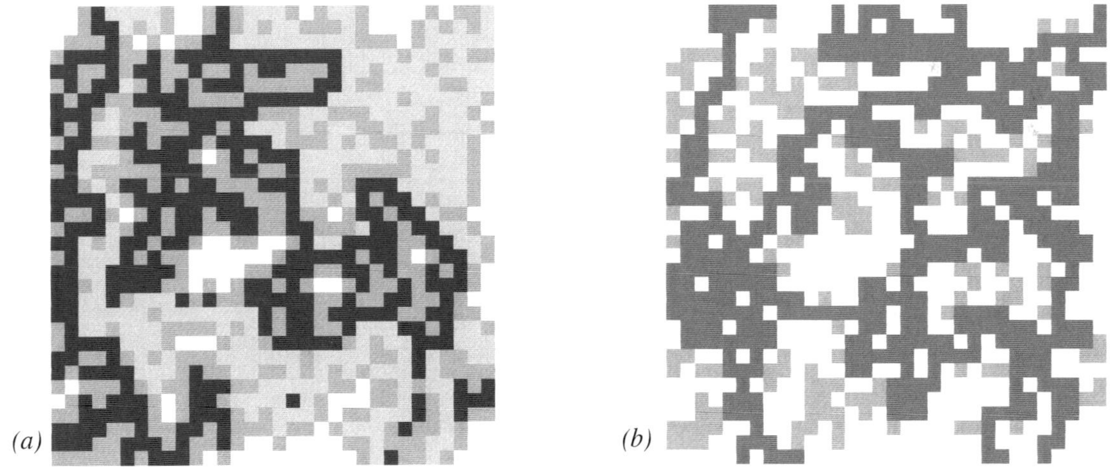


**Figure 4.** Example of application of "wall follower" algorithm [9] to the particular random network, $L = 32$, with bus-bar geometry. (*a*) Percolation cluster. Different directions of walks are shown in different colors. (*b*) Backbone and dead ends of the same percolation cluster.

The algorithm [10] deals with both original network and its dual. Its implementation consists of three steps. During the first step, the percolation cluster should be identified. All isolated clusters which, obviously, cannot participate in the conductivity should be removed. This step can be efficiently performed, e.g., by a single "burning" of the lattice [8]. The second step deals with the dual lattice. Notice, that the lattice of our primary interest, viz., a square lattice, is self-dual. The second step begins with identifying connected clusters in the dual lattice. Then, all bonds that are in between two neighbouring cells connected in the dual lattice should be removed. After this second step, the sites that used to be in the incipient infinite cluster after the first step are now either on the backbone or belong to disconnected clusters. It is finally straightforward to conclude the computation in a third step by extracting the incipient cluster which is now the backbone of the original RRN.

A rapid algorithm for identifying the current-carrying backbone in the 2D percolation model has been proposed by Yin and Tao [11, 12]. Its implementation has been recently presented [15]. Taking advantage of the modified Hoshen–Kopelman cluster labeling algorithm, this multistep algorithm identifies dead ends using their local properties (figure 5). The algorithm is based on Jordan's curve theorem: a loop in a 2D graph disconnects the area inside it from the area outside. In fact, the idea of the algorithm is rather similar to [10] although a dual lattice is not utilised explicitly. The algorithm can be easily parallelized and looks more promising for most tasks.

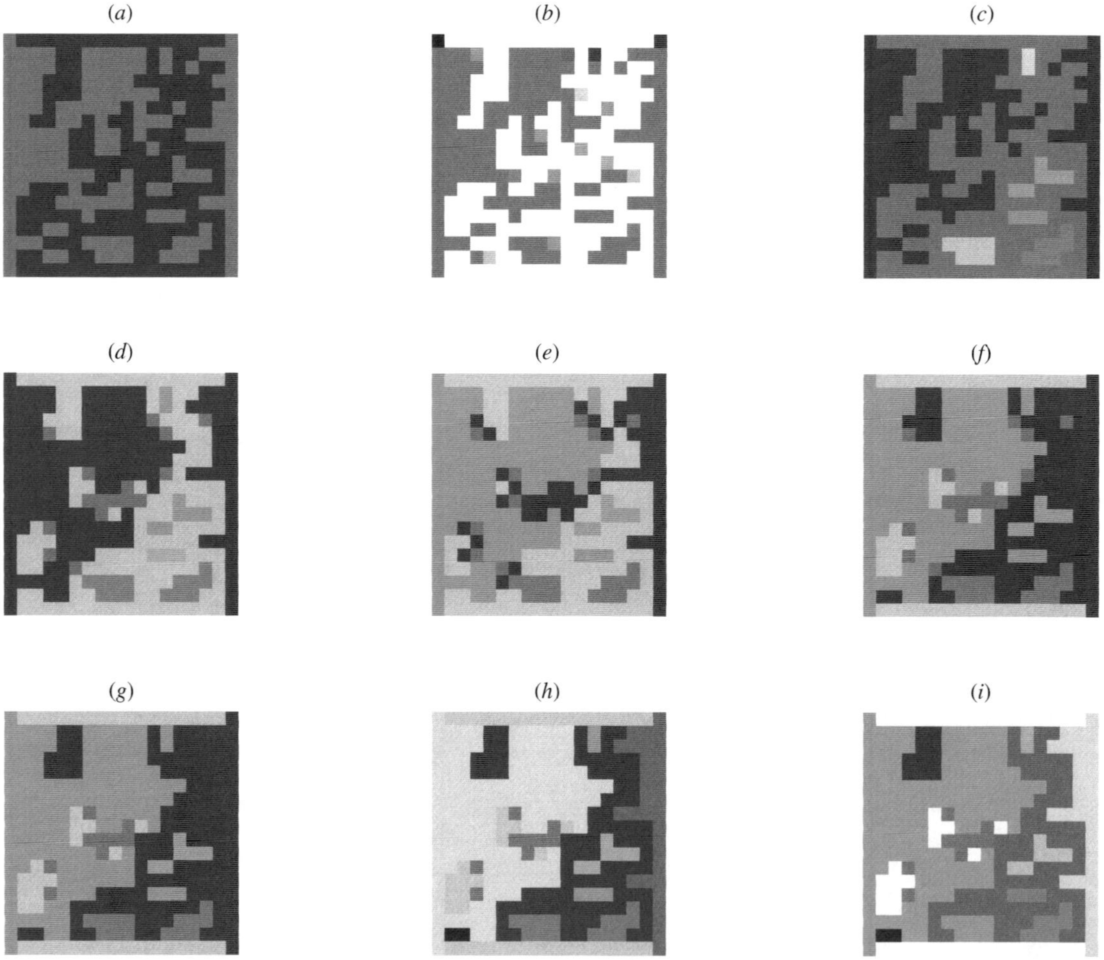


**Figure 5.** Steps of backbone identification with algorithm [11, 12] implemented as in [15].

## 4. Conclusion

We have implemented several different algorithms for backbone identification. The algorithms were applied to backbone identification for different system sizes and concentrations of conducting bonds.

Direct calculations of electrical potentials and currents are based on floating-point arithmetic and, hence, produce round-off errors. Some ghost currents may arise due to these round-off errors which impedes the backbone extraction. Moreover, these calculations deal with huge SLAE and require a lot of computer memory. Only relative small systems can be treated in these approaches because number of equations to be solved is proportional to the square of linear size of the system.

Graph theory algorithms are sometimes difficult to understand or/and to realize. Some of them require to store not only original network but its dual [10]. Moreover, some algorithms can produce stack overflow because of recursion. In fact, application of these algorithms also is restricted to the RRN of moderate size.

Our analysis suggests that a universal algorithm suitable for any possible problem is absent. Most likely, each particular task needs a specific algorithm. Nevertheless, the algorithms which allow parallelization (e.g., [15]) look very promising in many cases.

## Acknowledgments

The reported study was supported by the Ministry of Education and Science of the Russian Federation, Project No. 3.959.2017/4.6.

## References

[1] McLachlan D S and Sauti G 2007 *J. Nanomater.* **2007** 1–9 ISSN 1687-4110
[2] Wang M and Pan N 2008 *Mater. Sci. Eng.: R: Reports* **63** 1–30 ISSN 0927-796X
[3] Mutiso R M and Winey K I 2015 *Prog. Polym. Sci.* **40** 63–84 ISSN 0079-6700 Progress in Polymer Hybrid Materials
[4] Kirkpatrick S 1973 *Rev. Mod. Phys.* **45** 574–88 ISSN 0034-6861
[5] Li C and Chou T W 2007 *J. Phys. A: Math. Theor.* **40** 14679 ISSN 1751-8121
[6] Tarjan R 1972 *SIAM J. Comput.* **1** 146–60 ISSN 0097-5397
[7] Moukarzel C 1998 *Int. J. Mod. Phys. C* **9** 887–95 ISSN 0129-1831
[8] Herrmann H J, Hong D C and Stanley H E 1984 *J. Phys. A: Math. Gen.* **17** L261–6 ISSN 1361-6447
[9] Grassberger P 1992 *J. Phys. A: Math. Gen.* **25** 5475–84 ISSN 1361-6447
[10] Roux S and Hansen A 1987 *J. Phys. A: Math. Gen.* **20** L1281–5 ISSN 1361-6447
[11] Yin W G and Tao R 2000 *Physica B: Cond. Matt.* **279** 84–6 ISSN 0921-4526
[12] Yin W G and Tao R 2003 *Int. J. Mod. Phys. C* **14** 1427–37 ISSN 0129-1831
[13] Fattah M, Airola A, Ausavarungnirun R, Mirzaei N, Liljeberg P, Plosila J, Mohammadi S, Pahikkala T, Mutlu O and Tenhunen H 2015 *Proc. 9th Int. Symp. on Networks-on-Chip* NOCS '15 (New York, NY, USA: ACM) pp 18:1–18:8 ISBN 978-1-4503-3396-2
[14] Redner S 2009 *Encyclopedia of Complexity and Systems Science* ed Meyers R A (New York, NY: Springer New York) pp 3737–54 ISBN 978-0-387-30440-3
[15] Trobec R and Stamatovic B 2017 *Adv. Eng. Softw.* **103** 38–45 ISSN 0965-9978
[16] Bunde A and Havlin S 1991 *Fractals and Disordered Systems* ed Bunde A and Havlin S (Berlin, Heidelberg: Springer Berlin Heidelberg) pp 51–96 ISBN 978-3-642-51435-7
[17] Batrouni G G, Hansen A and Roux S 1988 *Phys. Rev. A* **38**(7) 3820–23 ISSN 2469-9926
[18] Kirkpatrick S 1971 *Phys. Rev. Lett.* **27**(25) 1722–5 ISSN 0031-9007
[19] Hoshen J and Kopelman R 1976 *Phys. Rev. B* **14** 3438–45 ISSN 1098-0121
[20] Lee C Y 1961 *IRE Trans. Electronic Comput.* **EC-10** 346–65 ISSN 0367-9950
[21] Herrmann H J and Stanley H E 1984 *Phys. Rev. Lett.* **53**(12) 1121–4 ISSN 0031-9007
[22] Mastorakos J and Argyrakis P 1993 *Phys. Rev. E* **48**(6) 4847–50 ISSN 2470-0045
[23] Porto M, Bunde A, Havlin S and Roman H E 1997 *Phys. Rev. E* **56**(2) 1667–75 ISSN 2470-0045
[24] Babalievski F 1998 *Int. J. Mod. Phys. C* **09** 43–60 ISSN 0129-1831
[25] Sheppard A P, Knackstedt M A, Pinczewski W V and Sahimi M 1999 *J. Phys. A: Math. Gen.* **32** L521–9 ISSN 1361-6447
[26] Mazer E, Ahuactzin J M and Bessiére P 1998 *J. Artif. Int. Res.* **9** 295–316 ISSN 1076-9757

# Asymptotics of N-dimensional tori in the generalized Korteweg – de Vries equation

**S D Glyzin[1,2], S A Kashchenko[1,3], M M Preobrazhenskaia[1,2]**

[1] P. G. Demidov Yaroslavl State University, Sovetskaya str., 14, Yaroslavl, 150000, Russia
[2] Scientific Center in Chernogolovka RAS, Lesnaya str., 9, Chernogolovka, Moscow region, 142432, Russia
[3] MEPhI National Research Nuclear University, Kashirskoe sh., 31, Moscow, 115409, Russia

E-mail: rita.preo@gmail.com

**Abstract.** We consider the generalized Korteweg – de Vries (KdV) equation with periodic boundary condition by space variable. The asymptotics of periodic solutions and invariant tori are considered in a sufficiently small neighbourhood of the zero equilibrium state. There are considered two cases: the coefficient of the quadratic term in the right part of generalized KdV equation is non-zero or zero. In the first case, the cycles and two-dimension tori are unstable. In the second, case it is proved that cycles are stable and tori are unstable.

## 1. Introduction

Let us consider the boundary problem (see [1, 2])

$$u_t + u_{xxx} + (\alpha u + \beta u^2)u_x = au + bu^2 + cu^3, \tag{1}$$

$$u(t, x + T) = u(t, x), \tag{2}$$

where $u(t,x) \in W_2^3$ for all $t > 0$, parameters $\alpha$, $\beta$, $a$, $b$, $c$, $T$ are constants.

Note that the position of the characteristic (wave) equation roots

$$\lambda_k = ik^3 + a \tag{3}$$

is important for problem of the zero solution stability. By (2), the value of $k$ in (3) is $2\pi n/T$, where $n$ is integer. If $a < 0$ then we have $\operatorname{Re}\lambda_k = a < 0$ and all solutions of (1), (2) from a sufficiently small neighbourhood of the zero equilibrium state tend to zero as $t \to \infty$. If $a > 0$ then $\operatorname{Re}\lambda_k > 0$ and the problem stops to be local. Suppose that the parameter $a$ is small enough:

$$a = \varepsilon a_0, \ 0 < \varepsilon \ll 1. \tag{4}$$

Let us investigate the question about a construction periodic solutions and invariant tori of different dimension in a small enough neighbourhood of zero equilibrium state. The condition (4) allows us to use the methods of bifurcation analysis [3, 4, 5]. Infinitely many roots of the characteristic equation (3) tend to imaginary axis as $\varepsilon \to 0$. Thus, the critical case of infinite dimension occurs in the problem of zero solution stability for (1), (2). We recall that well-known methods of local invariant integral manifolds (see, for example, [6, 7, 8]) and methods of normal forms (see [9, 10]) are used for investigation of the local dynamics of the equation in the finite-dimensional critical case. The methods permit us to study a special finite-dimensional non-linear system of ordinary differential equations (ODE) instead of the equation. It is impossible to pass boundary problem (1), (2), (4) to a finite-dimensional system of ODE. However it is possible to obtain the results about the existence of equilibrium states, periodic solutions, and invariant tori of various dimensions in a small neighbourhood of zero using the formal method of normal forms. It is possible to explore the question of their stability. We show this bellow.

Section 2 contains the formal problem statement. In sections 3 and 4, we construct the asymptotics of invariant tori. Moreover, in section 3, we study the general case $b \neq 0$ and give a result about the non-stability of cycles and two-dimension tori. Section 4 presents more simple situation $b = 0$. In this case, it is proved that there exists stable cycles and unstable tori.

## 2. Problem statement

Let $N$ be a fix natural number. We are interested in the construction of the $N$-dimension torus asymptotics in a sufficiently small neighbourhood of zero. We fix different natural $n_1, \dots, n_N$. Let

$$k_1 = 2\pi n_1/T, \dots, k_N = 2\pi n_N/T. \tag{5}$$

We find the solution of (1), (2) in the form of the series

$$u = \eta(t,\varepsilon) + \sum_{j=1}^{N} (\xi_j(t,\varepsilon)\exp(ik_jx + ik_j^3t) + c\bar{c}) + \dots, \tag{6}$$

where the symbol $c\bar{c}$ means the term which is complex conjugated to the previous one. The terms are trigonometric polynomials from $t$ and $x$. Here functions $\eta = \eta(t,\varepsilon)$, $\xi_j = \xi_j(t,\varepsilon)$ and their derivatives are small enough. The last ones are represented by $\eta$, $\xi_j$ and $\varepsilon$:

$$\begin{gathered}\dot{\eta} = \alpha_1\varepsilon\eta + \alpha_2\eta^2 + \sum_{j=1}^{N}\alpha_{3j}|\xi_j|^2 + \alpha_4\eta^3 + \eta\sum_{j=1}^{N}\alpha_{5j}|\xi_j|^2,\\ \dot{\xi}_j = \beta_{1j}\varepsilon\xi_j + \beta_{2j}\eta\xi_j + \xi_j\Big(\sum_{s=1}^{N}\beta_{3js}|\xi_s|^2 + \beta_{4j}\eta^2\Big),\ j = 1,\dots,N.\end{gathered} \tag{7}$$

Let the system (7) be called a normal form. Here we use the terminology of the classical theory of bifurcation analysis.

The following sequence of actions is standard for similar problems [11, 12, 13]. Put (6) to (1) instead of $u$. Change $\dot{\eta}$ and $\dot{\xi}_j$ to (7) (coefficients $\alpha_m$, $\alpha_{3j}$, $\alpha_{5j}$, $\beta_{mj}$, $\beta_{3js}$, $m = 1,2,4$, $j,s = 1,\dots,N$ are unknown). Then we equate the coefficients of the same powers of $\varepsilon$, $\eta$ and $\xi_j$. As a result we obtain values of $\alpha_m$, $\alpha_{3j}$, $\alpha_{5j}$, $\beta_{mj}$ from a resolvability condition. Then we substitute $\rho_j\exp(i\varphi_j)$, $\rho_j \geq 0$, for $\xi_j$, $j = 1,\dots,N$, in (7) to find automodel periodic solution. We obtain the amplitude system, which separates from (7). It is important whether the parameter $b$ is zero or not for learning the amplitude system.

## 3. Case $b \neq 0$

If $b \neq 0$, we obtain the following values of coefficients of normal form (7) for $j,s = 1,\dots,N$:

$$\begin{gathered}\alpha_1 = a_0,\ \alpha_2 = b,\ \alpha_{3j} = 2b,\ \alpha_4 = c,\ \alpha_{5j} = 6c,\\ \beta_{1j} = a_0,\ \beta_{2j} = 2b - i\alpha k_j,\ \beta_{3jj} = 3c + \frac{\alpha b}{2k_j^2} + i\Big(\frac{b^2}{3k_j^3} - \frac{\alpha^2}{6k_j} - \beta k_j\Big),\\ \beta_{3js} = 6c - \frac{4b\alpha}{3(k_j^2-k_s^2)} + i\Big(\frac{8b^2}{3k_j(k_j^2-k_s^2)} - 2\beta k_j\Big) \text{ where } s \neq j,\ \beta_{4j} = 3c - i\beta k_j.\end{gathered} \tag{8}$$

The substitution $\xi = \rho_j \exp(i\varphi_j)$ permits one to pass to the amplitude system

$$\begin{aligned}\dot\eta &= a_0\varepsilon\eta + 2b\sum_{s=1}^{N}\rho_s^2 + b\eta^2 + c\eta^3 + 6c\eta\sum_{s=1}^{N}\rho_s^2,\\ \dot\rho_j &= \rho_j(a_0\varepsilon + 2b\eta + \sum_{s=1}^{N}\operatorname{Re}\beta_{3js}\rho_s^2 + 3c\eta^2),\ j=1,\dots,N.\end{aligned} \tag{9}$$

If $c \neq 0$ then system (9) has three equilibrium points such that $\rho_1 = \ldots = \rho_N = 0$:

1) $\eta = \rho_1 = \ldots = \rho_N = 0$,
2) $\eta = -2a_0c\varepsilon/b + O(\varepsilon^2)$, $\rho_1 = \ldots = \rho_N = 0$,
3) $\eta = -b/c + O(\varepsilon)$, $\rho_1 = \ldots = \rho_N = 0$.

The equilibrium points in the first and second cases have different stability. If $a_0 > 0$, case 1) is unstable and 2) is unstable. (And conversely.) In the last case, the equilibrium state is not in a sufficiently small neighbourhood of zero. Learning this case is not a subject of our paper.

It is not difficult to prove the following statement for local case 2).

**Lemma 1.** *For all sufficiently small $\varepsilon > 0$, $a_0 > 0$, $b \neq 0$, $c \neq 0$, there exists an equilibrium stable state*

$$u(t,x) = -\frac{2a_0c\varepsilon}{b}\varepsilon + O(\varepsilon^2), \tag{10}$$

*which is asymptotic by the residual with solutions of (1), (2).*

Then let us consider the case $N = 2$. Suppose that $\eta$ and $\rho_j$ have $\varepsilon$ order. Thus, the equilibrium points of (9) are given at table 1.

**Table 1.** Equilibrium points of (9) in case $N = 2$.

| $\eta$ | $\rho_1 \geq 0$ | $\rho_2 \geq 0$ | Stability |
|---|---|---|---|
| 0 | 0 | 0 | U |
| $-\frac{a_0\varepsilon}{b} + O(\varepsilon^2)$ | 0 | 0 | S |
| $-\frac{a_0\varepsilon}{2b} + O(\varepsilon^2)$ | $\frac{a_0\varepsilon}{2\sqrt{2}\lvert b\rvert} + O(\varepsilon^2)$ | 0 | U |
| $-\frac{a_0\varepsilon}{2b} + O(\varepsilon^2)$ | 0 | $\frac{a_0\varepsilon}{2\sqrt{2}\lvert b\rvert} + O(\varepsilon^2)$ | U |
| $-\frac{a_0\varepsilon}{2b} + O(\varepsilon^2)$ | $\rho_{12*}\varepsilon + O(\varepsilon^2)$ | $\rho_{21*}\varepsilon + O(\varepsilon^2)$ | U |

Here (and bellow in the next section) U means "unstable", S means "stable" and

$$\rho_{js*} = \frac{a_0}{2\sqrt{2}|b|}\sqrt{\frac{3c - 4\alpha b/(3(k_j^2 - k_s^2)) - \alpha b/(2k_s^2)}{6c - \alpha b/(2k_j^2) - \alpha b/(2k_s^2)}}. \tag{11}$$

**Lemma 2.** *For all sufficiently small $\varepsilon > 0$, $a_0 > 0$, $k_1 = 2\pi n_1/T$, $k_2 = 2\pi n_2/T$, where $n_1 \neq n_2$ are natural, and $b$, $c$, $\alpha$ such that numbers*

$$3c - 4\alpha b/(3(k_1^2 - k_2^2)) - \alpha b/(2k_2^2), \quad 3c - 4\alpha b/(3(k_2^2 - k_1^2)) - \alpha b/(2k_1^2) \tag{12}$$

*are non-zero and have the same sign, there exists two-dimension unstable torus*

$$u(t,x) = -\frac{a_0}{2b}\varepsilon + \sum_{j=1}^{2}\left(\rho_{j*}\exp\left(ik_jx + i(k_j^3 + \varphi_{j*})t\right) + \bar{cc}\right) + O(\varepsilon^2), \tag{13}$$

*which are asymptotic by the residual with solutions of (1), (2). Here*

$$\rho_{1*} = \rho_{12*}\varepsilon + O(\varepsilon^2), \quad \rho_{2*} = \rho_{21*}\varepsilon + O(\varepsilon^2), \quad \varphi_{j*} = \frac{\alpha k_j a_0 \varepsilon}{2b} + O(\varepsilon^2), \ j = 1, 2, \tag{14}$$

$\rho_{js*}$ *is defined by (11).*

The similar result is proved in a more simple case when only one of $\rho_j$ is non-zero. Such cycles are also unstable.

## 4. Case $b = 0$

In the case, we obtain the following coefficients of normal form (7) for $j, s = 1, \ldots, N$:

$$\alpha_1 = \beta_{1j} = a_0, \ \alpha_2 = \alpha_{3j} = 0, \ \alpha_4 = c, \ \alpha_{5j} = 6c, \ \beta_{2j} = -i\alpha k_j,$$

$$\beta_{3js} = 6c - 2i\beta k_j \text{ where } s \neq j, \ \beta_{3jj} = 3c - i(\alpha^2/(6k_j) + \beta k_j), \ \beta_{4j} = 3c - i\beta k_j.$$

Thus, the substitution $\xi_j = \rho_j \exp(i\varphi_j t)$ leads the system (7) to the following amplitude system

$$\begin{aligned} \dot\eta &= \eta(a_0\varepsilon + 6c \sum_{s=1}^{N} \rho_s^2 + c\eta^2), \\ \dot\rho_j &= \rho_j(a_0\varepsilon + 6c \sum_{s=1}^{N} \rho_s^2 - 3c\rho_j^2 + 3c\eta^2), \ j = 1, \ldots, N. \end{aligned} \tag{15}$$

Let us find equilibrium states of (15). Suppose that $a_0 > 0$, $c < 0$.

Firstly, note that there exists zero equilibrium point and an equilibrium point such that $\rho_1 = \ldots = \rho_N = 0$ and $\eta$ is non-zero. It is easy to prove constructing the Jacobian matrix that the first one is unstable. If $\eta$ is non-zero, we have $\eta = \pm\sqrt{-a_0\varepsilon/(3c)}$. Both of the equilibrium states are stable.

Secondly, suppose that $\rho_{j_1}, \ldots, \rho_{j_L} > 0$, $L = 1, \ldots, N$, and $\rho_j = 0$ whenever $j \notin \{j_1, \ldots, j_L\}$. Let us consider separately two cases: $\eta = 0$ and $\eta \neq 0$.

If $\eta = 0$, we find $\rho_{j_s} = \sqrt{-a_0\varepsilon/(3(2L-1)c)}$, $s = 1, \ldots, L$. Such equilibrium states are stable just if $L = 1$. If $L > 1$, it is possible to prove that the Jacobian matrix of the system (15) has a positive eigenvalue.

If $\eta \neq 0$, we obtain equilibrium points of (15) such that $\eta = \pm\sqrt{-a_0\varepsilon/((4L+1)c)}$, $\rho_{j_s} = \sqrt{-2a_0\varepsilon/(3(4L+1)c)}$, $s = 1, \ldots, L$. All of them are unstable.

For example, we give the table 2.

As a result, we obtain the following

**Theorem.** *For all sufficiently small* $\varepsilon > 0$ *and for* $a_0 > 0$, $c < 0$, $k_1 = 2\pi n_1/T, \ldots, k_N = 2\pi n_N/T$ *(where* $n_1, \ldots, n_N$ *are different natural numbers), there exists stable equilibrium states*

$$u(t, x) = \pm\sqrt{-\frac{a_0\varepsilon}{c}} + O(\varepsilon),$$

*a stable cycles*

$$u(t, x) = \sqrt{-\frac{a_0\varepsilon}{3c}} \exp\left(ik_1 x + i(k_1^3 + \varphi_{j*})t\right) + c\overline{c} + O(\varepsilon),$$

*and unstable tori*

$$u(t, x) = \rho_{j*} \sum_{j=1}^{L} \left( \exp\left(ik_j x + i(k_j^3 + \varphi_{j*})t\right) + c\overline{c} \right) + O(\varepsilon), \ L = 2, \ldots, N,$$

**Table 2.** Equilibrium points of (15) in case $N = 2$.

| $\eta$ | $\rho_1 \geq 0$ | $\rho_2 \geq 0$ | Stability |
|---|---|---|---|
| 0 | 0 | 0 | U |
| $\pm\sqrt{-\frac{a_0\varepsilon}{c}}$ | 0 | 0 | S |
| 0 | $\sqrt{-\frac{a_0\varepsilon}{3c}}$ | 0 | S |
| 0 | 0 | $\sqrt{-\frac{a_0\varepsilon}{3c}}$ | S |
| 0 | $\sqrt{-\frac{a_0\varepsilon}{9c}}$ | $\sqrt{-\frac{a_0\varepsilon}{9c}}$ | U |
| $\pm\sqrt{-\frac{a_0\varepsilon}{5c}}$ | $\sqrt{-\frac{2a_0\varepsilon}{15c}}$ | 0 | U |
| $\pm\sqrt{-\frac{a_0\varepsilon}{5c}}$ | 0 | $\sqrt{-\frac{2a_0\varepsilon}{15c}}$ | U |
| $\pm\sqrt{-\frac{a_0\varepsilon}{9c}}$ | $\sqrt{-\frac{2a_0\varepsilon}{27c}}$ | $\sqrt{-\frac{2a_0\varepsilon}{27c}}$ | U |

$$u(t,x) = \eta_{j*\pm} + \rho_{j*}\sum_{j=1}^{L}\Big(\exp\left(ik_jx + i(k_j^3 + \varphi_{j*\mp})t\right) + c\overline{c}\Big) + O(\varepsilon),\ \ L = 1,\ldots,N,$$

*which are asymptotic by the residual with solutions of boundary problem (1), (2). Here*

$$\rho_{j*} = \sqrt{\frac{-2a_0\varepsilon}{3(4L+1)c}},\quad \eta_{j*\pm} = \pm\sqrt{\frac{-a_0\varepsilon}{(4L+1)c}},$$

$$\varphi_{j*} = \Big(\frac{\alpha^2}{6k_j} + 3\beta k_j\Big)\frac{a_0\varepsilon}{9c},\quad \varphi_{j*\mp} = \mp\alpha k_j\sqrt{-\frac{a_0\varepsilon}{9c}} + O(\varepsilon).$$

## 5. Conclusion

The local dynamics of generalized KdV equation is studied using the methods based on the classical bifurcation theory. The question about the existence, asymptotics and stability of periodic solutions and tori of the boundary problem with periodic boundary condition is considered in two cases: when the parameter $b$ of the quadratic non-linearity is zero and non-zero. In the first case, it is shown that, in condition of the positivity of the linear term in the right part ($a_0 > 0$), the cycles and two-dimension tori are unstable. In the second case, it is proved that, for $a_0 > 0$ and $c < 0$ (the condition guarantees the dissipativity of the problem), the cycles are stable and the tori are unstable.

## Acknowledgments

This work was supported by the Russian Science Foundation (project nos. 14-21-00158).

## References

[1] Korteweg D J, de Vries G 1895 On the change of form of long waves advancing in a rectangular canal and on a new tipe of long stationary waves *Phil. Mag.* **39** 422–43
[2] Kudryashov N A 2010 Metody nelineynoy matematicheskoy fiziki: Uchebnoye posobiye [in Russian] (Dolgoprudnyy: Intellekt) 368
[3] Kashchenko S A 2016 Normal form for the KdV – Burgers equation *Doklady Mathematics* **93** 331–33
[4] Bibikov Yu N 1980 Bifurcations of Hopf type for quasiperiodic motions *Differ. Uravn.* **16** 1539–44
[5] Bibikov Yu N 1990 Bifurcation of a stable invariant torus from an equilibrium *Mathematical notes of the Academy of Sciences of the USSR* **48** 15–19

[6] Hartman P 1964 Ordinary Differential Equations (John Wiley and Sons) 314
[7] Mitropolskii Y U, Lykova O B 1973 Integral'nyye mnogoobraziya v nelineynoy mekhanike [in Russian] (Moscow: Nauka) 512
[8] Bruno A D 1989 Local Methods in Nonlinear Differential Equations (Springer-Verlag Berlin Heidelberg) 348
[9] Hassard B D, Kazarinoff N D, Wan Y H 1981 Theory and Applications of Hopf Bifurcation (Cambridge Univ. Press) 320
[10] Guckenheimer J, Holmes P J 1983 Nonlinear Oscillations, Dynamical Systems, and Bifurcations of Vector Fields (Springer-Verlag New York) 462
[11] Glyzin S D, Kolesov A Yu, Rozov N Kh 2005 Chaotic Buffering Property in Chains of Coupled Oscillators *Diff. Eq.* **41** 41–49
[12] Glyzin S D, Kolesov A Yu, Rozov N Kh 2009 Extremal Dynamics of the Generalized Hutchinson Equation *Computational Mathematics and Mathematical Physics* **49** 71–83
[13] Glyzin S D, Kolesov A Yu, Preobrazhenskaia M M 2017 Existence and stability of periodic solutions of quasi-linear Korteweg – de Vries equation *Journal of Physics: Conference Series* **788** 012016

IOP Conf. Series: Journal of Physics: Conf. Series **955** (2017) 012023 doi:10.1088/1742-6596/955/1/012023

# Interfaces in evolutionary games

**Sergei Kolotev**[1,2], **Aleksandr Malyutin**[1,2], **Evgeni Burovski**[1,2], **Lev Shchur**[1,2,3]

[1] National Research University Higher School of Economics, 101000 Moscow, Russia
[2] Science Center in Chernogolovka, 142432 Chernogolovka, Russia
[3] Landau Institute for Theoretical Physics, 142432 Chernogolovka, Russia

**Abstract.** We investigate geometrical aspects of a spatial evolutionary game. The game is based on the Prisoner's dilemma. We analyze the geometrical structure of the space distribution of cooperators and defectors in the steady-state regime of evolution. We develop algorithm for the identification of the interfaces between clusters of cooperators and defectors, and measure fractal properties of the interfaces.

## 1. Introduction

In recent years much interest is paid to the investigation of physical and mathematical properties of interfaces in the plane and to identification of the origins of phase transitions. The interest is due to two main factors, the scientific factor and the practical factor. The first one is the development of the SLE (Stochastic Loewner Evolution [1], named later as Shcramm-Loewner Evolution) which connects properties of the random walk in the upper half-plane with the properties of Conformal Field Theory (CFT) [2]. It turns out to be the third alternative approach, complementing the algebraic solution [3] and the CFT solution [4], for solving critical properties of some two-dimensional systems of classical statistical physics. Among successful examples of this approach are the percolation problem [5, 6] and Ising model [7]. The second one is an interest to the investigation of physics and mathematics of the processes the planar geometry, connected to practicalities of the development of new materials for next generation of microelectronic devices [8].

In this paper we focus on the analysis of geometrical properties of interfaces emerging in the steady-state regime of the space-evolutionary game [9, 10]. The game is played by $L^2$ agents arranged on an $L \times L$ rectangular lattice with periodic boundary conditions. At each instant of time, an agent can be in a state of a cooperator $\mathcal{C}$, or a defector $\mathcal{D}$. Each agent computes their strategy to either keep the current state, or to change it depending on the payoff computed as the sum of the pair interaction with the neighbors. This is done for all players in parallel. The new state is assigned for all neighbors and time is incremented by one. It was found in [11] that starting with some initial distribution, system goes to the steady-state distribution of cooperators and defectors. They found also that starting from some regular distribution of players, regular self-similar structures appear, while random initial configurations lead to random steady-state structures, and the final steady-state stage depends on the payoff parameter.

It is natural to ask, what are the geometrical properties of the steady-state random structures? In particular, what is the fractal dimension of the interfaces between clusters of cooperators and defectors? The answer can be informative for the possible classification of the changes of the steady-state regimes while tuning payoff parameter.

In the theory of phase transitions [12], first- and second-order transitions are most investigated. For a second-order transition in two spatial dimensions, the interface length $\mathcal{L}$ scales with the systems size $L$ as a power law $\mathcal{L} \propto L^{\theta}$ [13]. For a first-order transition, the scaling of the interface of droplets is linear $\mathcal{L} \propto L$ [14]. In recent years, there is a discussion of the so-called mixed transitions, where some observables behave in a way expected for a second-order transition while other observables behave in a first-order way. So far, an analytic investigation of a mixed transition was done in a one-dimensional system [15]. In the usual percolation problem, the probability to belong to the infinite cluster behaves as power law of the parameter of the model. It was argued by some authors that in dynamical percolation, this probability can change sharply as a function of the parameter [16], which is is now known as discontinuous percolation [17]. We are not aware of any analysis of the interfaces in these novel phenomena.

In our model, the steady-state structures do percolate for some range of values of the payoff parameter, thus demonstrating a correlated behavior of players, which is similar to the case of a second-order phase transition. There are sharp changes of the observables while changing value of the payoff parameter, which is typical for first-order phase transition. At the same time, our model is not an equilibrium model of statistical mechanics, where one defines a thermodynamic equilibrium and fluctuations. The steady-state regime is achieved through a dynamic process, not some kind of an equilibration process. This way the steady-state structures are the self-organized.

The paper is organized as follows. In Section 2 we describe the model and review some previously known results. In Section 3 we describe algorithms we employ for simulations and analysis. Section 4 contains results of simulations and data analysis. In Section 5 we discuss results and future work.

## 2. Model of space evolution game

A prototypical model of the game theory is the so-called *Prisoner's dilemma* (PD), played by two players in discrete time steps. In each round of the game, each player uses one of two possible strategies, *cooperate*, $\mathcal{C}$, or *defect*, $\mathcal{D}$, and receives a payoff which depends on the strategies of the player and its opponent [18]. We use the following payoff structure [9, 11]: (i) If two players are $\mathcal{D}$ , they receive nothing; (ii) If both players $\mathcal{C}$, each of them receives a payoff of $S$, which we set to $S = 1$ without loss of generality; (iii) In the interaction of $\mathcal{C}$ and $\mathcal{D}$, the $\mathcal{D}$ receives a payoff $T > S$ and the $\mathcal{C}$ receives zero. This way, the payoff structure only depends on a single parameter, the payoff parameter $b = T/S$.

The game starts with some initial configuration of $\mathcal{C}$ and $\mathcal{D}$ placed in the vertices of the square lattice with periodic boundary conditions, i.e. it is square lattice on a torus with $L \times L$ vertices. Each agent computes the payoff of its "interaction" with 8 neighbors (in a way chess king's moves) and with itself using the above rule for pair interaction. Payoff is calculated as the sum of the pair-wise payoffs. In the next time step (next generation, on the language of the game [11]), an agent adopts the strategy of those of its neighbors (including itself) that received the highest payoff. Thus the future state of an agent depends on the state of the cell itself, its 8 neighbors and their neighbors, i.e., 25 agents in total.

The dynamics of the game depends on the value of the payoff parameter $b$. Possible payoffs of agents can take only discrete values of the form $i + bj$, with $i$ and $j$ being non-negative integers. Therefore, the switches between different dynamic regimes can only occur at some discrete set of $b$ values [11]. We concentrate our analysis for the narrow range of $3/2 < b < 2$, which we find the most interesting. This range includes several dynamic regimes, as seen in the left panel of Figure 1: the steady state average density of $\mathcal{C}$ changes abruptly at $b = 3/2, 8/5, 5/3, 7/4$

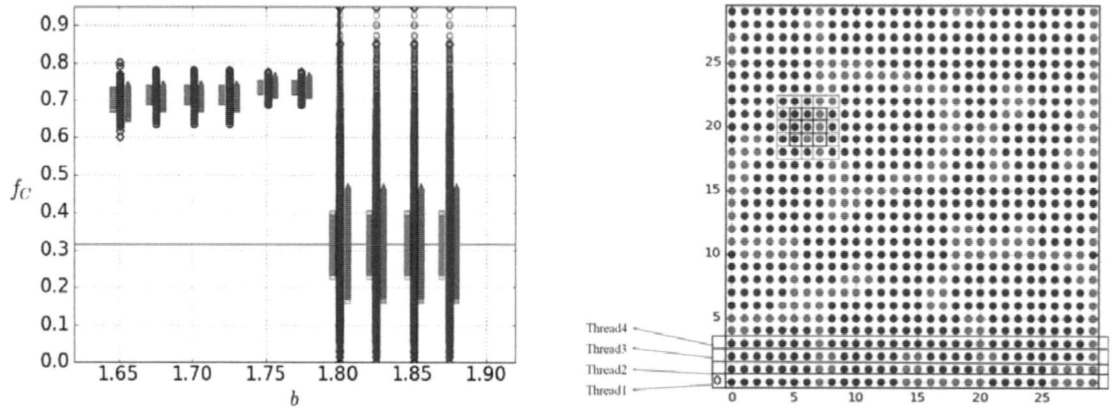


**Figure 1. Left**: Concentration of cooperators $\mathcal{C}$, $f_C$, as a function of the payoff parameter $b$ for the lattice sizes 20x20 (blue), 50x50 (green), and 100x100 (red). For clarity, green signs are shifted horizontally slightly to the right and red signs are shifted to the left. All simulations are performed at $b = 1.651, 1.675, 1.701, 1.725, 1.751, 1.775, 1.801, 1.825, 1.851, 1.875$. Each point is a single measurement in a simulation of $2\times10^4$ generations of 25 independent realizations of initial conditions. Blue line denotes the magic value $f_C = 12 \log 2 - 8 \approx 0.318$ [11].
**Right**: Illustration of the association of the lattice vertices and threads. We also show the vertex with coordinates (6,20) surrounded by the neighbors (inside the magenta box) which influence its decision on the next time step.

and 9/5. These values can be explained microscopically by analyzing probabilities of stable oscillations of small $\mathcal{C}$-clusters for $b < 9/5$ and probabilities of growing $\mathcal{D}$-clusters for $b > 9/5$ following [11].

Data for the Figure 1-left was obtained from simulations starting with sufficiently dense and random initial configuration of $\mathcal{C}$ and $\mathcal{D}$, f.e. for concentrations $f_C = 0.9$ of cooperators and $f_D = 0.1$ for defectors. The spread of the computed steady-state concentration $f_C$ in the figure can be clearly attributed to the finite size of the system, as red signs are spread less than green signs, and green signs is less than the blue signs, in accordance with the relation of the corresponding lattice sizes 100 > 50 > 20.

Figure 2 shows representative snapshots of steady-state structures in the game field. It can be seen that in the left panel there is a blue path (corresponding to $\mathcal{C}$) from one side of the lattice to another one. In other words, a blue cluster percolates, and in the "thermodynamic limit" ($L \to \infty$) will be infinite. At the right panel, the red cluster percolates. Clearly, something happens at the critical value $b = 9/5$. It is known [11] that for $b > 9/5$ a square with nine $\mathcal{D}$ can grow in the sea of the $\mathcal{C}$. Due to the random initial condition, system self organizes into a steady state similar to the one show in the right panel of Figure 2, and the average concentration is just the magic value $f_C = 12 \log 2 - 8 \approx 0.318$ calculated for the 9 $\mathcal{D}$ cluster in Ref. [11].

## 3. Algorithms

We use OpenMP to accelerate simulations. Each row of the lattice is processed by an OpenMP thread as illustrated in the right panel of Figure 1. This provides a significant speed-up and allows us to simulate larger system sizes, see Fig. 3

We use the Hoshen-Kopelman algorithm [19], which provides the exact decomposition of the game field into connected clusters of $\mathcal{C}$ neighbors and $\mathcal{D}$ neighbors. Left panel of Fig. 4 illustrates the algorithm. We start with the upper-left site (0,0) and assign it to the $\mathcal{C}$ cluster number 1.

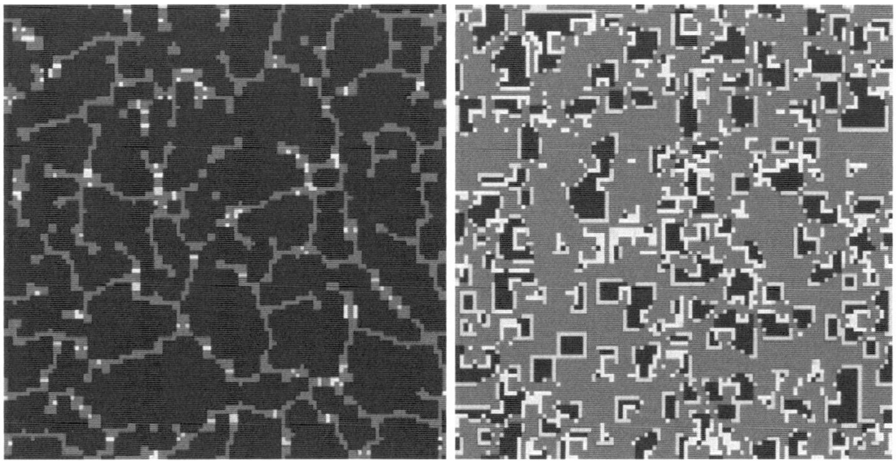

**Figure 2.** Snapshots of the game field for $b = 1.79$ (left) and $b = 1.81$ (right). The color coding is consistent with Ref. [9, 11]: blue is $\mathcal{C}$, red is $\mathcal{D}$, yellow is a $\mathcal{D}$ which was a $\mathcal{C}$ in the previous round, and green is a $\mathcal{C}$ which was a $\mathcal{D}$ in the previous round. See text for discussion.

Then we check the next site in the row (0, 1), which is connected (as a neighbor to the site (0,0)) and assign it to the same cluster number 1. The same happens with the third site (0,2). The fourth site (0,3) is a $\mathcal{D}$, and we assign it to the cluster number 2, and so on. Figure 4 shows what happens when the row-wise scan of the lattice reaches the site (6, 4). Its left-hand neighbor belongs to the cluster 1, and the upper site belongs to the cluster 3. Site (6, 4) is assigned to the cluster 1, as well as all sites which were assigned to the cluster 3, are reassigned to the cluster 1. Process of reassignment is generated by the red arrow. After the decomposition is finished, we calculate cluster "masses" or "sizes" (i.e., the number of sites which belongs to a given cluster), and update the histograms of the cluster sizes.

To extract the interface between $\mathcal{C}$-clusters and $\mathcal{D}$-clusters, we use the procedure of ref. [20]. It is defined on the dual lattice, which is plotted through the middle of the edges connecting sites, and perpendicular to them. An interface is shown by a black line in the right panel of Figure 4. We extract the external interface of the percolating cluster (it is $\mathcal{C}$-cluster in the figure), and calculate the average length $\mathcal{L}$ , which is the average length of the percolating interfaces in the steady-state regime. If the game field does not contain a percolating interface at this time step,

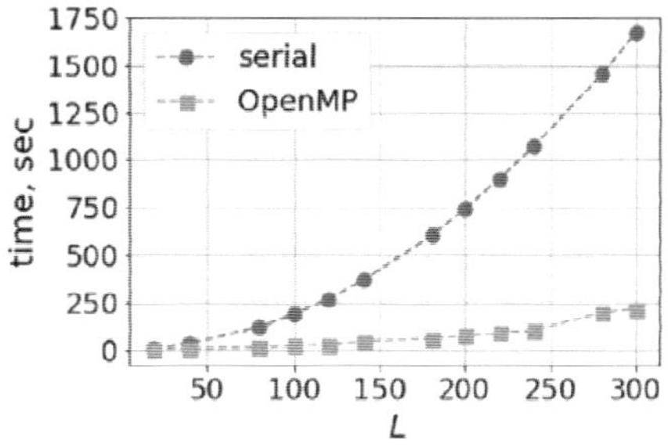


**Figure 3.** Computation time for a sequential and OpenMP implementations of the algorithm with 12 OpenMP threads. Each simulation is 50 independent realizations of initial conditions, simulated for 500 generations.

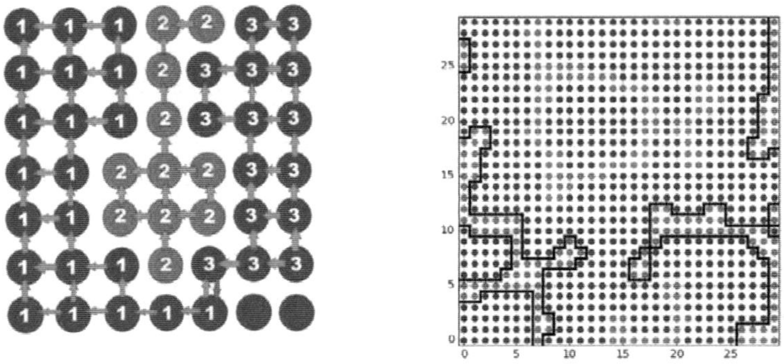


**Figure 4. Left**: An illustration of the Hoshen-Kopelman algorithm. See text for details. **Right**: An illustration on the interface extraction. (black lines) Color coding is the same as in the Figure 2: blue dots are $\mathcal{C}$ and red dots are $\mathcal{D}$.

we do not update the averages or histograms.

## 4. Results of simulations and data analysis

### *4.1. Phase transition*

The average density of $\mathcal{C}$ changes drastically from $f_c = 0.737(6)$ at $b = 1.775$ to $f_c = 0.315(6)$ at $b = 1.825$ (see the left panel of Figure 1). We stress that the latter value is consistent with the magic density [11], $f_c = 12\log 2 - 8 \approx 0.318$, which corresponds to the growth of $\mathcal{D}$ cluster. This way, we can attribute the transition from the $\mathcal{C}$-cluster percolation at $3/2 < b < 9/5$ (the blue one in the left panel of Figure 2) to the $\mathcal{D}$-cluster percolation (the red one in the right panel of Figure 2). In other words, the transition is from a random oscillatory regime of $\mathcal{C}$-clusters to the random growth regime of $\mathcal{D}$-clusters. The density changes abruptly at $b = b_c = 9/5$.

In both regimes, $b < 9/5$ and $b > 9/5$, the cluster size distribution demonstrates an exponential decay with size, although range of the distribution is much large in the case of proliferation of $\mathcal{D}$-clusters.

### *4.2. Scaling of the interface length*

We fit the interface length $\mathcal{L}$ with a power law form,

$$\mathcal{L} = AL^{\theta} + c\,. \tag{1}$$

The result for $b = 1.74$ is $A = 0.22(4)$, $c = 5(9)$, and $\theta = \mathbf{2.07(5)}$. For $b = 1.81$ it is $A = 0.35(1)$, $c = -20(5)$, and $\theta = \mathbf{1.99(1)}$. Estimates of the values of $c$ are rather rough, and we include then to extract some idea on the typical correlation length. What is very intriguing is that the interface of the percolating clusters in both regimes is close to **2**, which is not obvious while looking in the Figures 2. Value of $\theta = 2$ corresponds to the line which is space-filling in the thermodynamic limit of $L \to \infty$.

### *4.3. Minkowski dimension of the interface*

To check this finding we estimate the Minkowski dimension of the interface, i.e., we cover the interface with square boxes of linear size $l$, and count the minimum number of boxes $N(l)$ needed to completely cover the interface, and take the limit of $l \to 0$

$$d_M = \lim_{l \to 0} \frac{\log\, N(l)}{-\log\, l}\,. \tag{2}$$

This quantity is much more sensitive to the finite size of the lattice $L$. There are also visible deviations for $l \sim 1$, i.e., for the box size comparable to the lattice spacing. We fit data for $N(l)$ with the linear formula, $\log N(l) = B \log(1/l) + c$. Results of this fit procedure show a clear trend of $d_M$ to approaching the value of 2 in the thermodynamic limit $L \to \infty$.

## 5. Discussion and future work

We study an evolutionary game based on the Prisoner's Dilemma with agents arranged on a square grid in the plane. Despite its apparent simplicity, this deterministic, globally synchronous game displays surprisingly rich behavior in the long time limit. The steady state of the game features a series of dynamic regimes separated by sharp transitions at specific values of the payoff parameter. Transitions are characterized by jumps of the average density of cooperators, and non-trivial rearrangements of the game field. Interfaces between clusters of players with different strategies are stochastic fractals, which, however, are space-filling across the transitions, thus resembling specific Julia sets which are also space-filling [21].

It would be interesting to further characterize these emergent geometric structures. It would also be interesting to investigate the game for more complicated arrangements of players, such as non-square grids, complex networks and dynamic graphs.

## Acknowledgments

We thank Lev Barash for helpful discussions. This work was supported by RFBR grant 16-07-01122 (section 3) and grant 14-21-00158 from the Russian Science Foundation.

## References

[1] O. Schramm, *Scaling limits of loop-erased random walks and uniform spanning trees*, Isr. J. Math. **118**, 221 (2000).
[2] A. A. Belavin, A. M. Polyakov, and A. B. Zamolodchikov, *Infinite conformal symmetry in two-dimensional quantum field theory*, Nucl. Phys. B **241** 333 (1984).
[3] R.J. Baxter, *Exactly solvable models in statistical mechanics*, Academic Press, London, 1982.
[4] Vl.S. Dotsenko, V.A. Fateev, *Conformal algebra and multipoint correlation functions in 2D statistical models*, Nucl. Phys. B **240 (FS12)**, 312 (1984).
[5] O. Schramm and S. Sheffield, *Harmonic Explorer and Its Convergence to $SLE_4$*, The Annals of Probability **33**, 2127 (2005).
[6] S. Smirnov, *Critical percolation in the plane: conformal invariance, Cardy's formula, scaling limits*, Comptes Rendus de l'Académie des Sciences **333**, 239 (2001).
[7] S. Smirnov, *Conformal invariance in random cluster models. I. Holomorphic fermions in the Ising model*, Ann. Math., **172**, 1435 (2010).
[8] Y.H. Jung, H. Zhang, S.J. Cho, Z. Ma, *Flexible and Stretchable Microwave Microelectronic Devices and Circuits*, IEEE Trans. Elrctr. Dev. **64**, 1881 (2017).
[9] M.A. Nowak and R.M. May, *Evolutionary games and spatial chaos*, Nature **359**, 826 (1992).
[10] M.A. Nowak, *Evolutionary Dynamics: Exploring the equations of life*, The Belknap Press, (2006).
[11] M.A. Nowak and R.M. May, *The spatial dilemmas of evolution*, Int. J. Birurcation and Chaos **3**, 35 (1993).
[12] H.E. Stanley, *Introduction to phase transition and critical phenomena*, Oxford University Press 1971.
[13] M.N. Barber, *Finite size scaling*, in *Phase transitions and critical phenomena*, Eds. C. Domb and J Lebowitz, volume 8 (1983).
[14] K. Binder, *Theory of first-order phase transitions*, Rep. Prog. Phys. **50**, 783 (1987).
[15] A. Bar and D. Mukamel, *Mixed-Order Phase Transition in a One-Dimensional Model*, Phys. Rev. Lett. **112**, 015701 (2014).
[16] J. Adler, *Bootstrap percolation*, Physica A **171**, 453 (1991).
[17] H. Herrmann, *Discontinuous percolation*, J. of Phys: Conf. Series **681**,012003 (2015).
[18] see, *e.g.*, S. Tadelis, *Game Theory: An Introduction*, Princeton University Press, 2013, and references therein.
[19] J. Hoshen, R. Kopelman, *Percolation and cluster distribution. I. Cluster multiple labeling technique and critical concentration algorithm*, Phys. Rev. B. **14**, 8 (1976).
[20] A. Zatelepin and L. Shchur, *Duality of critical interfaces in Potts model: numerical check*, `arXiv`:1008.3573.
[21] M. Shishikura, *The Hausdorff dimension of the boundary of the Mandelbrot set and Julia sets*, Annals of Mathematics **147**, 225 (1998).

IOP Conf. Series: Journal of Physics: Conf. Series **955** (2017) 012024 doi:10.1088/1742-6596/955/1/012024

# Invariant characteristics of self-organization modes in Belousov reaction modeling

**S D Glyzin[1,2], V E Goryunov[1,2], A Yu Kolesov[1]**
[1] P.G. Demidov Yaroslavl State University, Sovetskaya str., 14, Yaroslavl, 150000, Russia
[2] Science Center in Chernogolovka RAS, Lesnaya str., 9, Chernogolovka, Moscow region, 142432, Russia

E-mail: glyzin@uniyar.ac.ru, salkar@ya.ru, kolesov@uniyar.ac.ru

**Abstract.** We consider the problem of mathematical modeling of oxidation-reduction oscillatory chemical reactions based on the mechanism of Belousov reaction. The process of the main components interaction in such reaction can be interpreted by a phenomenologically similar to it "predator-prey" model. Thereby, we consider a parabolic boundary value problem consisting of three Volterra-type equations, which is a mathematical model of this reaction. We carry out a local study of the neighborhood of the system's non-trivial equilibrium state and construct the normal form of the considering system. Finally, we do a numerical analysis of the coexisting chaotic oscillatory modes of the boundary value problem in a flat area, which have different nature and occur as the diffusion coefficient decreases.

## 1. Introduction

The reaction we are interested in was discovered by B.P. Belousov in 1951, but gained a wide popularity much later — a detailed article on this topic was published only 30 years later (see [1]). An experimental analysis of this reaction was done in the works of A.M. Zhabotinsky (see [2]).

We study a fairly simplified mathematical model in terms of the reaction components composition. The relations between the considering substances that enter into Belousov reaction (namely: bromo acid $HBrO_2$, cerium $Ce^{4+}$, and bromide $Br^-$ [2]) can be clearly described by the following scheme (see Fig. 1), where the sign "+" means that the presence of one substance contributes to the emergence of another, and the sign "−" indicates the oppression of one substance by another. A system of differential equations was proposed in [3] to describe this reaction on the assumption of phenomenological proximity of the substances interaction resulting scheme to the "predator-prey" problem. Later the system was modified in [4] for more accurately modeling of the ongoing chemical processes.

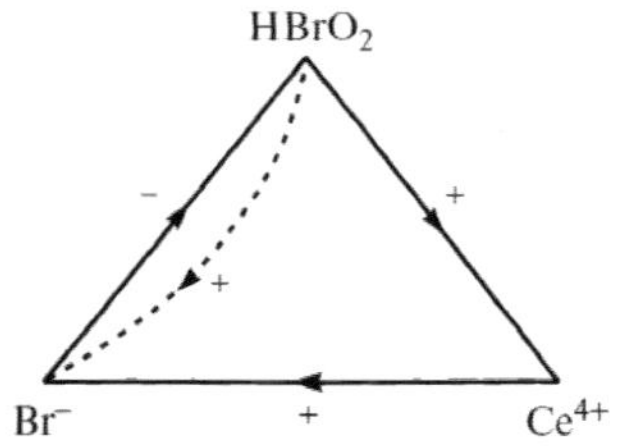


**Figure 1.** The nominal scheme of substances interaction in the modeling reaction.

## 2. Problem statement

We consider a boundary value problem of the form:

$$\begin{aligned} \dot{x} &= D_1\Delta x + r_1(1 + a(1 - z) - x)x, & \left.\frac{\partial x}{\partial \nu}\right|_{\partial\Omega} &= 0, \\ \dot{y} &= D_2\Delta y + r_2(x - y)y, & \left.\frac{\partial y}{\partial \nu}\right|_{\partial\Omega} &= 0, \\ \dot{z} &= D_3\Delta z + r_3(\alpha x + (1 - \alpha)y - z)z, & \left.\frac{\partial z}{\partial \nu}\right|_{\partial\Omega} &= 0, \end{aligned} \tag{1}$$

where $x(t, s)$, $y(t, s)$, and $z(t, s)$ stand for the concentration density of $HBrO_2$, $Ce^{4+}$, and $Br^-$ respectively, $s \in \Omega \subset \mathbb{R}^2$, $t > 0$; parameters $r_1$, $r_2$, $r_3$, $a$, $D_1$, $D_2$, $D_3$ are positive; $\alpha \in (0, 1)$; $(D_1, D_2, D_3) = d(D_1^0, D_2^0, D_3^0)$, $d > 0$; $\Delta$ is a Laplace operator; and $\nu$ is a direction of the outer normal to the border $\partial\Omega$ of bounded flat area $\Omega$ with measure equals to 1.

Our goal is to carry out the local study of limit cycle genesis in the neighborhood of the nontrivial equilibrium state of the system (1) at value of parameter $a$ close to the critical $a_{cr}$. In [4] for case of $a \gg a_{cr}$ the corresponding cycle was studied using the large parameter method. Moreover, it is required to consider the typical chaotic oscillatory modes with different structures at small diffusion value in a flat area using numerical analysis.

We introduce auxiliary notations: $p = r_1r_2r_3$, $q = r_1r_2 + r_1r_3 + r_2r_3$, $r = r_1 + r_2 + r_3$. A local analysis of the problem (1) will be performed near the equilibrium state $(1, 1, 1)$, since at

$$a > a_{cr} = (qr - p)/(p(1 - \alpha) - \alpha r_1r_3(r_1 + r_3)) \tag{2}$$

the system without diffusion terms loses stability in an oscillatory manner.

The main result of numerical analysis is as follows: the boundary value problem (1) has coexisting chaotic oscillations that retain the dynamic characteristics of spatially homogeneous cycle of the system, and self-organization type modes that are, in some way, more effective: the minima of the average value by area $\Omega$ increase and the maximum burst values decrease.

## 3. A local analysis of neighborhood of equilibrium state $(1, 1, 1)$ at $a$ close to critical

We clarify the problem statement in this case. We consider the problem for the spatially homogeneous solution of the system (1), in which we assume $a = a_{cr} + \varepsilon$, where $\varepsilon$ is a positive small parameter, $a_{cr}$ is determined by the formula (2), and the stability is getting lost in an oscillatory manner in the neighborhood of the equilibrium state $(1, 1, 1)$ at this value of $a$. To do this, we make replacements: $x = u + 1$, $y = v + 1$, $z = w + 1$. The resulting system has the form:

$$\begin{aligned} \dot{u} &= r_1(u + 1)(-u - (a_{cr} + \varepsilon)w), \\ \dot{v} &= r_2(v + 1)(u - v), \\ \dot{w} &= r_3(w + 1)(\alpha u + (1 - \alpha)v - w). \end{aligned} \tag{3}$$

In (3) we use the standard replacement of the normal forms method, and we search the solution in the form

$$(u, v, w)^T = \sqrt{\varepsilon}(he^{i\omega t}\xi(\tau) + \text{c.c.}) + \varepsilon U_1(t, \tau) + \varepsilon^{3/2}U_2(t, \tau) + \dots, \tag{4}$$

where c.c. is a complex conjugate to the expression in the same bracket, $\tau = \varepsilon t$ is a slow time, $U_j(t, \tau)$ $(j \geq 1)$ are the trigonometric polynomials by $t$, $h = (h_1, h_2, h_3)^T$ is the eigenvector of matrix of the linearized system (3) corresponding to eigenvalue $i\omega$, and $\omega = \sqrt{q + r_1r_3a\alpha}$.

After equalization of the coefficients by same powers of $\sqrt{\varepsilon}$, on the third step (at $\varepsilon^{3/2}$) we obtain the problem for $U_2$, from the solvability condition of which in the class of trigonometric polynomials we get the equation for complex amplitude:

$$\xi' = (\varphi_0 + i\psi_0)\xi + (d_0 + ic_0)|\xi|^2\xi,$$

which has a self-similar solution of the form $\xi = \rho e^{i\varphi}$, where $\rho(\tau)$ and $\varphi(\tau)$ satisfy the following equations: $\rho' = \varphi_0\rho + d_0\rho^3$, $\varphi' = \psi_0 + c_0\rho^2$. Here $\rho$ determines the amplitude, and $\varphi$ determines the correction to the oscillation frequency. Parameters $\varphi_0$ and $d_0$ have the form:

$$\varphi_0 = \frac{r_1 r_3 (r_2 - \alpha r)^2}{2(r_2(r^2+q) - \alpha(r^3+p))},$$

$$d_0 = \frac{(1-\alpha)^2 r_2^3 (r_1+r_2)^4 (r_1+r_3)^2 (r_2+r_3)^3 r((1+\alpha)r_1 + (\alpha-3)r_2 + (3\alpha-1)r_3)(q - r_1 r_3 \alpha)^2}{2r_3^2 (r_2-\alpha r)^3 (r_2(r^2+q) - \alpha(r^3+p))(r_2(r^2+4q) - \alpha(r^3+4p))}.$$

Hereinafter, we will only be interested in the amplitude of the oscillatory mode, therefore we do not give the form of $\psi_0$ and $c_0$ due to its cumbersomeness.

For the case $r_1 = 1$, $r_2 = 2$, $r_3 = 3$ the dependence of $\varphi_0$ and $d_0$ on the parameter $\alpha$ is represented on Fig. 2 and Fig. 3. Here, the solid line indicates a zone of positive values, and dotted line indicates a zone of negative values.

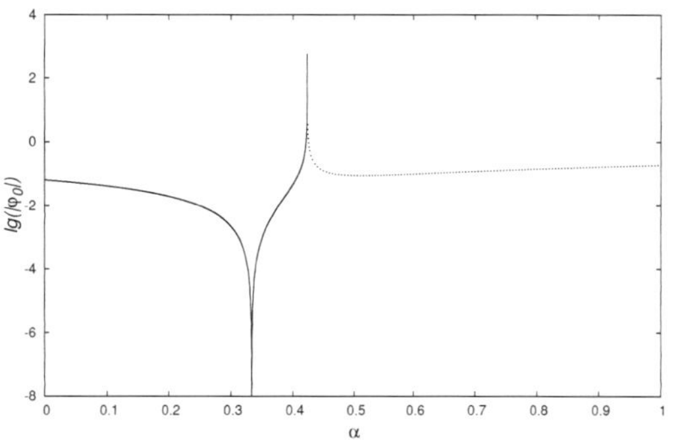


**Figure 2.** Dependence of $\varphi_0$ on $\alpha$ at $r_1 = 1$, $r_2 = 2$, $r_3 = 3$.

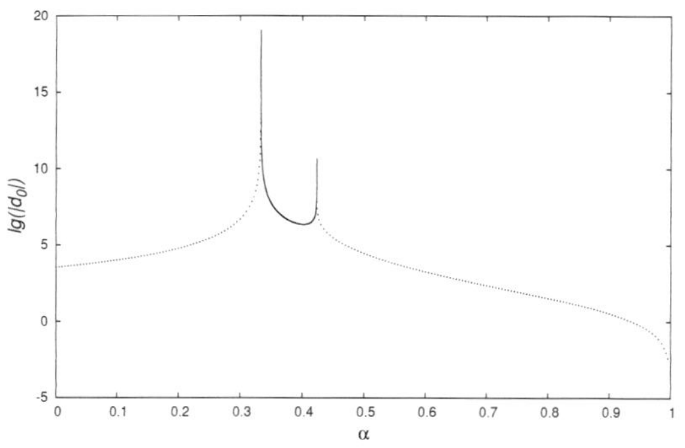


**Figure 3.** Dependence of $d_0$ on $\alpha$ at $r_1 = 1$, $r_2 = 2$, $r_3 = 3$.

Thereby, the following statement is proved.

**Theorem 1.** *Let these inequalities are satisfied:* $r_1, r_2, r_3 > 0$, $(r_2(r^2+q) - \alpha(r^3+p)) > 0$, $((\alpha+1)r_1 + (\alpha-3)r_2 + (3\alpha-1)r_3)(r_2(r^2+4q) - \alpha(r^3+4p)) < 0$, $r_2 \neq \alpha r$, *and* $a = a_{cr} + \varepsilon$, *where* $a_{cr}$ *is defined by the formula* (2). *Then there exists* $\varepsilon_0 > 0$ *such that for all* $\varepsilon \in (0, \varepsilon_0]$ *within some sufficiently small neighborhood of zero there exists an orbitally asymptotically stable limit cycle of the system* (3) *bifurcating from it with asymptotics* (4).

Note that a value of the oscillation amplitude of the system's limit cycle is determined by $\sqrt{-\varepsilon\varphi_0/d_0}$. To illustrate Theorem 1, we calculated the analytical and numerical values of the oscillation amplitude for a fixed set of parameters in order to determine the limits of applicability of this theorem. Based on Table 1, it is not difficult to see that the asymptotic formulas are satisfied within rather wide limits.

Suppose that $D_1 = \varepsilon d D_1^0$, $D_2 = \varepsilon d D_2^0$, $D_3 = \varepsilon d D_3^0$. After that, using the methods of asymptotic analysis we can find the value $d$, at which the constructed by us spatially homogeneous cycle with the asymptotics (4) loses stability. By analogy with the assumptions of the article [5], we can find such a critical value $d$ and the asymptotics of a spatially

**Table 1.** Values of the cycle oscillation amplitudes at $r_1 = 1$, $r_2 = 2$, $r_3 = 3$.

| $\alpha$ | $\varepsilon$ | analytical result | numerical result |
|---|---|---|---|
| 0.1 | $10^{-2}$ | 0.22090794 | 0.22116161 |
| 0.1 | $10^{-1}$ | 0.69869911 | 0.69937442 |
| 0.3 | $10^{-2}$ | 0.08908736 | 0.08855718 |
| 0.3 | $10^{-1}$ | 0.28086436 | 0.28004238 |

inhomogeneous cycle bifurcating from a homogeneous one. Unfortunately, asymptotic methods do not allow us to study the problem by further decreasing the parameter $d$ and increasing the parameter $a$. Thereby, we have to apply numerical methods. It should also be noted that asymptotic formulas for spatially homogeneous and spatially inhomogeneous cycles were obtained in [4] assuming that $r$ is a large parameter.

Spatially inhomogeneous modes bifurcating from the homogeneous cycle of the system (3) usually inherit the properties of a spatially homogeneous cycle relatively to the average value, minima and maxima of the oscillatory modes. In a numerical experiment, it is interesting to find spatially inhomogeneous modes of another structure, which we conventionally call self-organization modes. The next section is devoted to a numerical analysis of the boundary value problem (1) and to finding such modes.

## 4. Numerical analysis of the boundary value problem (1)

We study the boundary value problem (1) using numerical methods in the square area $\Omega$ ($\Omega = \{(x_1, x_2) \mid 0 \le x_1 \le 1,\ 0 \le x_2 \le 1\}$). For the numerical analysis of the boundary value problem (1) we replace Laplace operator with its difference analogue. Suppose the number of partition points $m$ gets such greater, than the diffusion coefficient gets smaller. Also we assume that the values in the appropriate rectangles of the region $\Omega$ are the same. Such approach allows us to apply the methods of parallel calculations.

Numerical analysis showed that the cycle constructed by us with the increasing of parameter $a$ becomes relaxational and turns into the modes described in [4]. Therefore, we further consider the case of finite $r_1, r_2, r_3$ and a relatively large $a$. Based on this, we set in (1)

$$d = 0.002,\ D_1^0 = 0.01,\ D_2^0 = 0.08,\ D_3^0 = 0.01,\ r_1 = 1,\ r_2 = 2,\ r_3 = 3,\ \alpha = 0.2,\ a = 100.$$

As the diffusion coefficient $d$ decreases, the oscillating modes of distributed model become more complicated. Coexisting chaotic modes of different nature arise including chaotic oscillations with relaxation properties and self-organization modes.

- The solutions of the first type are similar in their properties to the homogeneous stable relaxation cycle of the system (deep minima and high $\delta$-like bursts of the average by space value $x(t, s)$) (see Fig. 4 a), Fig. 5 a), and Fig. 6 a)).
- In self-organization modes the minima of the average by space value $x(t, s)$ are separated from zero, and the bursts are not so large by amplitude as in other solutions (see Fig. 4 b), Fig. 5 b), and Fig. 6 b)).

A comparison of their dynamic characteristics is given in Table 2. Here $\widehat{x}(t)$ is the average by space defined by the formula $\widehat{x}(t) = 1/(\text{mes}\,\Omega) \int\limits_{\Omega} x(t, s)ds$ and $M(\widehat{x}(t))$ is the average by time. The largest Lyapunov exponents (see [6, 7, 8, 9]) of these modes are calculated for various number of partition points (see Fig. 7). These graphs clearly show the difference in the chaotic structure of considered modes.

**Table 2.** Comparison of modes dynamic characteristics.

| Mode type | $M(\widehat{x}(t))$ | $\min \widehat{x}(t)$ | $\max \widehat{x}(t)$ | $\min x(t,s)$ | $\max x(t,s)$ |
|---|---|---|---|---|---|
| Relaxation mode | 0.9540 | $10^{-11}$ | 3.44373 | $10^{-12}$ | 29.81307 |
| Self-organization mode | 0.9722 | 0.79906 | 1.15436 | $10^{-12}$ | 18.03019 |

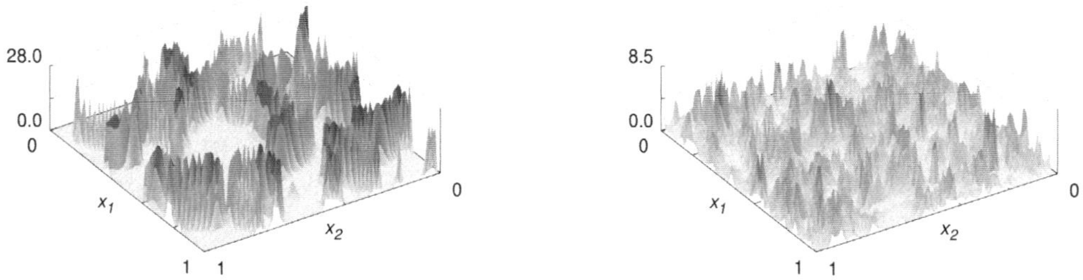


a) b)

**Figure 4.** Distribution of $x(t, s)$ by area $\Omega$ at $t = T_*$ in case of mode with a) relaxation properties; b) self-organization.

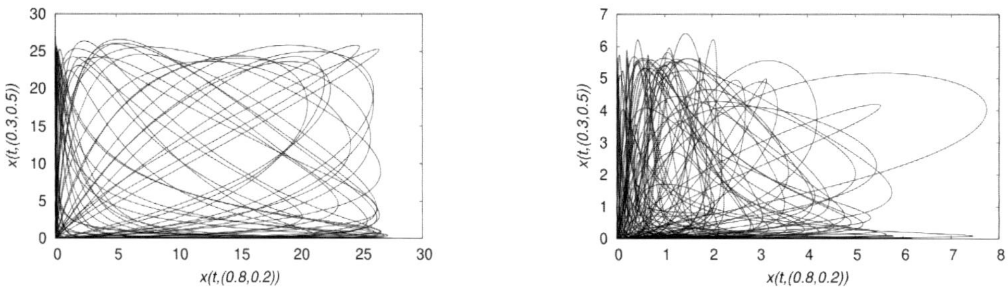


a) b)

**Figure 5.** Phase portrait of $x(t, s)$ (built on $x(t, (0.8, 0.2))$ and $x(t, (0.3, 0.5))$) in case of mode with a) relaxation properties ($0 \leq t \leq 200$); b) self-organization ($0 \leq t \leq 50$).

## 5. Conclusion

We summarize obtained results. First of all, we note the found for a point model asymptotic formulas for a stable periodic solution. For this solution, which is a spatially homogeneous cycle of a model system, we determine the conditions on which a spatially inhomogeneous cycle bifurcates from it. These asymptotics are used as a starting point for a numerical experiment. The main result of numerical modeling of the boundary value problem is as follows: it has more complex oscillatory modes (self-organization modes) along with spatially inhomogeneous modes bifurcating from spatially homogeneous ones. These modes are unordered both by time and space. The largest Lyapunov exponent, calculated for them, turned out to be positive, and that just represents its disorder. At the same time, these solutions have a more explicit biophysical meaning. This allows us to consider the offered approach for oscillatory chemical reactions modeling as fairly adequate.

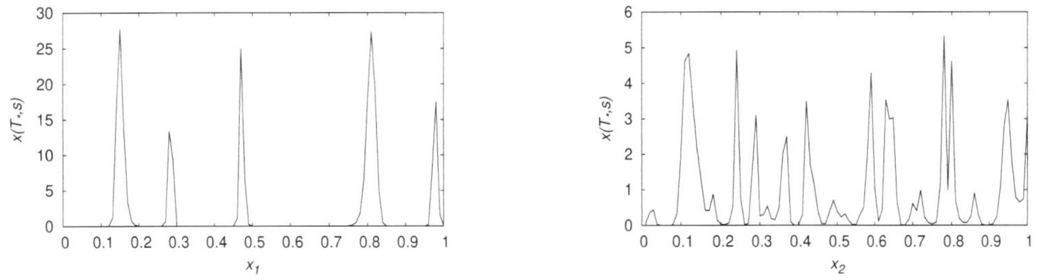


a) b)

**Figure 6.** Cross-section of $x(t,s)$ along one of area $\Omega$ axes at $t = T_*$ in case of mode with a) relaxation properties; b) self-organization.

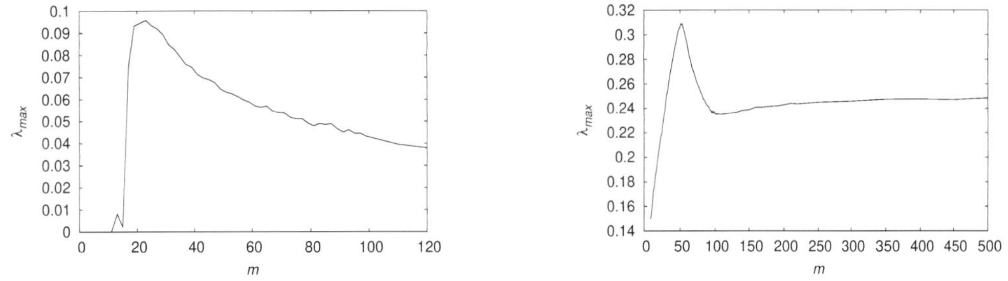


a) b)

**Figure 7.** Dependence of largest Lyapunov exponent on the number of partition points in case of mode with a) relaxation properties; b) self-organization.

## Acknowledgments

This work is supported by the Russian Science Foundation (project no. 14-21-00158).

## References

[1] Belousov B P 1981 Periodically acting reaction and its mechanism *Autowave Processes in Systems with Diffusion (Institute of Applied Physics, USSR Academy of Science, Gorky)* 176–86
[2] Zhabotinsky A M and Zaikin A N 1973 Autowave processes in a distributed chemical system *Journal of Theoretical Biology* **40:1** 45–61
[3] Kolesov Yu S 1985 Adequacy problem for ecological equations *Dep. VINITI 1901-85*
[4] Glyzin S D, Kolesov A Yu and Rozov N Kh 2011 Relaxation oscillations and diffusion chaos in the Belousov reaction *Computational Mathematics and Mathematical Physics* **51:8** 1307–24
[5] Glyzin S, Goryunov V and Kolesov A 2017 Spatially inhomogeneous modes of logistic differential equation with delay and small diffusion in a flat area *Lobachevskii Journal of Mathematics* **38:5** 898–905
[6] Glyzin S D, Kolesov A Yu and Rozov N Kh 2005 Chaotic buffering property in chains of coupled oscillators *Differential Equations* **41:1** 41–9
[7] Glyzin S D, Kolesov A Yu and Rozov N Kh 2010 Finite-dimensional models of diffusion chaos *Computational Mathematics and Mathematical Physics* **50:5** 816–30
[8] Glyzin S D 2013 Dimensional characteristics of diffusion chaos *Automatic Control and Computer Sciences* **47:7** 452–69
[9] Glyzin S D and Shokin P L 2016 Diffusion chaos in the reaction – diffusion boundary problem in the dumbbell domain *Automatic Control and Computer Sciences* **50:7** 625–35

# Properties of the Tent map for decimal fractions with fixed precision

**V M Chetverikov**

Department of Applied Mathematics of Moscow Institute of Applied Mathematics, National Research University Higher School of Economics, Moscow, Russia

E-mail: vchetverikov@hse.ru

**Abstract**. The one-dimensional discrete Tent map is a well-known example of a map whose fixed points are all unstable on the segment [0,1]. This map leads to the positivity of the Lyapunov exponent for the corresponding recurrent sequence. Therefore in a situation of general position, this sequence must demonstrate the properties of deterministic chaos. However if the first term of the recurrence sequence is taken as a decimal fraction with a fixed number "k" of digits after the decimal point and all calculations are carried out accurately, then the situation turns out to be completely different. In this case, first, the Tent map does not lead to an increase in significant digits in the terms of the sequence, and secondly, demonstrates the existence of a finite number of eventually periodic orbits, which are attractors for all other decimal numbers with the number of significant digits not exceeding "k".

## 1.Introduction

Interest in deterministic chaos among mathematicians and physicists was accompanied by a huge number of publications after the 70s of the last century [1-5]. A large survey of only one of the important areas of these studies [6] contained 216 references.

Significant interest of cryptologists towards deterministic chaos began, apparently from work [7], where the author gave a symmetric encryption scheme based on ergodic property of chaos.

Deterministic chaos is interesting for cryptologists because on the one hand it looks like a random process, but on the other hand it allows to restore the process accurately if the generating model and the initial conditions are known. A detailed discussion of the prospects and problems of new cryptosystems is devoted to the work [8-11]. Statistical analysis of the Tent map properties by constructing a four-level discrete model is devoted to work [12].

It is known that when chaotic systems are realized with a finite precision in digital computers, their dynamical properties are often found to be entirely different from the original versions in the continuous setting. To emphasize the essential difference between continuous chaos and digital chaos, the latter is also called pseudo chaos [13]. Similarly, digital chaotic orbits are also called pseudo (chaotic) orbits [14].

## 2.Mathematical definitions and assertions

The paper [15] reports some findings on a new series of dynamical indicators, which can quantitatively reflect the degradation effects on digital 1D piecewise linear chaotic maps realized with a fixed-point finite precision. To describe the processes taking place with limited accuracy of the representation of numbers, the authors used n-bit fixed-point binary decimals in the form $0.b_1b_2\ldots b_n \in [0,1),\ b_i \in \{0,1\}$

In this paper we consider orbits of a recurrent sequence

$$x_{n+1} = T(x_n) \quad n = 0,1,2\ldots \quad x_0 \in [0,1)\,, \tag{1}$$

where

$$T(x) = 2 \cdot \min\{x, 1-x\} = 1 - 2 \cdot |x - 1/2| \tag{2}$$

is a simplest Tent map without any parameter which maps [0,1] into [0,1]. In contrast to [15], it is assumed that the initial point of the orbit can be represented as a decimal fraction containing a given number of digits after the decimal point, and that all calculations related to the mapping (2) are carried out exactly without rounding. This possibility is implemented in mathematical packages in the form of symbolic calculations (for example, in the package "Mathematica").

We introduce several definitions that make it possible to formulate the results most simply.

Definition1. The set $I_0$ is represented by one number 0. The set $I_k$ (k=1,2, ...) is a collection of numbers of the form $0.\alpha_1\alpha_2\alpha_3\ldots,\alpha_k = \sum_{i=1}^{k}\alpha_i \cdot 10^{-i} \in I_k$, where $\alpha_i$ are integers from 0 to 9, and $\alpha_k \neq 0$. The set $\tilde{I}_k \subset I_k$ (k=1,2, ...) is a collection of numbers in $I_k$ for which $\alpha_k \neq 5$.

Definition2. The set $J_k$ is defined as the union of sets $J_k = \bigcup_{q=0}^{k} I_q$.

Obviously, for symbolic computations, if $x_0 \in J_k$ then $T(x) \in J_k$ since T (x) is a 1D piecewise linear mapping. It is easy to prove that the number of elements of this set is $NI_k = 9 \cdot 10^{k-1}$ if $k > 0$ and $NI_0 = 1$. The number of elements of the set $\tilde{I}_k$ is $N\tilde{I}_k = 8 \cdot 10^{k-1}$. Then the number of elements of the set $J_k$ equals to $NJ_k = 10^k$.

Definition 3. The sequence of numbers defined by the recurrence relation (1) is called the orbit of the map (2) induced by the quantity $x_0$.

Definition 4. We call an eventually periodic orbit with period $m$ an orbit that has the structure such as $x_0, x_1, x_2, \ldots, x_p, x_{p+1}, x_{p+2}, \ldots, x_{p+m}$, where $x_{p+m} = x_{p+1}$.

Definition 5.We call a cycle $S_k(m_k)$ of length $m_k$ of the mapping $T(x)$ the set of elements in $\in I_k$ for which the following property holds: for any $x \in S_k(m_k)$ the mapping $T(x) \in S_k(m_k)$. In other words, the cycle is an invariant mapping set.

The above definitions allow us to formulate the main assertions.

Assertion 1. Let the orbit of the mapping be induced by an element $x_0 \in I_k$. Then:
1). if $x_0$ looks like $x_0 = 0.\alpha_1\alpha_2\alpha_3\ldots,\alpha_{k-1}5$, then $x_1 = T(x_0) \in I_{k-1}$ and all further elements of the orbit will belong to the set $J_{k-1}$ ;
2). if $x_0 \in \tilde{I}_k$, then all further elements of the orbit will be eventually periodic with the elements $\in I_k$. The only attractor on this set is a cycle $S_k(m_k)$.

Assertion 2. The length of a cycle $S_k(m_k)$ is determined by a number $k$ by the formula.

$$m_k = 2 \cdot 5^{k-1} \tag{3}$$

The minimum and maximum elements of a cycle $S_k(m_k)$ are defined by the formulas:

$$\min S_k(m_k) = 2^{k+1} \cdot 10^{-k}; \qquad \max S_k(m_k) = 1 - 2^k \cdot 10^{-k} \tag{4}$$

If you arrange the elements of the cycle $S_k(m_k)$ in ascending order, then the differences between two adjacent elements will be equal to either $2^{k+1}\cdot 10^{-k}$ or $2^{k+2}\cdot 10^{-k}$. In this case, the following rigid order of alternation is realized: at the beginning of the ordered series, a smaller difference is repeated three times, and then a large difference appears once. Then the situation repeats.

Assertion 3. Since the number of elements in the set $\tilde{I}_k$ is $N\tilde{I}_k$ and the cycle $S_k(m_k)$ attracting these elements contains $m_k$ elements, the information compression ratio is $N\tilde{I}_k \cdot (m_k)^{-1} = 2^{k+2}$.

The union of all cycles contained in a set $J_k$ is the set of numbers $G_k(M_k)$ occurring in all cycles $G_k(M_k) = 0 \cup \bigcup_{q=1}^{k} S_q(m_q)$. The number of elements in the collection of cycles is $M_k = 1 + \sum_{q=1}^{k} m_q = \frac{1}{2}(5^k + 1)$ (odd number). In this case, all possible orbits induced by all the different elements $x_0 \in J_k$ are attracted by the $M_k$ elements of the $k$ cycles and, consequently, the degree of information compression is $NJ_k \cdot (M_k)^{-1} = 2\cdot 10^k \cdot (5^k + 1)^{-1}$.

Figure 1 shows a diagram of all possible orbits initiated by different values $x_0 \in I_1$. Bold lines indicate the links between the two elements of the cycle $S_1(2) = \{0.4, 0.8\}$. There are ten elements from the initial set $x_0 \in I_2 \setminus \tilde{I}_2$ which go to the level $I_1$: $x_1 = T(x_0) \in I_1$, and only the element 0.5 leaves the level $I_1$ to the level $I_0$.

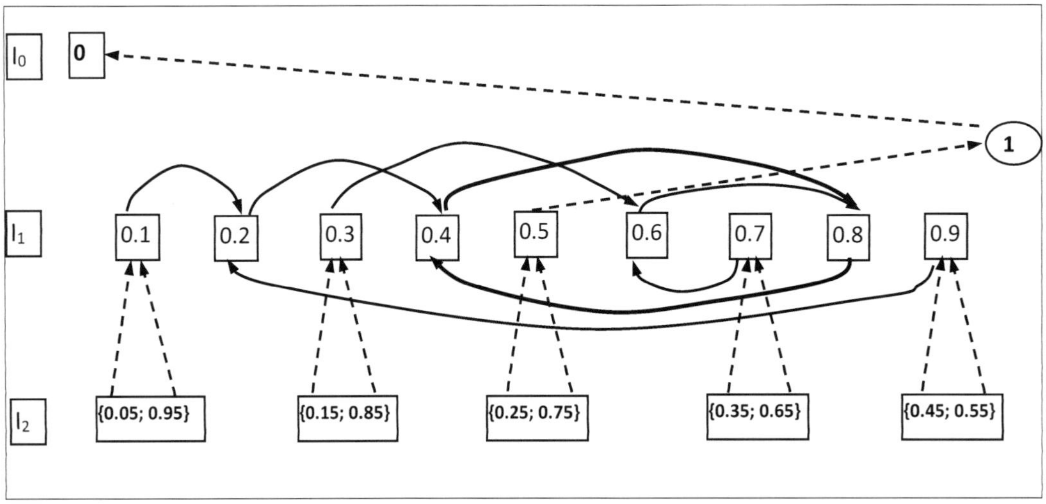


Figure 1. Orbits passing through the elements $x \in I_1$, as well as transitions from $I_2$ to $I_1$ and from $I_1$ to $I_0$ as a result of the transformation (1).

Each element in the set $I_k$ of the orbit has two incoming lines and one output line.

This means that there are two different elements $\hat{x}$, $\breve{x} \in I_k \cup I_{k+1}$, for each element $x_i$ of the orbit $x_i \in I_k$ such that

$$T(\hat{x}) = x_i, \quad T(\check{x}) = x_i \tag{5}$$

Equations (5) imply the irreversibility of motion along the orbit and, consequently, the inability to completely restore the orbit in reverse time. For orbits from the set $J_k$ the ergodicity condition is not satisfied. This condition is essential for the encryption scheme proposed in [7], using the chaotic sequence of the logistic mapping.

However, it can be suggested to use the scheme proposed in [7] for the deterministic sequence (1), (2) having a definite cycle system if the calculations (1), (2) are carried out symbolically.

In addition to the above cycles built on decimals, there are a large number of cycles built on various other denominators. We introduce the notation for cycles based on a fraction with the denominator $q$, where the number $m$ denotes the cycle length (the number of elements in the cycle), and l is the number of the cycle of length $m$. In case the cycle of the given period is the only one, then the superscript is not mentioned. Some of them are listed below.

Three of these cycles are often encountered in the literature

$C_7(3) = \{2/7, 4/7, 6/7\}$, $\quad C_9(1) = \{2/9, 4/9, 8/9\}$,

$C_{11}(5) = \{2/11, 4/11, 6/11, 8/11, 10/11\}$.

Five others do not attract great care of the researchers:

$C_{31}^{(1)}(5) = \{2/31, 4/31, 8/31, 16/31, 30/31\}$, $\quad C_{31}^{(2)}(5) = \{6/31, 12/31, 14/31, 24/31, 28/31\}$,

$C_{31}^{(3)}(5) = \{10/31, 18/31, 20/31, 22/31, 26/31\}$,

$C_{41}^{(1)}(10) = \{2/41, 4/41, 8/41, 10/41, 16/41, 18/41, 20/41, 32/41, 36/41, 40/41\}$,

$C_{41}^{(2)}(10) = \{6/41, 12/41, 14/41, 22/41, 24/41, 26/41, 28/41, 30/41, 34/41, 38/41\}$.

## 3.Conclusions

The Tent map (1) has the following two important properties. First, this transformation does not increase the number of significant digits for each subsequent transformation. Secondly, this transformation has attractors, which are determined by the number of digits after the decimal point. These properties are proposed to be used in the Baptista's encryption scheme instead of logistic mapping.

The advantage of the proposed mapping is that there is no need to partition the interval (0,1) into discrete subsets to fix a number. The drawbacks of the proposed scheme include the need for symbolic calculations.

## Acknowledgment

In conclusion, the author expresses his gratitude to professor E A. Krouk for his consultations on cryptography.

## References

[1] Lorenz E N 1963 Deterministic nonperiodic flow, *J. Atmos. Sci.* **20** 130
[2] Sparrow C 1982 *The Lorenz equations: bifurcations, chaos and strange attractors* Appl. Math.Sci.V.41 (Berlin: Springer)
[3] Schuster H G 1984 *Deterministic Chaos* (Weinheim: Physik-Verlag)
[4] Shapiro A P and Luppov S P 1983 *Recurrent equations in the theory of population biology* (Moscow: Nauka)

[5] Nedostupov E V, Sarancha D A, Chigerov E N and Yurezanskaya Yu S 2010 On some properties of one-dimensional unimodal mappings. *Reports of the Academy of Sciences* **430** No 1 pp 23–28
[6] Marwan N, Carmen Romano M, Thiel M and Kurths J 2007 Recurrence plots for the analysis of complex systems *Physics Reports* **438** 237 – 329
[7] Baptista M S 1998 Cryptography with chaos *Physics Letters* A, 240 (12) 50-54
[8] Alvarez G, Montoya F, Romera M and Pastor G 2003 Cryptanalysis of a chaotic secure communication system *Physics Letters*, A **306** 200–205
[9] Alvarez G, Montoya F, Romera M and Pastor G 2003 Cryptanalysis of a discrete chaotic cryptosystem using external key *Physics Letters*, A **319** 334–339
[10] Alvarez G and Li S Some Basic 2006 Cryptographic Requirements for Chaos-Based Cryptosystems *International Journal of Bifurcation and Chaos* **16** no 8 pp 2129-2151
[11] Nitharwal B, Rani M and Saini H C Improving security of the Baptista's cryptosystem using two-step logistic map 2015 *I J Computer Network and Information Security* **5** 34-40
[12] Diaconu A V 2012 Multiple bitstreams generation using chaotic sequences. *The annals of "Dunărea de Jos" University of galati fascicle* **35**, no 1, 1221
[13] Chirkikov B V and Vivaldi F 1999 An algorithmic view of pseudochaos *Physica* D **129**, 223
[14] Levy Y E 1982 Some remarks about computer studies of dynamical systems *Physics Letters* A **88** 1–3
[15] Li S, Chen G and Mou X 2005 On the dynamical degradation of digital piecewise linear chaotic maps *International Journal of Bifurcation and Chaos* **15** no 10 pp 3119-3151

# Research of the effectiveness of parallel multithreaded realizations of interpolation methods for scaling raster images

**A A Vnukov**[1], **M B Shershnev**[1]

[1]National research University Higher school of Economics, 20 Myasnitskaya ulitsa, Moscow 101000 Russia

**Abstract.** The aim of this work is the software implementation of three image scaling algorithms using parallel computations, as well as the development of an application with a graphical user interface for the Windows operating system to demonstrate the operation of algorithms and to study the relationship between system performance, algorithm execution time and the degree of parallelization of computations. Three methods of interpolation were studied, formalized and adapted to scale images. The result of the work is a program for scaling images by different methods. Comparison of the quality of scaling by different methods is given.

## 1. Introduction

These Scaling of images is used in a wide range of applied problems. This is one of the most demanded operations in computer graphics. [1]

Interpolation methods are widely used to solve various mathematical problems. There are many methods of interpolation. The most popular are the following image interpolation methods: bicubic interpolation, bilinear interpolation, stepwise interpolation (nearest-neighbor interpolation). These methods are listed in order of decreasing computational complexity and, accordingly, in order of decreasing the quality of the resulting image. [2]

The speed of execution of interpolation algorithms directly depends on the possibilities of their practical application. Often you need to scale images in real time, for example, when playing video. To effectively use the computing resources of modern multiprocessor computers, you need to use multi-threaded computing. [3]

## 2. Overview of existing methods of interpolation of images

When interpolating images, the image itself plays the role of an interpolated function, the value of which is the color of the pixel. Image pixels are points where the value of the function (pixel color) is known. Interpolation contains intermediate values of the function (colors of new pixels), the image does not necessarily increase, the interpolation methods allow to arbitrarily change the image size and aspect ratio.

### *2.1 Nearest neighbor method*

Stepwise interpolation (also known as the nearest neighbor method) is the simplest method of multidimensional interpolation. The nearest neighbor algorithm chooses the nearest point value and does not take into account the values of neighboring points at all. The algorithm is very simple to implement and widely used.

In Figure 1, the red dots represent the input data, the points at which the value of the interpolated function is known. The lines of blue denote the interpolation function. The stepwise interpolation algorithm is the simplest, both in terms of implementation and in terms of computational complexity. However, when the image is enlarged by this method, the so-called staircase effect is observed.

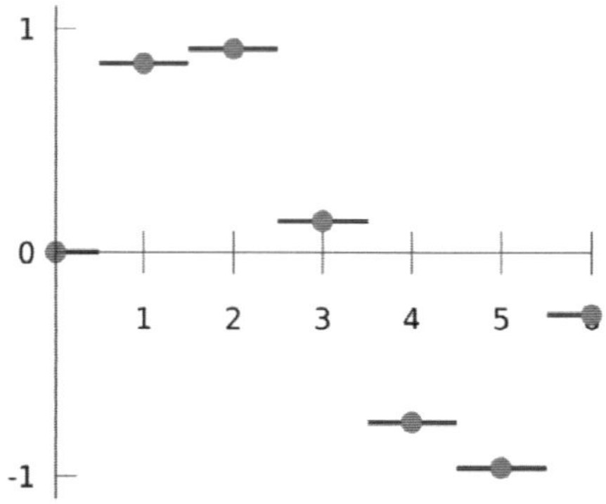


**Figure 1.** Interpolant function for nearest neighbour method.

### 2.2 *Bilinear interpolation*

In the case of linear interpolation, in order to find intermediate values, a piecewise linear function is constructed from known values. Bilinear interpolation is used for functions of two variables: first, linear interpolation is performed for each of the measurements, thus two auxiliary points are calculated, then linear interpolation between these auxiliary points is performed.

The formulas for linear and bilinear interpolation are:

$$f_{linear}(p_0, p_1, x) = x(p_1 - p_0) + p_0$$

$$f_{bilinear}(p_{00}, p_{01}, p_{10}, p_{11}, x, y) = f_{linear}(f_{linear}(p_{00}, p_{01}, x), f_{linear}(p_{10}, p_{11}, x), y)$$

$p_{00}, \ldots, p_{11}$ – the colors of the four nearest pixels.

$x, y$ – the relative coordinates of the point at which the color is calculated.

The bilinear interpolation algorithm is computationally more complex than the stepwise interpolation algorithm, but the resulting image looks much smoother than the result of using the stepwise interpolation method. This algorithm is often used in graphic editors, as it allows to obtain acceptable in most cases quality, without requiring a large computing power. However, if the original image is enlarged many times with this algorithm, unnatural sharp color transitions can occur in some places of the resulting image.

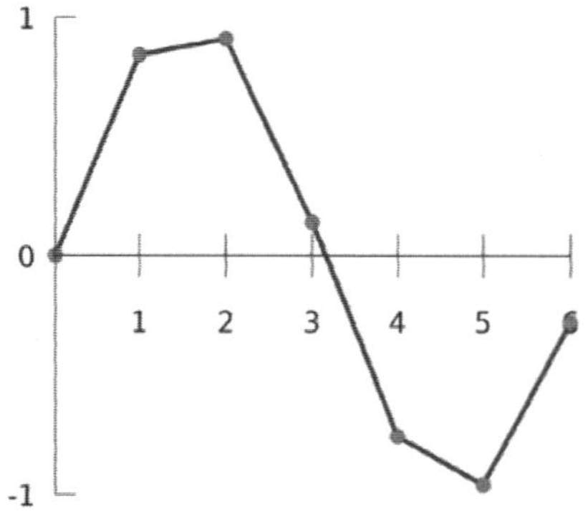


**Figure 2.** Interpolant function for bilinear interpolation.

### *2.3 Bicubic interpolation*

Bicubic interpolation is the extension of one-dimensional interpolation by polynomials of the third degree to the case of a function of two variables. The surface obtained as a result of bicubic interpolation is a smooth function, unlike surfaces obtained as a result of bilinear interpolation or interpolation by the nearest neighbor method.

Bicubic interpolation assumes that the interpolated function is given not only by its values at certain points, but also by the values of the derivative at these points.

In the case of image interpolation, it plays the role of an interpolated function, whose known values are the colors of the pixels of the original image, the values of the derivative are not known. You can assume the derivative value to be zero, but you can use the slope of the line between the previous and next point as a derivative to get a smoother look.

The formulas for cubic and bicubic interpolation are:

$$f_{cubic}(p_0, \dots, p_3, x) =$$
$$= x^3\left(-\frac{1}{2}p_0 + \frac{3}{2}p_1 - \frac{3}{2}p_2 + \frac{1}{2}p_3\right) + x^2\left(p_0 - \frac{5}{2}p_1 + 2p_2 - \frac{1}{2}p_3\right)$$
$$+ x\left(-\frac{1}{2}p_0 + \frac{1}{2}p_2\right) + p_1$$

$$f_{bicubic}(p_{00}, \dots, p_{33}, x, y) =$$
$$= f(f_{cubic}(p_{00}, \dots, p_{03}, x), f_{cubic}(p_{10}, \dots, p_{13}, x), f_{cubic}(p_{20}, \dots, p_{23}, x), f_{cubic}(p_{30}, \dots, p_{33}, x), y)$$

Where $p_{00}, \dots, p_{33}$ – the color values of the 16 nearest pixels surrounding the calculated value.
$x, y$ – the relative coordinates of the point at which the color is calculated.

The bicubic interpolation algorithm is computationally more complex than the bilinear interpolation algorithm, however the resulting image looks smoother.

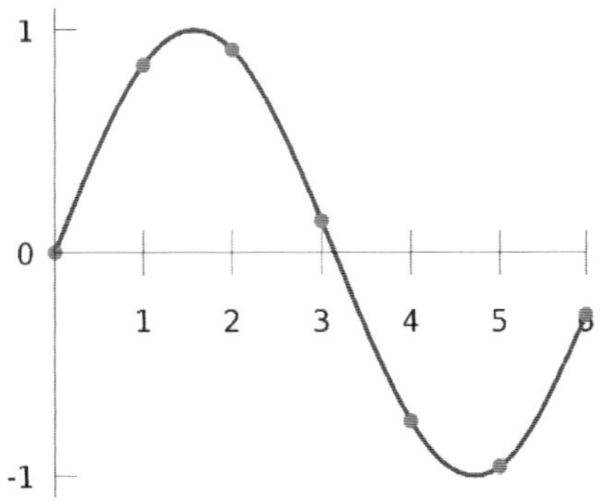


**Figure 3.** Interpolant function for bicubic interpolation.

## 3. Parallelizing an algorithm

Raster images often consist of a large number of pixels of the order of several million. Interpolation of such images is associated with a large number of calculations. At the same time, new pixels can be calculated independently of each other, so that the efficiency of the algorithm can be significantly increased by parallelizing the computations.

Algorithms of interpolation considered in this paper can be parallelized according to the following scheme: the image is divided into segments by rows. Each thread processes its part of the image, then the parts are combined to form the resulting image. Segmentation by line is important, since in the computer's memory the image is stored as a string sequence, and sequential memory access operations are faster than random access operations.

## 4. Create an application

The Main features:

- Open and save files with images in BMP format
- Display an editable image in the application window
- Scale images using the user-specified parameters: interpolation type, degree of scale change for each direction, number of streams used.

The main task of the application is to demonstrate the work of realized image interpolation algorithms. The application should simplify the process of testing algorithms and make it more visible. The interface is represented by the following elements: the main window in which the image is displayed, the main menu for accessing functions, the dialogs for opening and saving a file, and the scaling dialog in which the user can specify the scaling parameters. Also, the program displays the amount of time spent processing the image.

## 5. Performance Comparison and Scalability Research

Graph 1 shows the results of measurements of the image processing time when the algorithm is parallelized to a different number of streams for each of the interpolation methods under consideration. On the vertical axis there is time, on the horizontal axis - the number of streams. The graph shows that the bilinear interpolation algorithm works on average 2.4 times longer than the stepwise interpolation algorithm, and the bicubic interpolation algorithm is 4.5 times longer than the bilinear interpolation algorithm. Parallelization has resulted in a 300% - 400% increase in productivity.

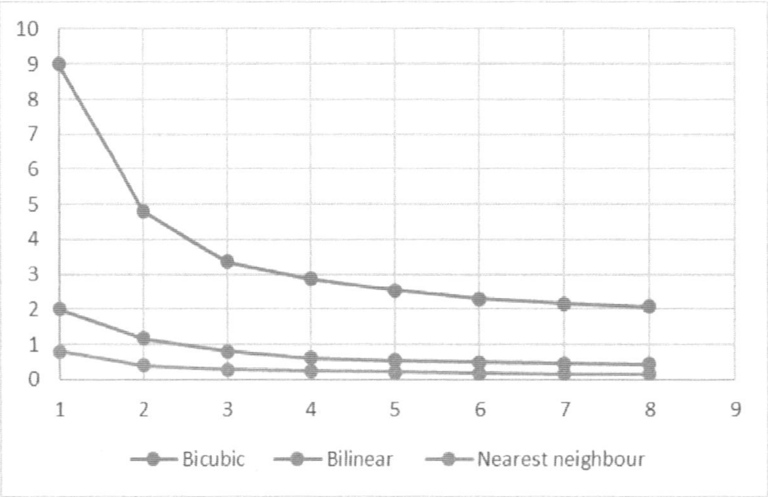


**Graph 1.** Performance.

## 6. Examples of image scaling

### *6.1 Example 1*

The original image has a size of 2x2 pixels. It was increased 100 times on both sides. The result of the three methods is shown in Figure 4.

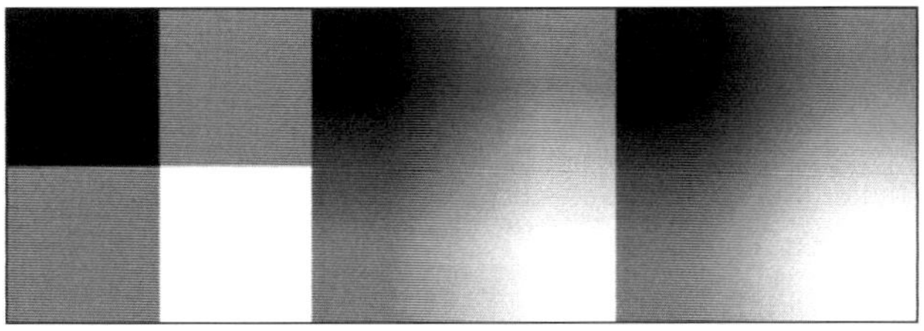

**Figure 4.** From left to right: the method of the nearest neighbour, bilinear and bicubic interpolation.

Stepwise interpolation does not change the structure of the image. Four areas are clearly visible, each of which corresponds to one of the four pixels of the original image. Bilinear interpolation makes the color transitions smooth, but in some places unnatural vertical and horizontal bands are noticeable. Bicubic interpolation gives an absolutely smooth image with smooth color transitions.

### *6.2 Example 2*
The original image was enlarged 100 times on both sides. The difference in interpolation quality is clearly visible if you look at the yellow dot in the upper left corner. In the case of bilinear interpolation, the vertical and horizontal lines in the center are expressed inside the point. With the increase using bicubic interpolation, one can see an almost ideal circle with smooth color transitions at the boundary.

**Figure 5.** From left to right: the method of the nearest neighbour, bilinear and bicubic interpolation.

## 7. Conclusion
In the course of this work, three parallel algorithms for scaling raster images by different interpolation methods were implemented. As a result of the work, an application with a graphical user interface was created, which allows to scale images by different interpolation methods and to save the result. The application allows you to compare different interpolation methods for the quality of image processing.

In the course of testing, the assumption of the feasibility of parallelizing the bicubic interpolation algorithm was confirmed. The use of several calculation flows provides a significant acceleration of the algorithm.

## 8. References
[1] G.Keys Robert, "Cubic Convolution Interpolation for Digital Image Processing", IEEE *Transactions on Acoustics, Speech and Signal Processing*, vol.29, no.6, 1981, pp.1153-1160.

[2] Meijering Erik, Unser Michael, "A Note on Cubic Convolution Interpolation", IEEE *Transactions on Image Processing*, vol. 12, no. 4, April 2003, pp. 477–479.

[3] Vnukov A. A., Egorov I. V. "Investigation of the effectiveness of software implementation of the multithreaded scaling algorithm by the method of bilinear interpolation", *Mining Information and Analytical Bulletin* (scientific and technical journal). 2014. № 9. C. 264-271.

IOP Conf. Series: Journal of Physics: Conf. Series **955** (2017) 012027 doi:10.1088/1742-6596/955/1/012027

# Transitional circuitry for studying the properties of DNA

**N Trubochkina**[1]

[1] National Research University Higher School of Economics, 121354, Russia

ntrubochkina@hse.ru

**Abstract**. The article is devoted to a new view of the structure of DNA as an intellectual scheme possessing the properties of logic and memory. The theory of transient circuitry, developed by the author for optimal computer circuits, revealed an amazing structural similarity between mathematical models of transition silicon elements and logic and memory circuits of solid state transient circuitry and atomic models of parts of DNA.

## 1. Introduction
Knowledge of DNA is the most important for humanity. DNA plays a determining biological role in the development of living beings, as it encapsulates their genetic information as a certain development program. A new look at the known object of research will help to a biophysicists to better understand the nature of this object, and use this knowledge for the benefit of man.

## 2. What is DNA?
There are two ways of looking at what exactly is DNA.

### *2.1. Overview of the history*
It is believed that in 1953 James Watson and Francis Crick unraveled the structure of deoxyribonucleic acid (DNA), a molecule of heredity (Crick, 1981) [1]. In Search of the Double Helix: Quantum Physics and Life. United States of America: McGraw-Hill Book Company. DNA has the form of a swirling rope ladder (spiral), the crossbeams of which are represented by pairs of nucleobases. Adenine is combined with thymine, and cytosine - with guanine.
The main value of DNA is the ability to carry information about the protein and the ability to double.
A sequence of three linked nucleotides is a code for a particular amino acid. From the sequence of amino acids proteins that control the body's biochemical mechanisms of development and metabolism appear. DNA is a replicator molecule. After the DNA molecule is divided into two chains, each nucleic base attracts complimentary to itself. Thus, adenine attracts a new thymine, guanine - a new cytosine, etc. At the end, two new DNA molecules are obtained, which are exact replicas of the original DNA. This [2-7] is a vision of the essence of DNA by biochemists and biophysicists.

### *2.2. A new look at the structure of DNA*
The author is the developer of the optimal computer transitional circuitry [8-13], which allows you to look at DNA as a certain biochemistry, built on the principles of transitional circuitry. And if we understand how this bioscheme works at the level of intellectual functions of logic and memory, we have the possibilities of its circuit redesign, optimization, and elimination of failures.

So the new view is that DNA is an intellectual scheme with all the consequent opportunities of circuit analysis, synthesis and optimization.

## 3. Fundamentals of the theory of transitional circuitry

A significant difference between transitional circuit technology and transistor circuitry is the determination of the minimum part for the synthesis of a complex circuit.

### *3.1. Mathematical model of an element of transitional circuitry*

The mathematical model of a functionally integrated element (FIE) is an undirected graph G (X, A, T), where: X = (x1, x2, ... xN) is the vertex set, A = (a1, a2, ... aM) is the set of edges . The predicate Γ is a three-place predicate and is described by a logical statement Γ (xi, ak, xj), which means that the edge ak joins the vertices xi and xj. The element of the set of vertices xi corresponds to the part of the integral structure in which Ti determines the qualitative composition of the part of the integral structure, Fi is the element of the functional set. The set consists of subsets: Tt - subsets of materials of solid-state circuitry, Tb - subsets of atoms of biochemistry. A set of materials with different properties Tt = { } (i = 1, ... n) = (p, n, p+, n+, ... SiO2, ..., Al, Ga ...) = SUDUM corresponds to the subset Tt, where p is the semiconductor region p-type, n is an n-type semiconductor region, SiO2 is a region of silicon dioxide, Al is an aluminum region, Ga is a region of a halide, etc., S is a subset of semiconductor regions, D is a subset of dielectric regions, M is a subset of conductors. The functional set F = Fc U Fp consists of two subsets: Fc = { } = (E1, ..., Ek1, I1, ..., Ik2, φ1, ..., φk3 ...): subsets of control actions in the form of voltage Ei, current Ij, of light φk and Fp = { } = (in1, ..., inm, out1, ..., outn) of the destination subset, which specifies the input and output functions of the area of the set T, relative to which transfer characteristics of the elements that determine their performance. N is the number of regions of the integral structure, the dimension of the element. The elements of the set of edges ak, ai correspond to transitions between different parts of the integral structure that perform certain functions, and there exist xi, xj (xi ≠ xj) & Γ (xi, ak, xj) & Γ (xj, ak, xi). Examples of transitions of solid-state circuitry (graphs of dimension N = 2) are: Si-Sj transition - transition between semiconductors, for example, p-n junction - transition between p and n semiconductors performing a diode (barrier) function; Si - Dj transition - transition between a semiconductor and a dielectric; Si-Mj transition - transition between a semiconductor and a metal (Schottky diode), transitions between transparent and opaque layers in optoelectronic elements, membranes in biological elements, etc. By analogy, in biosystems [14-15], a simple group (R-Z) - a combination of two atoms of element-organogens (C, N, O, P, S), containing a bond, is a system of components of biochemistry.

## 4. What gives the new circuitry for computers and DNA

On the basis of the new circuitry for solid-state microcircuits, optimal mathematical models, logical nanostructures and memory nanostructures protected by patents were developed. The new elements contained fewer areas and connections and their mathematical models were surprisingly similar to some molecules and parts of DNA with similar functions of logic and memory. The results of many years of work on the synthesis of intellectual elements in the developed transitional circuitry were brought into the system of elements of various dimensions [16-18].

### *4.1. Results of comparison of models of transition solid-state nanostructures and organic molecules*

Because of the surprising structural similarity of some models of solid-state intelligent circuitry for computers and organic molecules (parts of DNA), they were compared for further in-depth study by biochemists and biophysicists of DNA as some intellectual scheme of bio-transitional circuitry. The results of the comparison are given in [8]. Let us compare the models of transition schemes and parts of bio molecules

*4.1.1. Components.* The components of transitional circuitry of dimension 1 N = 1 (N is the dimension - the number of minimum parts) from which various structures are built, • for solid-state

IOP Conf. Series: Journal of Physics: Conf. Series **955** (2017) 012027 doi:10.1088/1742-6596/955/1/012027

silicon TC there are materials (for example, p-type semiconductor ——p, oxide ——ox), • for bio carbon TC, these are atoms (for example, hydrogen ——H or groups of atoms, for example ——$CH_3$). for components of dimension 2 (N = 2): • for a solid-state silicon TC, the component (N = 2) is a transition (the boundary between two materials, for example, a p——n transition). • for a bio carbon TC - this is a valence bond or some other interaction (energy, information, chemical), for example: 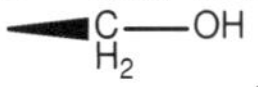 .

*4.1.2. Switching circuits – transistors.* From dimension 3 (N = 3), the circuits of transient circuitry (TC) begin. A tree-like model of dimension 3 under certain conditions can be a transistor model. Examples of models of a solid state silicon TC transistor are bipolar transistor models 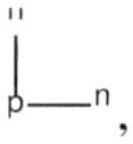, 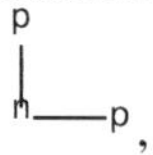, for a bio carbon TC, it is part of the amino acids 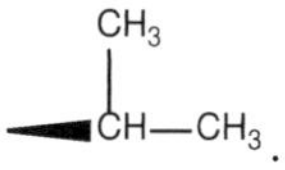.
Transistors (binary devices capable of transmitting or not passing current, charge, electron, proton, information in any representation) can be schemes of dimension 4 (N = 4). For a solid-state TC this is, for example, MOS transistors 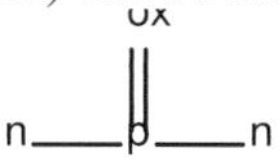 and 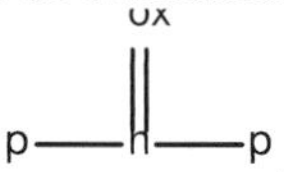, for bioplasmics, parts of amino acids 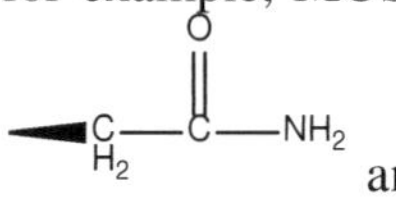 and 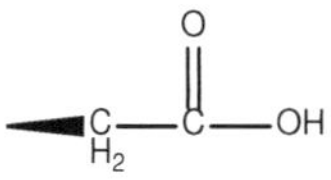 .

*4.1.3. Logical structures.* Realization of logic functions begins from the dimension 4 (N = 4).
For solid-state TC, the graph model of the structure of the inverter has the form p——n——p——n .
Principal for the implementation of any functions is the availability for the structure of the management system. *Any structure without control is dead.*
There are, for example, parts of amino acids similar in structure to a semiconductor inverter, for example 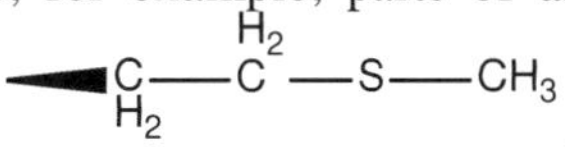, but the implementation of inversion on them has not yet been considered. As was shown, the inversion in bio TC has been proved for more complex structures.
Using the parallelization of inputs in a computer TC leads to the possibility of implementing logical functions AND, OR, for example, 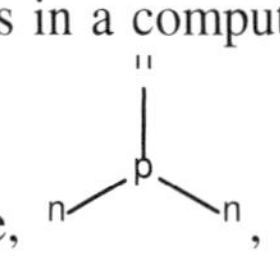, 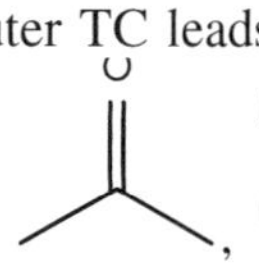, 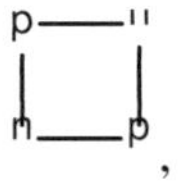, 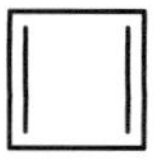. And parallelization of the outputs, allows you to work with the load in the macro circuit. Examples of parallelizing the output (N = 5) for a solid-state TC is a model 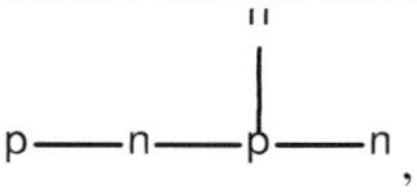, and for a bio TC a model 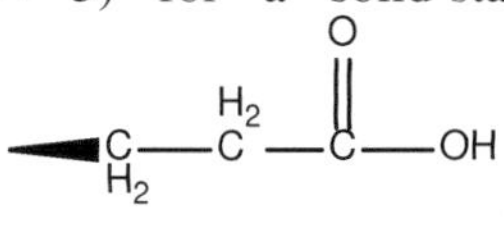. An increase in the dimensionality of the circuit model to 6 (N = 6) gives inverters with a structure p——n——p——n——p——n for a solid state TC. It can put in correspondence a part of amino acids 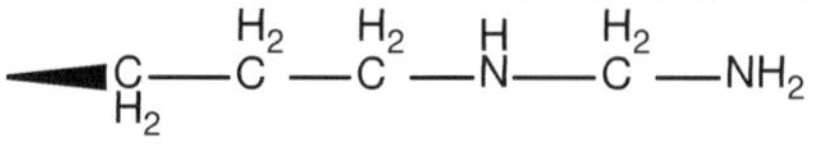 with an uncertain yet logical function.

With a further increasing in the dimensionality of the models (N = 7), a Bi-MOS inverter 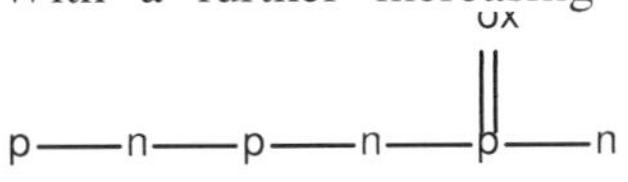 and an AND circuit with a power 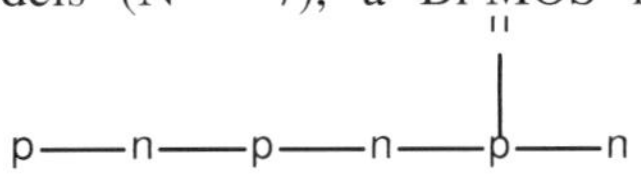 supply appear in the solid state TC; in the bio TC, the analog of the structure may be a part of the amino acids

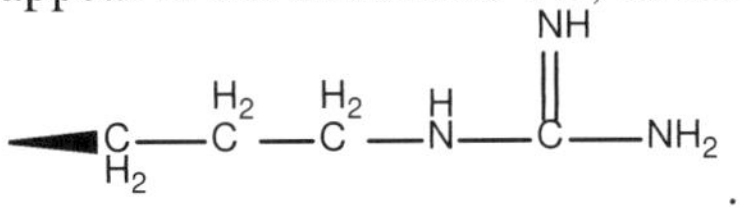

*4.1.4. Memory Structures.* Among the structures of dimension 7 (N = 7) there are structures with a cycle of dimension 6.

In a solid state TC, it is a model of a bistable cell 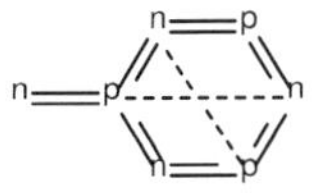 (another name for a bistable cell is an RS trigger) [19]. The synthesis and modeling of the physical nanostructure with the model obtained in the equation was performed in Fig. 1: In the physical simulation of this model in the form of a nanostructure, a system of parameters was found in which the nanostructure was efficient and performed the memory function. In Figure 2 shows the electron density distributions for two states (logical 1 (S = 1, R = 0) and logical zero (S = 0, R = 1) of a bistable cell with a cyclic mathematical model in Figure 1.

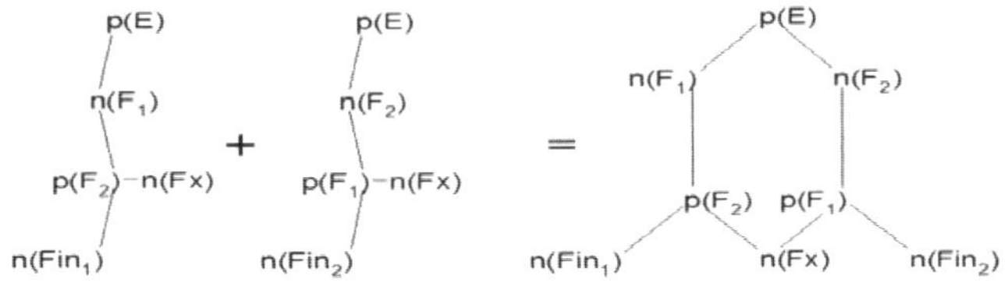


**Fig. 1.** The equation of synthesis of a model of a disjunctive bistable cell (BC) in a solid state TC

Simulation of current through a direct and inverted output, the results of which are shown in Fig. 3, proves the existence of a memory function for the nanostructure and confirms the execution of the logical "zero" and "unity" recording functions (the asymmetry of physical states).

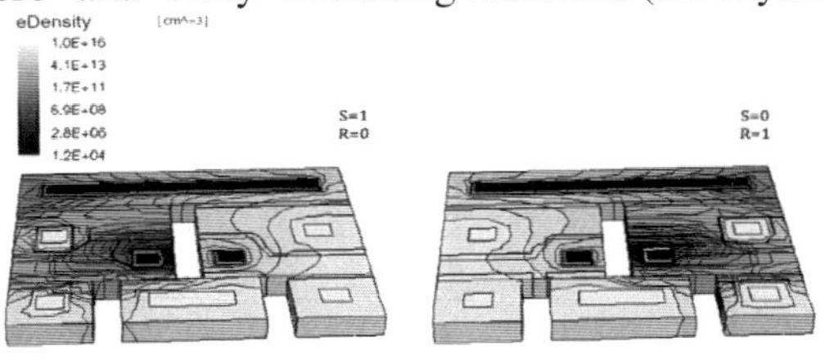


**Fig. 2.** The distribution of the electron density for two states (logical 1 (S = 1, R = 0) and the logical zero (S = 0, R = 1) of a bistable cell with a cyclic mathematical model in Fig. 1

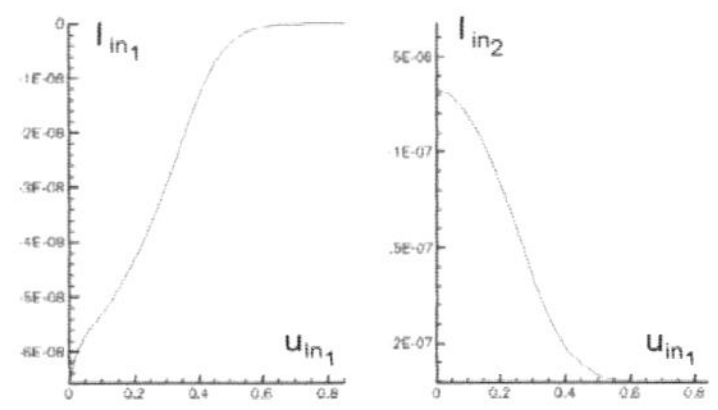


**Fig.3.** Current curves for direct and inverted outputs

BC (RS-trigger) is the basis of a huge number of different types of memory in a solid state TC. A necessary condition for the realization of the memory function is the presence of cross feedbacks between the inputs and outputs of a pair of logic gates that form the memory structure (they are denoted by dash-dotted lines). If, for example, in the benzene ring, or another hexagonal structure, information is transmitted on such bonds, hexagonal structures can be considered as memory cells in bio-TC under certain conditions (controls). With a further increase in the dimensionality of the structure, models of a functionally complete logical basis for AND-NOT appear in the solid state TC, for example, 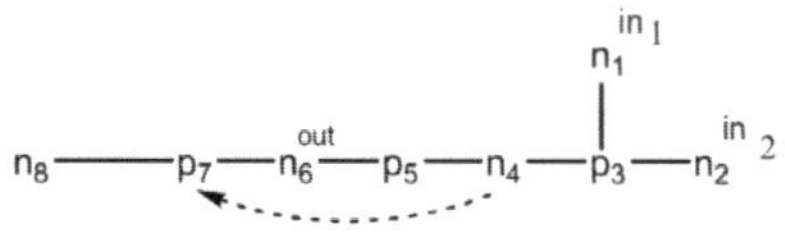. Here the dash-dot line indicates the internal connection between the regions of the structure, which ensures the equivalence of the control functions (in this case, the equality of potentials). In the bio TC reserve there is a structural analogue 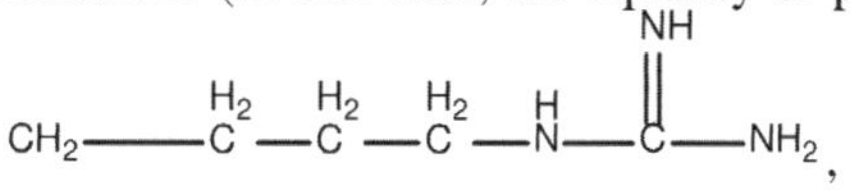, the function of which has not yet been determined. Interesting 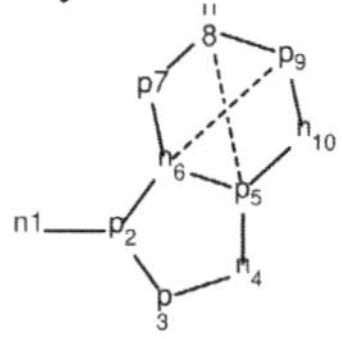 are the model of the structure of the D-trigger in the solid-state TC and the bioanalog, 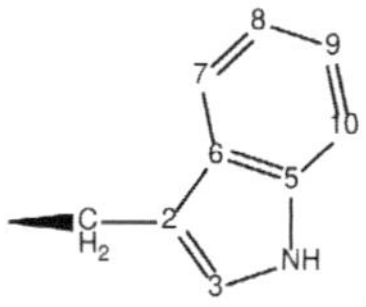 which is constantly encountered in the structure DNA. Further complication of the TC leads to the generation of models of single-digit memory circuits with control logic circuits (Fig. 4 (a) - adenine in DNA, (b) - D-trigger, driven by a logic circuit in a solid state TC).

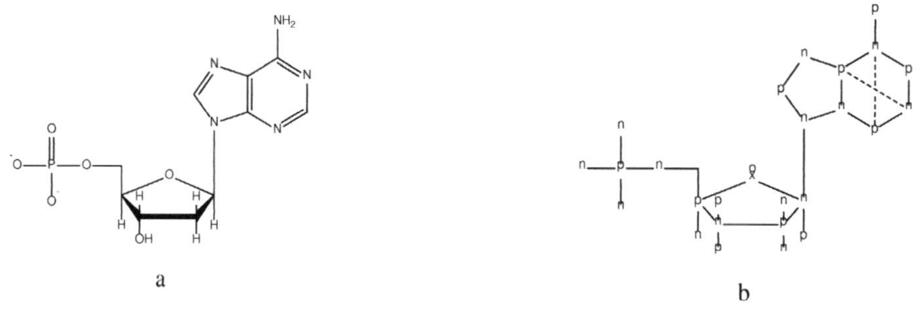

**Fig. 4.** Comparison of transient models of nanostructures: (a) - adenine in DNA, (b) - D-trigger, driven by a logic circuit in a solid state TC

*4.2. Principal differences*

With the similarity of the models structures of the intelligent circuits of the solid-state TC and the bio TC, the principal difference, apart from the components (materials and connections), is the various structure management systems and the differences in the representation of the processed information. Despite the fundamental differences, the surprising repetitive similarity of the models of solid-state TC and bio-TC circuitry makes us think.

## 5. Conclusion

In continuation to the history of the study of DNA, a new view of the structure of DNA as an intellectual scheme of transient circuitry (TC) was suggested.

The fundamentals of transitional circuitry theory are considered.

A mathematical model of an element of transitional circuitry is proposed.

Advantages of using transitional circuitry for computers and DNA are considered.

The results of models comparison of transitive solid-state nanostructures and organic molecules (parts of DNA) on components, switching, logical structures and memory structures and their fundamental differences are presented. Knowledge of transitional circuitry and a new view of DNA as an intellectual transitional scheme of bio-transitional circuitry can help researchers look at their objects of research in a new way and reveal secrets that have been closed to them for some time.

**References**

[1] Gribbin, J. (1985). In Search of the Double Helix: Quantum Physics and Life. United States of America: McGraw-Hill Book Company
[2] Watson, James, and Francis Crick. (1953). Molecular Structure Of Nucleic Acids. Nature: n. pag. Print.
[3] Berg, Paul. (1972). Biochemical Method For Inserting New Genetic Information Into DNA Of Simian Virus 40: Circular SV40 DNA
[4] Brown, T. A. (2011). Introduction to genetics: A molecular approach. New York: Garland Publishing.
[5] Crick, F. (1988). What Mad Pursuit: A Personal View of Scientific Discovery. New York: Basic Books
[6] Wilkins, M. (2004). The Third Man of The Double Helix. Oxford University Press.
[7] Jaenisch, Rudolf, and Arnold Levine. (1971) DNA Replication In SV40-Infected Cells. Virology 44.3: 480-493. Web.
[8] Trubochkina N.K. Modelirovanie 3D nanoskhemotekhniki. M.: BINOM. Laboratoriya znanij, 2012 — ISBN 978-5-9963-0291-8 — s.499
[9] Trubochkina N.K. Razrabotka i modelirovanie kachestvenno novoj 3D nano-struktury KMOP invertora s proektnoj normoj 20 nm — osnovy novoj ehlementnoj bazy ehnergosberegayushchih SBIS. // Kachestvo. Innovacii. Obrazovanie, 2011. № 6. C. 58—63
[10] Trubochkina N.K. Nanotekhnologicheskie proekty MIEM: ot matematicheskih modelej do prakticheskih razrabotok. // Rossijskie nanotekhnologii, 2010. T. 5. № 11-12. C. 12—15
[11] Trubochkina N.K. Sravnenie modelej nanostruktur chetyrekh tipov perekhodnoj skhemotekhniki. V kn.: «Novye informacionnye tekhnologii». Doklady XVIII mezhdunarodnoj studencheskoj konferencii-shkoly-seminara, 2010. C. 40—50
[12] Trubochkina N.K. Teoriya perekhodnoj skhemotekhniki kak instrument razrabotki intellektual'nyh nanostruktur i nanosistem. V kn.: Sbornik trudov po tematicheskomu napravleniyu NNS «Nanoehlektronika», 2010. C. 53—79
[13] Trubochkina N.K.Interactive 3D in Science and Art: From Visual Analysis jf 3-D Intelligent nano-Circuitry to 3-D Art Galleries on the Web. International Scientific Seminar “Visual Computing in Fundamental Science and Academic Research”, 19-25, 2009
[14] Karasev V.A, Luchinin V.V. Vvedenie v konstruirovanie bionicheskih nanosistem. – M.: FIZMATLIT, 2009. – 464 s.
[15] Karasev V.A. Arhitektura, principy organizacii i funkcionirovaniya bioorganicheskih nanostruktur // V kn. Nanotekhnologiya. Fizika, processy, diagnostika, pribory / Red. V.V. Luchinin i YU.M.Tairov. – M:. FIZMATLIT, 2006. S. 65-97.
[16] Integral'nyj logicheskij ehlement I-NE na osnove sloistoj tryohmernoj nanostruktury Patent RF na izobretenie № 2010119061 ot 14.05.2010. Federal'naya sluzhba po intellektual'noj sobstvennosti patentam i tovarnym znakam (ROSPATENT)Vorob'ev V.V, Sosnin A.A., Trubochkina N.K.
[17] Poluprovodnikovaya struktura logicheskogo ehlementa I-NE. Patent RF na izobretenie № 2010119553 ot 17.05.2010 Federal'naya sluzhba po intellektual'noj sobstvennosti patentam i tovarnym znakam (ROSPATENT) Belyj A.V., Luk'yanchikov N.I., Trubochkina N.K.
[18] Yachejka pamyati na baze tonkoslojnoj nanostruktury Patent RF na izobretenie № 2010119999 ot 19.05.2010 Federal'naya sluzhba po intellektual'noj sobstvennosti patentam i tovarnym znakam (ROSPATENT) Orlov P.V., Popovich I.P., Trubochkina N.K.

# On the mixing time in the Wang-Landau algorithm

**Marina Fadeeva**[1,2], **Lev Shchur**[1,2,3]

[1] Science Center in Chernogolovka, 142432 Chernogolovka, Russia
[2] National Research University Higher School of Economics, 101000 Moscow, Russia
[3] Landau Institute for Theoretical Physics, 142432 Chernogolovka, Russia

**Abstract.** We present preliminary results of the investigation of the properties of the Markov random walk in the energy space generated by the Wang-Landau probability. We build transition matrix in the energy space (TMES) using the exact density of states for one-dimensional and two-dimensional Ising models. The spectral gap of TMES is inversely proportional to the mixing time of the Markov chain. We estimate numerically the dependence of the mixing time on the lattice size, and extract the mixing exponent.

## 1. Introduction

Wang-Landau algorithm is the way of the direct estimation of the density of states (DOS) [1, 2]. It can be applied to any system with a defined partition function. The idea is based on the representation of the partition function as a sum over the energy levels

$$Z = \sum_{k=1}^{N_E} g(E_k) e^{-E_k/k_B T} , \tag{1}$$

where $g(E_k)$ is the number of states (density of states) with the energy $E_k$, $N_E$ is the number of energy levels, $k_B$ is the Boltzmann constant, and $T$ is the temperature.

It is proven to work quite well for many systems, with the examples in different areas of statistical physics and mechanics. There are more than 15 hundreds of papers on the application of the algorithm and its improvements. In the paper we focus on the discussion of the mixing time in the final stage of the Wang-Landau algorithm. It is connected to the rate of convergence of the estimated DOS.

At the final stage of the correct Wang-Landau algorithm, the density of states is close to the true values [3, 4]. Accordingly, the random walk in the energy space with the Wang-Landau probability of acceptance of the new energy

$$P_{WL}(E_k, E_m) = \min\left(1, \frac{g(E_k)}{g(E_m)}\right) \tag{2}$$

becomes a reversible Markov process. A random walk in the energy space is characterized by the transition matrix in the energy space (TMES) [3]

$$T(E_k, E_m) = P_{WL}(E_k, E_m) P(E_k, E_m) \tag{3}$$

where $P(E_k, E_m)$ is a probability of one step of the random walk to move from a configuration with energy $E_k$ to any configuration with energy $E_m$. The TMES matrix is the stochastic symmetric matrix if computed with the true density of states $g(E_k)$.

We are interested in the mixing properties of the Markov random walk in the energy space generated by transitions from one energy level to a different one with the Wang-Landau probability. It is known (see, f.e. [5]) that mixing time of the Markov process is inversely proportional to the spectral gap $G$, which is the difference of the two largest eigenvalues $\lambda_1 > \lambda_2$ of the transition matrix

$$G = \lambda_1 - \lambda_2 \ . \tag{4}$$

We present here a preliminary estimation of the spectral gap for one- and two-dimensional Ising models. The Ising model plays in statistical physics the same role as the fruit flies in biology. Ising model [7] is the simplest model of magnetics, and is the first model for which a problem of a continuous phase transition was solved in detail [8].

In section 2 we present the estimation of the spectral gap and scaling of the mixing time for a one dimensional Ising model, for which we know exactly both the density of states and the elements of transition matrix for any length of the Ising spin chain. In section 3 we calculate the transition matrix numerically for the 2D Ising model for which we know exactly the density of states [6]. We estimate the gap dependence on the system size, and present a numerical estimate of the mixing time scaling with the system size. We compare our results with the known estimates of the tunnelling time in a 2D Ising model. In the final section 4 we discuss future work.

## 2. Mixing time for one-dimensional Ising model

Consider a one-dimensional Ising model with $L$ spins $\sigma_i$ $(i = 1, 2, ..., L)$ placed on the string with periodic boundary conditions. The energy is given by

$$E = -\sum_{i=1}^{L} J\sigma_i\sigma_{i+1} \tag{5}$$

and $\sigma_{L+1} = \sigma_1$. Without loss of generality and for the sake of simplicity we can put $J = 1$ (in other words, we measure energy $E$ in units of $J$). Energy (5) can be parameterized by $k$, the number of pairs of the domain walls, $E_k = -L + 4k$. Occupations of energy levels are nothing but binomial coefficients, therefore the density of states is $g(E_k) = 2C_L^{2k}$. Therefore, off-diagonal matrix elements of TMES are [3]

$$T(E_k, E_{k+1}) = T(E_{k+1}, E_k) = \frac{C_{L-2}^{2k}}{\max\left(C_L^{2k}, C_L^{2k+2}\right)}. \tag{6}$$

More details can be found in the paper [3].

Using this expression, we compute elements of TMES for the number of spins $L$ ranging from 4 to 100, and compute two largest eigenvalues using the `dgev` function from the LAPACK [9] library. Dependence of the gap on $L$ is shown in the left panel of Figure 1 in the logarithmic scale for both axes, which clearly demonstrates the power law decay of the spectral gap. The linear fit gives the slope value $-2.19(2)$. Therefore, the mixing time $T_m = 1/G$ in the Wang-Landau algorithm for one-dimensional Ising model scales with the system size $L$ as $T_m \propto L^{2.19}$.

## 3. Mixing time for two-dimensional Ising model

Matrix elements of TMES for the two-dimensional Ising model can be computed using the Wang-Landau probabilities (2) with the exact values of DoS [6]. We compute TMES for the

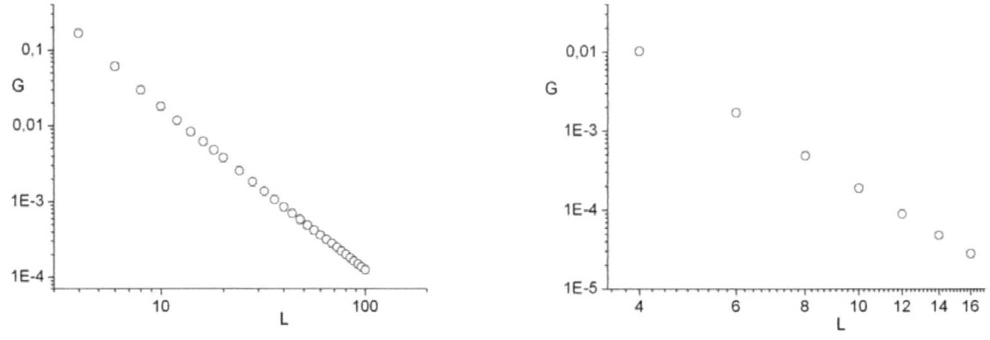


**Figure 1.** Spectral gap $G$ (circles) as a function of the linear lattice size $L$. Dotted line is the linear fit to $G$. Left: one-dimensional Ising model. Right: two-dimensional Ising model. Note the log-log scale.

system sizes $L$ from 4 to 16. We have to stress that convergence of the TMES spectrum is very slow, and is slowing down with the system size $L$. For example, the eigenvalue $\lambda_2$ reaches its stationary value after approximately $10^6$ WL-steps for $L = 4$, after $10^8$ steps for $L = 8$, and after $10^{12}$ steps for $L = 16$. The linear fit to the $G(L)$ data in the log-log scale gives slope of the line $-4.28(4)$. Therefore, $T_m \propto L^{4.28}$ for the two-dimensional Ising model.

In the paper [10], the average *tunnelling time* $\tau$ to move from a configuration state with the lowest energy to the configuration with the highest energy has been measured for the two-dimensional Ising model, and was found to scale as $\tau \propto L^{4.743(7)}$. The number $N_E$ of energy levels for this model is $N_E \propto L^2$. The tunnelling time for the unbiased random walk in the energy space will be $\tau_0 \propto N^2$. The bias due to the WL probability for the random walk does increase this time.

## 4. Discussion

In the paper we estimate the mixing time of the random walk in the energy spectrum generated by the Wang-Landau probability. Using the approach proposed in the paper [3], we compute the transition matrix in the energy spectrum (TMES) for one-dimensional and two-dimensional Ising models using the exact knowledge of the density of states. We estimate the spectral gap of TMES, and obtain the exponents of the mixing time $T_m$. This time can be viewed as the characteristic time at the final stage of a $1/t$ Wang Landau algorithm [11, 12] (or a stochastic approximation algorithm [13, 14]).

The connection of the mixing time we measure in the paper and the tunnelling time [10] for the Markov process is not obvious. Future work is necessary for the analysis of performance limitations of the Wang-Landau algorithm.

## Acknowledgments

This work was supported by grant 14-21-00158 from the Russian Science Foundation.

## References

[1] F. Wang, D. P. Landau, *Efficient, multiple-range random walk algorithm to calculate the density of states*, Phys. Rev. Lett. **86**, 2050 (2001).

[2] F. Wang, D. P. Landau, *Determining the density of states for classical statistical models: A random walk algorithm to produce a flat histogram*, Phys. Rev. E **64**, 056101 (2001).

[3] L. Yu. Barash, M. A. Fadeeva, and L. N. Shchur, *Control of accuracy in the Wang-Landau algorithm*, Phys. Rev E **96**, 043307 (2017).
[4] C. Zhou, J. Su, *Optimal modification factor and convergence of the Wang-Landau algorithm*, Phys. Rev. E **78**, 046705 (2008).
[5] S. Boyd, P. Diaconis, L. Xiao, *Fastest mixing Markov chain on a graph*, SIAM Review **46**, 667 (2004).
[6] P. D. Beale, *Exact distribution of energies in the two-dimensional Ising model*, Phys. Rev. Lett. **76**, 78 (1996).
[7] E. Ising, *Beiträg zür theorie des ferromagnetismus*, Z. Physik **31** 253 (1925).
[8] L. Onsager, *Crystal statistics. I. A two-dimensional model with an order-disorder transition*, Phys. Rev. **65** 117 (1944).
[9] LAPACK, http://www.netlib.org/lapack/
[10] P. Dayal, et al, *Perfomance limitations of flat-histogram method*, Phys. Rev. Lett. **92**, 097201 (2004).
[11] R. E. Belardinelli and V. D. Pereyra, Phys. Rev. E **75**, 046701 (2007). [
[12] R. E. Belardinelli and V. D. Pereyra, J. Chem. Phys. **127**, 184105 (2007).
[13] F. Liang, C. Liu, and R. J. Carroll, J. Am. Stat. Assoc. **102**, 305 (2007).
[14] F. Liang, J. Stat. Phys. **122**, 511 (2006).

IOP Conf. Series: Journal of Physics: Conf. Series **955** (2017) 012029 doi:10.1088/1742-6596/955/1/012029

# Application of grid-characteristic method to some seismic exploration problems in the Arctic

**D I Petrov**[1]
[1]Moscow Institute of Physics and Technology
diapetr@gmail.com

**Abstract**. The aim of this work is simulation of wave propagation in the Arctic with the presence of different ice structures, using high-performance computing. In the given paper grid-characteristic method is applied, which provides correctly describing the contact and boundary conditions.

## 1. Introduction

The question of developing the Arctic in Russia is of current importance as there are eight hydrocarbon fields, and their supplies are estimated approximately 2,7 trillion $m^3$. The considerable barrier on way of oil extraction in north seas is presence of different ice formations, particularly, ice ridges, icebergs. One of the main stages of planning geological survey works is mathematical modeling, that allows significantly bring down the cost of carrying out seismic exploration. In this work numerical experiments on solving problems of seismic exploration in the conditions of the Arctic shelf were carried out. and applied to calculate the fluxes in the finite volume discretization of the governing equations.

In this paper we present the results of a numerical experiment on the modeling of wave processes in water-soil systems by the grid-characteristic method. A form was obtained from clique from a reservoir with a hydrocarbon deposit at various initial conditions at significant depths. The experiment is performed for various configurations of geological sediments and two types initial perturbations. Also, wave processes in "ice-water-soil", the soil contains a non-uniform oil inclusion. .

## 2. Mathematical model of meduim

A complete system of equations describing the state continuous linear-elastic medium and a complete system of equations describing sonic acoustic field are being solved. The components of the velocity $v$ and symmetric Cauchy stress tensor $\boldsymbol{\sigma}$ in a linearly elastic medium are described by the following system of equations [1]:

$$\rho \frac{d\vec{v}}{dt} = -\nabla p \tag{1}$$

$$\frac{\partial \sigma}{\partial t} = \lambda(\nabla \cdot \vec{v}) I + \mu\left(\nabla \otimes \vec{v} + (\nabla \otimes \vec{v})^T\right) \tag{2}$$

For numerical modeling of sea water and oil-containing inclusions, the ideal fluid approximation was used and the complete system of equations:

$$\rho \frac{\partial \vec{v}}{\partial t} = -\nabla p$$
$$\frac{\partial p}{\partial t} = -c^2 \rho (\nabla \cdot \vec{v}) \quad (3)$$

where λ, μ – Lamé parameters, which determine the properties of the elastic material, $c$ denotes the speed of sound in an ideal fluid. The velocity of longitudinal waves in a linearly elastic medium can be found from

$$c_p = \sqrt{\frac{\lambda + 2\mu}{\rho}}$$

and the velocity of transverse waves is calculated in accordance with

$$c_s = \sqrt{\frac{\mu}{\rho}}$$

## 3. Numerical method

For the numerical solution of the given systems, a grid-characteristic method is used on curvilinear structural grids, which allows to build correct numerical algorithms for calculating boundary points and points lying on the interfaces of two media with different Lamé parameters and/or densities [2].

The system in the two-dimensional case can be represented in the form

$$\frac{\partial \vec{q}}{\partial t} + \mathbf{A}_1 \frac{\partial \vec{q}}{\partial x_1} + \mathbf{A}_2 \frac{\partial \vec{q}}{\partial x_2} + \mathbf{A}_3 \frac{\partial \vec{q}}{\partial x_3} = 0$$

In the given expression, the vector $q$ is understood to be the vector formed by of two velocity components and three components of a symmetric tensor of stresses

$$\vec{q} \, \text{O} \{v_1, v_2, \sigma_{11}, \sigma_{22}, \sigma_{12}\}'$$

First, the spatial splitting method is applied coordinates, as a result of which we have two one-dimensional systems:

$$\frac{\partial \vec{q}}{\partial t} = \mathbf{A}_\mathbf{j} \frac{\partial \vec{q}}{\partial x_j}$$

Each of these systems is hyperbolic and has a complete set of eigenvectors with real eigenvalues, therefore each of the systems can be rewritten in the form

$$\frac{\partial \vec{q}}{\partial t} = \Omega^{-1}{}_\mathbf{j} \Lambda \Omega_j \frac{\partial \vec{q}}{\partial x_j}$$

where $\mathbf{\Omega}$ is a matrix composed of eigenvectors, $\mathbf{\Lambda}$ is a diagonal matrix, a matrix whose elements are eigenvalues (index $j$ is ommited here and later). The general view of the matrix $\mathbf{\Lambda}$ will be

$$\Lambda = diag\{c_p, -c_p, c_s, -c_s, 0\}$$

IOP Conf. Series: Journal of Physics: Conf. Series **955** (2017) 012029 doi:10.1088/1742-6596/955/1/012029

After changing the variables

$$\vec{p} = \mathbf{\Omega}\vec{q}$$

each of the systems decomposes on five independent scalar advection equations:

$$\frac{\partial \vec{p}}{\partial t} + \mathbf{\Lambda}\frac{\partial \vec{p}}{\partial x} = 0$$

One-dimensional avdection equations are solved by the method of characteristics or by ordinary finite-difference schemes. After all the components have been transferred, the solution may be restored:

$$\vec{q}^{\,n+1} = \mathbf{\Omega}^{-1}\vec{p}^{\,n+1}$$

The solution of system in the two-dimensional case is carried out similarly

The program implements the application of TVD-difference schemes of the $2^{nd}$ order of accuracy, 15 different limiters, the superbee limiter and grid-characteristic schemes of 2-4 order of accuracy [3].

**4. Boundary conditions**

Let in a part of the integration region *a* a system of equations (1), (2), and in the part *b* of the integration region b the system of equations (3) are being solved. Consider the contact condition between them. System (1), (2) has three outgoing characteristics in the three-dimensional case, and system (3) in the three-dimensional case has one outgoing characteristic. Thus, in order to find all four outgoing characters, it is necessary to fulfill the following contact conditions:

$$p^{(b,n+1)} = -\left(\sigma^{(a,n+1)} \cdot \vec{p}\right) \cdot \vec{p} \tag{4}$$

$$\sigma^{(a,n+1)} \cdot \vec{p} - \left(\sigma^{(a,n+1)} \cdot \vec{p}\right) \cdot \vec{p} = 0 \tag{5}$$

$$\overrightarrow{v^{(a,n+1)}} \cdot \vec{p} = \overrightarrow{v^{(b,n+1)}} \cdot \vec{p} \tag{6}$$

The condition (4) is the equality of the normal component surface density of forces on the part of a solid body to the pressure inideal fluid, condition (5) ensures that the tangentialcomponent of the surface density of forces on the side of of the body, and expression (6) gives the equality of the normal components velocities in an ideal fluid and a solid. In (4) - (6) for *p* is noted the outer normal to the solid body, which is the inner normal to the liquid. The velocity vector *v* such that the conditions (4)-(6) are fulfilled. Further, for a rigid body, the boundary condition is used with a given velocity, and for a liquid, a boundary condition with given by the normal component of velocity.

**5. Result of the numerical simulation**

We consider a region for integration 1200 m wide and 600 m depth, representing multilayered geological medium under conditions of the Arctic shelf. The medium consists of ice, sea water, rock and carbohydrate reservoir. The density of water was assumed equal to 1000 kg/m$^3$, the speed of sound is 1500 m/s. The ground has a density of 2500 kg/m$^3$, the longitudinal propagation velocity of the disturbance is 4500 m / s, the transverse velocity is 2250 m/s. The following parameters of the hydrocarbon layer were used: density 2000 kg/m$^3$, longitudinal speed the propagation of sound is 4000 m/s, the transverse speed - 1250 m/s. The source of the initial Rieker wavelet with dominant frequency equals 40 Hz is placed in the center of the ice field.

The following grid parameters are taken: spartial step is 0.2 m, time step is $3 \cdot 10^{-5}$s. 15 000 iterations were calculated. Absorbing conditions are set at the sides and at the bottom of the region. Free boundary condition is set on the top side of the region.

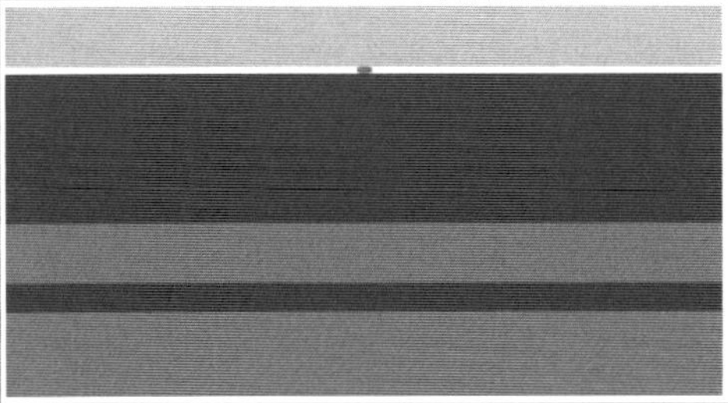

**Figure 1.** A model of geological meduim. From the top to bottom: air, ice, water, rock, reservoir, rock.

**Figure 2.** A model of geological meduim. From the top to bottom: air, ice, water, rock. No layer of carbohydrates.

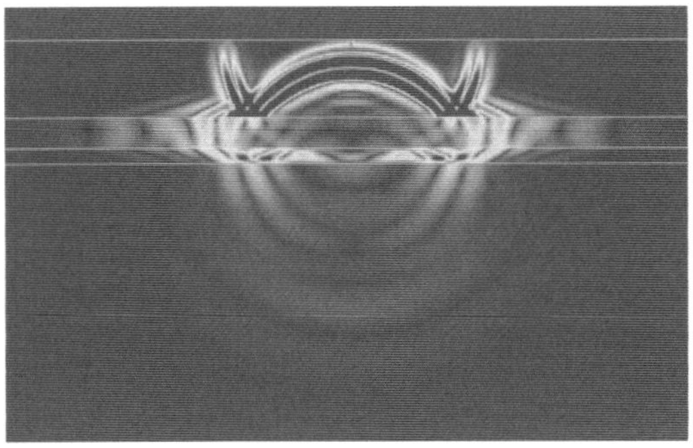

**Figure 3.** The wave field in the medium after initial perturbation. A response of reservoir is being observed.

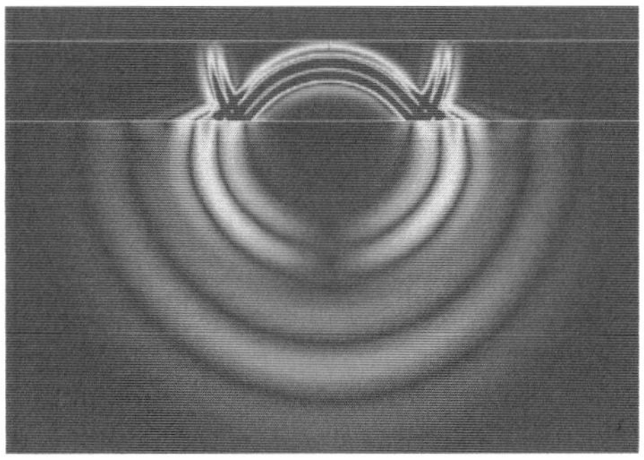

**Figure 4.** he wave field in the medium after initial perturbation.

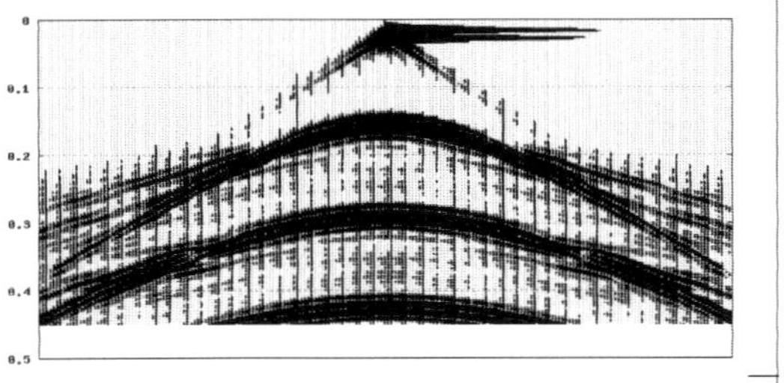

**Figure 5.** Synthetic seimogram corresponding the first definition.

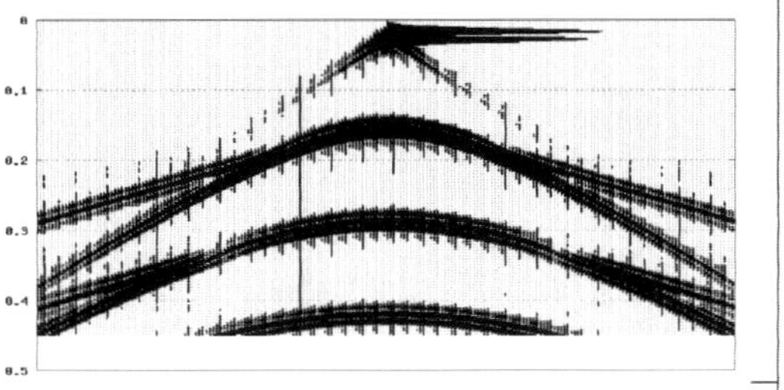

**Figure 6.** Synthetic seimogram corresponding the second definition.

## 6. Conclusion

In this paper, we considered the solution of elastic-acoustic problems of shelf seismic exploration using grid-characteristic method. A boundary condition was developed between a part of the integration in which a system of equations is solved that describes the state of an infinitesimal volume of a continuous linearly elastic and the part of the region of integration in which of the equations describing the acoustic field.

**Acknowledgements**

This work was supported by the Russian Science Foundation, grant no. 14-11-00263.

This work has been carried out using computing resources of the federal collective usage center Complex for Simulation and Data Processing for Mega-science Facilities at NRC “Kurchatov Institute”, http://ckp.nrcki.ru/.

## References

[1] Novatskiy V K *Teoriya uprugosti* [Elasticity theory], Moscow, Mir Publ, 1975. 872 p.

[2] Kvasov I E, Pankratov S A and Petrov I B 2011 Numerical simulation of seismic responses in multilayer geologic media by the grid-charcteristic method. *Mathematical Models and Computer Simulations* ***3***. No 2 pp 196–204.

[3] Harten Ami. High resolution schemes for hyperbolic conservation laws 1997 *Journal of Computational Physics* **135(2)** pp 260–278.

[4] Favorskaya A V, Petrov I B, Khokhlov N I and Petrov D I 2014 Numerical modeling of arctic problems by grid-characteristic method *News UFU. Technical science*. U.D.C. 519.6.

[5] Petrov I B, Favorskaya A V, Kvasov I E and Sannikov A V 2013 Grid-characteristic method with interpolation of high orders on tetrahedral hierarchical grids with divisible time step *Mathematical modelling* **25,** No2 pp 42–52.

# Singular solutions for vibration control problems

**Larisa Manita**[1] **and Mariya Ronzhina**[2]

[1] National Research University Higher School of Economics, Moscow Institute of Electronics and Mathematics, Russia

[2] Lomonosov Moscow State University, Faculty of Mechanics and Mathematics, 119991, GSP-1, Moscow, Russia

E-mail: lmanita@hse.ru, maryaronzhina@gmail.com

**Abstract.** Optimal control problem for the system of partial differential equations of hyperbolic type is considered. By using the Fourier method this problem is reduced to the optimal control problem for the corresponding Fourier coefficients. For some special initial data we prove the existence of optimal solutions with a countable number of switchings on a finite time interval and optimal spiral-like solutions which attain the origin in a finite time making a countable number of rotations. The problem of controlling the vibrations of the Timoshenko beam is considered as an example of the optimal control problem for linear system of PDE.

## 1. Optimal control problem for linear system of PDE

We consider a control system

$$\frac{\partial^2}{\partial t^2}p(x,t) + A_x p(x,t) = g(x,t), \quad x \in G, \quad t \in \mathbb{R}_+ \tag{1}$$

subject to initial

$$p(x,0) = v_0(x), \quad p_t(x,0) = v_1(x) \tag{2}$$

and boundary conditions

$$B_x p|_{\partial G} = 0 \tag{3}$$

Here, $p(x,t)$ and $g(x,t)$ are $m$-dimensional vector-valued functions defined on $G \times [0,\infty)$, $G$ is a compact subset of $\mathbb{R}^n$ with smooth boundary $\partial G$, $A_x$ is a linear differential operator of order 2 in $G$, $B_x$ is a linear differential operator of order 1 in a neighborhood of $\partial G$. Suppose that $p(x,t) \in L_2(G \times [0,\infty), \mathbb{R}^m)$, $g$ is a measurable function in $t$, $g(\cdot,t) \in L_2(G,\mathbb{R}^m)$, $p(\cdot,t), \frac{\partial^2}{\partial t^2}p(\cdot,t), v_0(x), v_1(x) \in H^2(G,\mathbb{R}^m)$. The external force $g(x,t)$ is considered as a control function. We assume that $g(\cdot,t)$ is bounded in the following sense

$$\|g(\cdot,t)\|^2_{L_2(G,\mathbb{R}^m)} \leq 1 \tag{4}$$

We study the following control problem: to find the admissible control such that the corresponding solution of (1)-(3) minimize the functional

$$\int_0^\infty \|p(\cdot,t)\|^2_{L_2(G,\mathbb{R}^m)}\, dt \to \inf \tag{5}$$

Assume that $A_x$ with domain $D_A = \left\{y \in H^2(G, \mathbb{R}^m), \quad B_x p|_{\partial G} = 0\right\}$ is a self-adjoint operator, has the discrete spectrum $\{\lambda_j\}_{j=1}^{\infty}$ and the orthonormal complete system of eigenfunctions $\{h_j(x)\}_{j=1}^{\infty}$ that are smooth in $G$ and satisfying (3). We seek a solution of (1)-(5) in the form

$$p(x,t) = \sum_{j=1}^{\infty} q_j(t) h_j(x) \tag{6}$$

where $q_j(t) = (p, h_j)_{L_2(G,\mathbb{R}^m)}, \quad j = 1, 2, \ldots,$ are Fourier coefficients. Expand the functions $g, v_0, v_1$ in the basis $\{h_j(x)\}_{j=1}^{\infty}$

$$g(t,x) = \sum_{j=1}^{\infty} u_j(t) h_j(x), \quad v_0(x) = \sum_{j=1}^{\infty} s_j h_j(x), \quad v_1(x) = \sum_{j=1}^{\infty} r_j h_j(x) \tag{7}$$

Substituting (6) and (7) into (1)-(5) we get an optimal control problem for the Fourier coefficients in the space $l^2$:

$$\int_0^{\infty} \sum_{j=1}^{\infty} q_j^2(t) dt \to \min \tag{8}$$

$$\ddot{q}_j(t) + \lambda_j q_j(t) = u_j(t), \quad \sum_{j=1}^{\infty} u_j^2(t) \le 1 \tag{9}$$

$$q_j(0) = s_j, \quad \dot{q}_j(0) = r_j, \quad j = 1, 2, \ldots \tag{10}$$

Problem (1)-(5) with the external force in the form $g(x,t) = u(t) f(x)$ was considered in [1]. In this case the corresponding control problem for the Fourier coefficients has the same form but (9) is replaced by

$$\ddot{q}_j(t) + \lambda_j q_j(t) = C_j u(t), \quad -1 \le u(t) \le 1 \tag{11}$$

where $C_j \in \mathbb{R}$. In [1] for the problem (8), (10), (11) it was constructed the optimal synthesis containing singular extremals and extremals with accumulation of switchings.

In the present paper for some initial conditions (2) we prove that the behavior exhibited by the solutions of (8)-(10) is similar to that of the problem with scalar control.

Note that the system (1)-(3) governs vibrations of beams, strings and other mechanical models. At the end of this paper we will consider, as an example, the problem of controlling the vibrations of the Timoshenko beam.

## 2. Problem with $k$-dimensional control

Let $h_1(x), h_2(x), \ldots, h_k(x)$ be the orthonormal eigenfunctions of the operator $A_x$, and $\lambda_1, \lambda_2, \ldots, \lambda_k$ be the corresponding eigenvalues. Consider the $k$-dimensional subspace $\mathcal{X}$ in $L_2(G, \mathbb{R}^m)$ spanned on $h_1, h_2, \ldots, h_k$. Assume $v_0(x), v_1(x) \in \mathcal{X}$. Then one can prove that the subspace $\mathcal{X}$ is an integral manifold of the flow of optimal trajectories of the problem (1)-(5). It means that for any $t > 0$ the corresponding optimal solution $p(x,t)$ and the optimal control $g(x,t)$ can be written in the form

$$p(x,t) = \sum_{j=1}^{k} q_j(t) h_j(x), \quad g(x,t) = \sum_{j=1}^{k} u_j(t) h_j(x)$$

where $q_j(t)$ and $u_j(t)$ satisfy (9)-(10). Thus we can reduce (8)-(10) to the following problem with $k$-dimensional control:

$$\int_0^\infty \|q(t)\|^2\, dt \to \min \tag{12}$$

$$\dot{q} = z, \quad \dot{z} = \Lambda q + u, \quad \|u(t)\| \le 1 \tag{13}$$

$$q(0) = s, \quad z(0) = r. \tag{14}$$

Here $q,\ z,\ u \in \mathbb{R}^k$ , $\|\cdot\|$ means the Euclidean norm of $\mathbb{R}^k$, $\Lambda = \operatorname{diag}\{-\lambda_1, -\lambda_2, \ldots, -\lambda_k\}$.

To construct optimal solutions for the problem (12)-(14) we use the Pontryagin Maximum Principle. Define the Hamiltonian

$$H(q, z, \phi, \psi) = -\frac{1}{2}\langle q, q\rangle + \langle z, \phi\rangle + \langle \Lambda q, \psi\rangle + \langle u, \psi\rangle$$

where $\phi$, $\psi$ are adjoint variables. We have the system of equations of the Pontryagin maximum principle

$$\dot{q} = z, \quad \dot{z} = \Lambda q + \hat{u}, \quad \dot{\phi} = \frac{1}{2}q - \Lambda\psi, \quad \dot{\psi} = -\phi. \tag{15}$$

The optimal control $\hat{u}(t)$ satisfies the *maximum condition*:

$$H(q(t), z(t), \phi(t), \psi(t), \hat{u}(t)) = -\frac{1}{2}\langle q, q\rangle + \langle z, \phi\rangle + \langle \Lambda q, \psi\rangle + \max_{\|u(t)\|\le 1}\langle u, \psi\rangle$$

The maximum condition gives: if $\psi \ne 0$, then $\hat{u}(t) = \psi/\|\psi\|$; if $\psi = 0$, then any control $\|u(t)\| \le 1$ meets the maximum condition.

A solution $\xi(t) = (u(t), z(t), q(t), \phi(t), \psi(t))$ of the system (15) is called a *singular* one on the interval $(t_1, t_2)$ if $\psi(t) = 0$ for all $t \in (t_1, t_2)$. It was proved [2] that in the problem (12)-(14) the origin $(z, q, \phi, \psi) = 0$ with $\hat{u}(t) = 0$ is the only singular solution. For sufficiently small initial data the optimal solutions attain the origin in a finite time $T$ and the corresponding optimal control function $\hat{u}(t)$ does not have a limit as $t \to T$. In the case $k = 1$ it was shown that the behaviour of the optimal solutions of (12)-(14) in the vicinity of the origin is similar to the optimal synthesis in the two-dimensional Fuller problem with a scalar control. Namely, the optimal control $u(t)$ has an infinite number of switchings in a finite time interval (chattering mode). In this paper we study the case $k = 2$.

## 3. Model problem

Consider the following optimal control problem with two-dimensional control which is called a *model problem*.

$$\int_0^\infty \|q(t)\|^2\, dt \to \inf$$

$$\ddot{q} = u, \quad \|u(t)\| \le 1, \quad q(0) = s^0, \quad \dot{q}(0) = r^0, \quad q,\ u \in \mathbb{R}^2$$

Note that the only difference between the model problem and (12)–(14) is the summand $\Lambda q$ in (13).

The Hamiltonian system (15) for the model problem has a group of symmetries $G = \mathcal{O}(2)\times\mathbb{R}_+$ that acts on $\xi(t) = (u(t), z(t), q(t), \phi(t), \psi(t))$ as follows.

Let $\zeta \in \mathcal{O}(2)$ then $\zeta \circ \xi \stackrel{def}{=} (\zeta u,\ \zeta z,\ \zeta q,\ \zeta\varphi,\ \zeta\psi)$. Let $\lambda \in \mathbb{R}_+$ then $g_\lambda(\xi(t)) \stackrel{def}{=} \left(u(t/\lambda),\ \lambda z(t/\lambda),\ \lambda^2 q(t/\lambda),\ \lambda^3\varphi(t/\lambda),\ \lambda^4\psi(t/\lambda)\right)$. The action of the group $G$ respects the set of solutions of the model problem. This allows to find some explicit solutions in the model problem.

**Proposition 1** ( [3]). *There exists a one-parameter family of optimal solutions of the model problem that contains logarithmic spirals*

$$q^*(t) = A_1 t^2 e^{i\alpha \ln|T-t|},\ z^*(t) = A_2 t e^{i\alpha \ln|T-t|},\ u^*(t) = -e^{i\alpha \ln|T-t|}, \qquad (16)$$
$$\alpha = \pm\sqrt{5},\ A_1 = \frac{1}{126}(4+i\alpha)(3+i\alpha),\ A_2 = -A_1(2+i\alpha)$$

*and all its possible rotations and reflections. They approach the origin in a finite time $T$ with countable number of rotations.*

Unlike trajectories of a focus for linear differential equations, these optimal logarithmic spirals approach the origin in a finite time, while making a countable number of rotations.

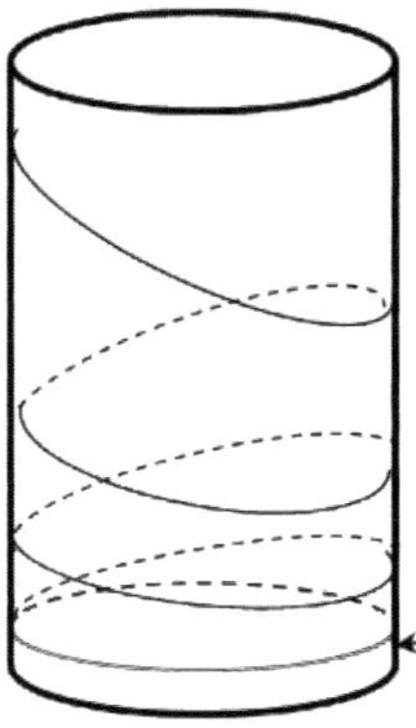


**Figure 1.** A blowup

## 4. Blowing up the singularity

Unfortunately, the problem (12)-(14) does not have a symmetry group. However, using the blowing up the singularity at the origin we prove that in the neighborhood of the origin the behavior of optimal solutions for (12)-(14) is determined by optimal solutions of the model problem. Consider the blowing up [4] the singularity at the origin by the map $B : (q, z, \phi, \psi) \mapsto (\mu, \tilde{q}, \tilde{z}, \tilde{\phi}, \tilde{\psi})$:

$$\tilde{z} = \frac{z}{\mu},\quad \tilde{q} = \frac{q}{\mu^2},\quad \tilde{\phi} = \frac{\phi}{\mu^3},\quad \tilde{\psi} = \frac{\psi}{\mu^4},\quad \mu = \left(|z|^{24} + |q|^{12} + |\phi|^8 + |\psi|^6\right)^{\frac{1}{24}}$$

where $\mu \in \mathbb{R}_+$, and $\left(\tilde{q},\ \tilde{z},\ \tilde{\phi},\ \tilde{\psi}\right) \in \mathbb{R}^8$ lies on the manifold $C_0 = \left\{|\tilde{z}|^{24} + |\tilde{q}|^{12} + |\tilde{\phi}|^8 + |\tilde{\psi}|^6 = 1\right\}$.

Compare the blowup for the model problem with the blowup for system (15).

| *Problem* (12)–(14) | *Model problem* |
|---|---|
| $\begin{aligned} \mu' &= \mu\mathcal{M}(\tilde{q}, \tilde{z}, \tilde{\phi}, \tilde{\psi}) \\ \tilde{\psi}' &= -\tilde{\phi} - 4\tilde{\psi}\mathcal{M} \\ \tilde{\phi}' &= \tilde{q} - 3\tilde{\phi}\mathcal{M} \\ \tilde{q}' &= \tilde{z} - 2\tilde{q}\mathcal{M} \\ \tilde{z}' &= \Lambda\mu^2\tilde{q} + u - \tilde{z}\mathcal{M} \\ u &= \tilde{\psi}/\lvert\tilde{\psi}\rvert \end{aligned} \qquad (17)$ | $\begin{aligned} \mu' &= \mu\mathcal{M}_0(\tilde{q}, \tilde{z}, \tilde{\phi}, \tilde{\psi}) \\ \tilde{\psi}' &= -\tilde{\phi} - 4\tilde{\psi}\mathcal{M}_0 \\ \tilde{\phi}' &= \tilde{q} - 3\tilde{\phi}\mathcal{M}_0 \\ \tilde{q}' &= \tilde{z} - 2\tilde{q}\mathcal{M}_0 \\ \tilde{z}' &= u - \tilde{z}\mathcal{M}_0 \\ u &= \tilde{\psi}/\lvert\tilde{\psi}\rvert \end{aligned} \qquad (18)$ |
| $\mathcal{M} = \frac{1}{24}\big(24\lvert\tilde{z}\rvert^{22}\langle\tilde{z}, u\rangle + 12\lvert\tilde{q}\rvert^{10}\langle\tilde{q}, \tilde{z}\rangle + 8\lvert\tilde{\phi}\rvert^6\langle\phi, q\rangle - 6\lvert\tilde{\psi}\rvert^4\langle\psi, \phi\rangle + 24\mu^2\lvert\tilde{z}\rvert^{22}\langle\tilde{z}, \tilde{q}\rangle\big)$ | $\mathcal{M}_0 = \frac{1}{24}\big(24\lvert\tilde{z}\rvert^{22}\langle\tilde{z}, u\rangle + 12\lvert\tilde{q}\rvert^{10}\langle\tilde{q}, \tilde{z}\rangle + 8\lvert\tilde{\phi}\rvert^6\langle\phi, q\rangle - 6\lvert\tilde{\psi}\rvert^4\langle\psi, \phi\rangle\big)$ |

Note that if we put $\mu = 0$ into the system (17) then we get the system (18).

We get that the blowup (18) for the model problem coincides with the blowup (17) for the system (17) as $\mu = 0$. One can prove that there is the contraction in the $\mu$-direction and the expansion in the other directions (see figure 1).

Summarizing the above, we have that the spirals (16) represent two circles in the quotient space $(\tilde{q}, \tilde{z}, \tilde{\phi}, \tilde{\psi})$ depending on the direction of rotation. Both of the circles are hyperbolic periodic trajectories and their stable manifolds are woven in logarithmic spirals. Moreover these circles exist in the Hamiltonian system (15) and the eigenvalues for the model problem circles and for the general Hamiltonian system circles coincide. So we can reconstruct the stable manifold in the Hamiltonian system (15) and it is also woven of the analogues of the logarithmic spirals. Thus the following proposition is proved.

**Proposition 2.** *There exists the optimal spiral-like solution of (12)–(14) which attains the singular point in finite time making a countable number of rotations.*

## 5. Timoshenko beam

Consider a linear model of a rotating Timoshenko beam [5–7]

$$\begin{aligned} w_{tt}(x,t) - \frac{1}{\gamma} w_{xx}(x,t) + \frac{1}{\gamma} \varsigma_x(x,t) &= g_1(x,t) \\ \varsigma_{tt}(x,t) - \varsigma_{xx}(x,t) + \frac{1}{\gamma}\varsigma(x,t) - \frac{1}{\gamma} w_x(x,t) &= g_2(x,t) \end{aligned}$$

Here, the $X$-axis coincides with the beam when it's at rest, $w(x,t)$ is the displacement of the center line of the beam in the direction that is perpendicular to the $X$ -axis and $\varsigma$ is the rotation angle of the cross-section area at the location $x \in [0,1]$ and time $t$. The controls are introduced as forcing terms $g_1(x,t)$ and $g_2(x,t)$ in the right sides of both equations. Assume that $\int_0^1 \left(g_1^2(x,t) + g_2^2(x,t)\right) dx \le 1$ for all $t \ge 0$. We impose the following boundary conditions:

$$w(0,t) = \varsigma(0,t) = 0, \quad w_x(1,t) - \varsigma(1,t) = 0, \quad \varsigma_x(1,t) = 0$$

These boundary conditions mean that the beam is clamped at the left end and free at the right end. The initial state of the beam is

$$w(x,0) = w_0(x),\ w_t(x,0) = w_1(x),\ \varsigma(x,0) = \varsigma_0(x),\ \varsigma_t(x,0) = \varsigma_1(x)$$

We study the minimization problem of the deviation of the beam from the equilibrium state in the sense of the functional

$$\int_0^\infty \int_0^1 \left(w^2(x,t) + \varsigma^2(x,t)\right) dx\, dt$$

The control problem for the Timoshenko beam refers to the class of problems (1)-(5). Indeed, define a linear operator $A_x : D \to H$ by $A_x \begin{pmatrix} \eta_1 \\ \eta_2 \end{pmatrix} = \begin{pmatrix} -\frac{1}{\gamma}\eta_1'' + \frac{1}{\gamma}\eta_2' \\ -\frac{1}{\gamma}\eta_1' - \eta_2'' + \frac{1}{\gamma}\eta_2 \end{pmatrix}$ where $H = L^2([0,1], \mathbb{C}^2)$ and

$$D = \left\{ \begin{pmatrix} \eta_1 \\ \eta_1 \end{pmatrix} \in H \,\middle|\, \eta_1(0) = \eta_2(0) = 0, \eta_1'(1) - \eta_2(1) = 0, \eta_2'(1) = 0 \right\}$$

It was proved [5,6] that $A_x$ is self-adjoint in $H$ and positive. Hence all the above results on the singularities of optimal solutions are applied to the problem of controlling the vibration of the Timoshenko beam.

IOP Conf. Series: Journal of Physics: Conf. Series **955** (2017) 012030 doi:10.1088/1742-6596/955/1/012030

## 6. Conclusion

We have considered a control system of partial differential equations that governs vibrations in many mechanical systems. Vibration control problems are widely used in many areas of engineering (robots with flexible links, spacecrafts with flexible antennas, etc). We have studied the problem of minimizing the deviation from the rest position. This problem was reduced to the optimal control problem for the corresponding Fourier coefficients. For some initial data we have proved the existence of optimal solutions with a countable number of switchings on a finite time interval and optimal spiral-like solutions which attain the origin in a finite time making a countable number rotations. We have shown that optimal solutions contained the singular arc and the chattering mode. Thus, optimal controls have a complicated structure and are difficult to implement in control devices. We believe that knowledge of the behavior of exact optimal solutions can be useful in numerical modeling of mechanical processes.

This research was supported in part by the Russian Foundation for Basic Research under grant no. 17-01-00805.

## References

[1] Borisov V F, Zelikin M I and Manita L A 2010 Optimal synthesis in an infinite-dimensional space *Proceedings of the Steklov Institute of Mathematics* **271**(1) 34–52

[2] Zelikin M I 1997 One-parameter families of solutions to a class of pde optimal control problems *Contemporary Mathematics* **209** 339–349

[3] Zelikin M I and Borisov V F 1994 *Theory of Chattering Control with Applications to Astronautics, Robotics, Economics and Engineering* (Birkhäuser, Boston)

[4] Manita L, Lokutsievskiy L and Ronzhina M 2017 Optimal solutions in a neighborhood of a singular extremal for a problem with two-dimensional control *Proceedings of the International Conference on Mathematical Theory of Optimal Control* (Steklov Mathematical Institute, Moscow) pp. 94–96

[5] Gugat M 2001 Controllability of a slowly rotating Timoshenko beam *ESAIM: Control, Optimisation and Calculus of Variations* **6** 333–360

[6] Krabs W and Sklyar G 2000 On the stabilizability of a slowly rotating Timoshenko beam *Zeitschrift für Analysis und ihre Anwendungen* **19**(1) 131–146

[7] Shubov M A 2000 Exact controllability of damped Timoshenko beam *IMA Journal of Mathematical Control and Information* **17**(4) 375–395

# Algorithms for computing solvents of unilateral second-order matrix polynomials over prime finite fields using lambda-matrices

**Filipp Burtyka**

Junior Research Fellow, Institute of Computer Technologies and Information Security of Southern Federal University, 2 Chekhova St., Taganrog, Russia

E-mail: `bbfilipp@yandex.ru`

**Abstract.** The paper considers algorithms for finding diagonalizable and non-diagonalizable roots (so called *solvents*) of monic *arbitrary* unilateral second-order matrix polynomial *over prime finite field*. These algorithms are based on polynomial matrices (lambda-matrices). This is an extension of existing general methods for computing solvents of matrix polynomials over field of complex numbers. We analyze how techniques for complex numbers can be adapted for finite field and estimate asymptotic complexity of the obtained algorithms.

## 1. Introduction

Let $\mathbb{K}$ be a field, $\mathsf{M}_n(\mathbb{K})$ be the ring of $n \times n$ matrices over $\mathbb{K}$ (it can be, for example, the field of complex numbers $\mathbb{C}$ or finite field $\mathbb{F}_q$, where $q = p^t$, $p \in \mathbb{N}$ is prime). The *unilateral matrix polynomial (UMP)* of $n$-th order and $d$-th degree over $\mathbb{K}$ is the following expression:

$$\mathcal{F}(X) = \mathbf{F}_d \cdot X^d + \mathbf{F}_{d-1} \cdot X^{d-1} + \cdots + \mathbf{F}_2 \cdot X^2 + \mathbf{F}_1 \cdot X + \mathbf{F}_0, \quad (1)$$

where $\mathbf{F}_i \in \mathsf{M}_n(\mathbb{K})$, $0 \le i \le d$ are coefficients, $\mathbf{F}_d \ne \mathbf{0}$, $X \in \mathsf{M}_n(\mathbb{K})$ is variable.

Let $(\mathsf{M}_n(\mathbb{K}))[X]$ be the set of all UMP for given $n$ and $\mathbb{K}$. The *root* (or *solvent*) of UMP $\mathcal{F}(X)$ over $\mathbb{K}$ is a matrix $\mathbf{S} \in \mathsf{M}_n(\mathbb{K})$ such that $\mathcal{F}(\mathbf{S}) = \mathbf{0}$ (we denote by $\mathbf{0}$ the zero matrix of appropriate order). The solvents set of $\mathcal{F}$ we denote by $\mathsf{Ker}(\mathcal{F}(X))$.

A *Monic UMP (MUMP) is such UMP that* $\mathbf{F}_{\deg(\mathcal{F})} = \mathbf{I}$, *where* $\mathbf{I}$ *is identity* $n \times n$ *matrix.*

In this work we consider only MUMP, more particular *left* MUMP. *Right* UMP looks like $\mathcal{F}(X) = X^d + X^{d-1} \cdot \mathbf{F}_{d-1} + \cdots + X^2 \cdot \mathbf{F}_2 + X \cdot \mathbf{F}_1 + \mathbf{F}_0$. All below analysis is true and for them.

*An important task for* $\mathcal{F}(X)$ *is computing* $\mathsf{Ker}(\mathcal{F}(X))$. Difficulties occur because of noncommutativity and the presence of zero divisors. Unlike scalar polynomials when $\mathbb{K}$ is $\mathbb{C}$ it can be true that $|\mathsf{Ker}(\mathcal{F}(X))| = \infty$. And when $\mathbb{K}$ is $\mathbb{F}_p$ one may get $|\mathsf{Ker}(\mathcal{F}(X))| > \deg(\mathcal{F}(X))$. For example, solvents of $\mathcal{F}(X) = X^2 - \begin{pmatrix} 0 & 0 \\ 0 & 0 \end{pmatrix} \in (\mathsf{M}_2(\mathbb{K}))[X]$ are all nilpotent $2 \times 2$ matrices. So, for $\mathbb{K} = \mathbb{C}$ we have $|\mathsf{Ker}(\mathcal{F}(X))| = \infty$, and for $\mathbb{K} = \mathbb{F}_p$ there is $|\mathsf{Ker}(\mathcal{F}(X))| = p^2$.

In this paper we consider algorithms for computing solvents of $\mathcal{F}(X) \in \mathsf{M}_2(\mathbb{F}_p)[X]$, where $p$ is prime, that belong to the class of matrices $Eig_2(\mathbb{F}_p) \subset \mathsf{M}_2(\mathbb{F}_p)$. These matrices have a Jordan normal form (JNF) over $\mathbb{F}_p$. From the paper [11] we get $|Eig_2(\mathbb{F}_p)| = \frac{1}{2} \cdot p^4 + p^3 - \frac{1}{2} \cdot p^2$. Since $|\mathsf{M}_2(\mathbb{F}_p)| = p^4$, we have $|Eig_2(\mathbb{F}_p)| > \frac{1}{2}|\mathsf{M}_2(\mathbb{F}_p)|$. So, the class of held matrices is quite large.

Algorithms for computing solvents of MUMP over $\mathbb{F}_p$ having arbitrary form and degree have not been considered in literature, even for $n = 2$. We undertook an analysis of existing works [2, 8, 4, 5] devoted to MUMP over $\mathbb{C}$. These works describe a general approach to compute solvents. They use the following fact: any matrix in $\mathsf{M}_2(\mathbb{C})$ has JNF. However this approach was not presented in algorithmic form. *This paper analyzed how it can be generalized for $\mathbb{F}_p$ and described the algorithms.*

Our research has important practical application in cryptography. Complexity analysis of computing $\mathsf{Ker}(\mathcal{F}(X))$ is necessary to estimate security level of one recently proposed homomorphic cryptosystem based on MUMP over $\mathbb{F}_p$ [9, 10].

*The known works devoted to UMPs over $\mathbb{C}$*

The major part of works dealing with matrix polynomials is devoted to calculating solvents of UMP $\mathcal{F}(X) \in (\mathsf{M}_n(\mathbb{C}))[X]$. Several researchers [1, 2, 5, 8, 6] presented the result formalized in Theorem 1.

**Theorem 1.** *(i) There exist unresolvable UMPs $\mathcal{F}(X) \in (\mathsf{M}_n(\mathbb{C}))[X]$.*

*(ii) There exist UMPs $\mathcal{F}(X) \in (\mathsf{M}_n(\mathbb{C}))[X]$, having infinity number of solvents.*

*(iii) If $\mathcal{F}(X) \in (\mathsf{M}_n(\mathbb{C}))[X]$ has finite number of solvents, then $|\mathsf{Ker}(\mathcal{F}(X))| \leq C^n_{n\cdot d}$.*

This result was obtained using polynomial matrix theory [7] (see section 3). The detailed case study for second-order matrices was done in [4, 3].

In [8] the author thoroughly studies the properties of block-companion matrix. It can be composed by coefficients of UMP $\mathcal{F}(X) \in (\mathsf{M}_n(\mathbb{C}))[X]$ and gives an insight how matrix polynomials solvents are built inside. Also the conditions for the number of solvents to be infinite are stated and some cases of finite (positive) solvents number are considered.

Theorem 1 states if $\mathcal{F}(X) \in (\mathsf{M}_n(\mathbb{C}))[X]$ has the finite number of solvents $t$, then $t \leq C^n_{n\cdot d}$. But for arbitrary $n$ it's unknown if $\mathcal{F}(X) \in (\mathsf{M}_n(\mathbb{C}))[X]$ exists having exactly $t$ roots. In [5] the author solves this problem for $n = 2$. He shows for $n = 2$ and $\forall d, t \leq C^2_{2\cdot d}$ the $d$-th degree UMP exists with exactly $t$ solvents. To achieve this result the author builds UMP having only diagonalizable (recall that a square matrix $\mathbf{A}$ is called *diagonalizable* if there exists an invertible matrix $\mathbf{P}$ such that $\mathbf{P}^{-1}\mathbf{A}\mathbf{P}$ is a diagonal matrix) solvents, and their number is exactly $t$.

Finally there is a plenty of works proposing numerical algorithms for computing solvents of UMP $\mathcal{F}(X) \in \mathsf{M}_n(\mathbb{C})[X]$. For example, in [1, 12, 13, 14, 15] the authors propose how to adapt numerical methods for computing matrix polynomials solvents.

## 2. Theoretical background

For each $\mathcal{F}(X) \in (\mathsf{M}_n(\mathbb{F}_p))[X]$ there is $\lambda$-matrix $\mathcal{F}(\lambda)$. We get it by replacing matrix variable $X \in \mathsf{M}_n(\mathbb{F}_p)$ by scalar variable $\lambda \in \mathbb{F}_p$, i.e.

$$\mathcal{F}(\lambda) = \lambda^d + \mathbf{F}_{d-1} \cdot \lambda^{d-1} + \cdots + \mathbf{F}_0 \in (\mathsf{M}_n(\mathbb{F}_p[\lambda])).$$

Below definitions were given in [8] for $\mathbb{C}$. We rewrote them for $\mathbb{F}_p$.

**Definition 1** ([8]). *Derivative of $\mathcal{F}(\lambda)$ is $\lambda$-matrix $\mathcal{F}'(\lambda)$ such that $\mathcal{F}'(\lambda) = d \cdot \mathbf{I} \cdot \lambda^{d-1} + (d - 1) \cdot \mathbf{F}_{d-1} \cdot \lambda^{d-2} + \cdots + \mathbf{F}_1$.*

**Definition 2** ([8]). *Eigenvalue (EVal) of $\mathcal{F}(\lambda)$ is such $\alpha \in \mathbb{F}_p$ that $\mathsf{det}(\mathcal{F}(\alpha)) = 0$.*

**Definition 3** ([8]). *Eigenvector (EVec) of $\mathcal{F}(\lambda)$, corresponding to EVec $\alpha$, is such $\vec{v} \in \mathbb{F}_p^n$ that $\mathcal{F}(\alpha) \cdot \vec{v} = \vec{0}$.*

**Definition 4** ([8]). *Let $\alpha \in \mathbb{F}_p$ be EVal and $\vec{v} \in \mathbb{F}_p^n$ be corresponding EVec of $\mathcal{F}(\lambda)$. The pair $(\alpha, \vec{v})$ is eigenpair (EP) of $\mathcal{F}(\lambda)$.*

**Definition 5** ([8]). *Let there be $\mathcal{F}(\lambda) \in (\mathsf{M}_2(\mathbb{F}_p))[\lambda]$ having EVal $\alpha \in \mathbb{F}_p$. Vectors $\vec{v_1}, \vec{v_2} \in \mathbb{F}_p^n$, $\vec{v_1} \neq \vec{0}$ satisfying the system*

$$\begin{cases} \mathcal{F}(\alpha)\, \vec{v_1} = \vec{0} \\ \mathcal{F}(\alpha)\, \vec{v_2} + \mathcal{F}'(\alpha)\, \vec{v_1} = \vec{0}, \end{cases} \tag{2}$$

*are called a Jordan chain of length 2 $\mathcal{F}(\lambda)$ corresponding to EVal $\alpha$.*

## 3. The general form of solvents, having JNF

According to analysis done for $\mathbb{C}$ [2, 8, 4] we came to conclusion that there are two types of solvents of $\mathcal{F}(X) \in (\mathsf{M}_2(\mathbb{F}_p))[X]$ from class $Eig_2(\mathbb{F}_p)$:

- *Diagonalizable solvents*
$$\mathbf{S} = \mathbf{V} \cdot \begin{pmatrix} \alpha_1 & 0 \\ 0 & \alpha_2 \end{pmatrix} \cdot \mathbf{V}^{-1}, \tag{3}$$
where $\mathbf{V} = [\vec{v_1}\ \vec{v_2}] \in \mathsf{M}_2(\mathbb{F}_p)$ – matrix with columns $\vec{v_1}$ and $\vec{v_2}$ respectively, $(\alpha_i, \vec{v_i}), i = 1, 2$ – EP of $\mathcal{F}(\lambda)$, $\vec{v_1}, \vec{v_2}$ are linear independent.
- *Non-diagonalizable solvents*
$$\mathbf{S} = \mathbf{V} \cdot \begin{pmatrix} \alpha & 1 \\ 0 & \alpha \end{pmatrix} \cdot \mathbf{V}^{-1}, \tag{4}$$
where $\mathbf{V} = [\vec{v_1}\ \vec{v_2}] \in \mathsf{M}_2(\mathbb{F}_p)$, $\vec{v_1}, \vec{v_2}$ – Jordan chain $\mathcal{F}(\lambda)$ satisfying (2), EVal $\alpha \in \mathbb{F}_p$ has algebraic multiplicity $\geq 2$, $\vec{v_1}, \vec{v_2}$ are linear independent.

## 4. Algorithms of computing solvents of MUMP based on $\lambda$-matrices

### *4.1. Algorithm for computing diagonalizable solvents*

In this section we consider solvents (3). Let $\ker_{\mathcal{D}}(\mathcal{F}(X))$ be the set of all such solvents. We need algorithm ConstructEigenValues, computing all EVal of $\mathcal{F}(\lambda)$. It uses algorithm FindRoots, whose input is $d(\lambda) \in \mathbb{F}_p[\lambda]$. It computes all different roots of $d(\lambda)$.

**Algorithm 1:** ConstructEigenValues($\mathcal{F}(\lambda)$)

**Input:** $\lambda$-matrix $\mathcal{F}(\lambda) \in (\mathsf{M}_2(\mathbb{F}_p))[\lambda]$
**Output:** $\mathcal{A} = \{\alpha_1, ..., \alpha_r\}$, $\alpha_i \in \mathbb{F}_p$, $\alpha_i \neq \alpha_j$

```
1 d(λ) := det(F(λ));
2 A := FindRoots(d(λ));
3 return A;
```

Further we need to compute $\ker(\mathcal{F}(\alpha_i))$ for $\forall \alpha_i \in \mathcal{A}$. For $\mathsf{M}_2(\mathbb{F}_p)$ there are only two cases: 1) $\ker(\mathcal{F}(\alpha_i)) = \mathbb{F}_2^p$, 2) $\ker(\mathcal{F}(\alpha_i)) = Lin(\vec{v_i})$, where $Lin(\vec{v_i})$ is a linear subspace with basis $\vec{v_i} \in \mathbb{F}_2^p \setminus \{\vec{0}\}$. We divide eigenvalues into two types in accordance with $\dim(\ker(\mathcal{F}(\alpha_i)))$. It helps us to build algorithm for finding $\ker_{\mathcal{D}}(\mathcal{F}(X))$ in such a way that all repetitive solvents will be excluded. It will decrease the complexity.

Algorithm ConstructEigenPairs computes EP of $\mathcal{F}(\lambda)$. Its output is a set $\mathcal{A}_0 = \{\alpha_{0,1}, ..., \alpha_{0,t_0}\} \subseteq \mathcal{A}$, $\alpha_{0,i} \neq \alpha_{0,j}$, containing all EVal such that $\dim(\ker(\mathcal{F}(\alpha_{0,i})))= 2$. Also it computes pairs $(\vec{v_1}, \mathcal{A}_1), ..., (\vec{v_s}, \mathcal{A}_s)$, $\dim(Lin(\vec{v_i}, \vec{v_j}))= 2$. The set $\mathcal{A}_k = \{\alpha_{k,1}, ..., \alpha_{k,t_k}\} \subseteq \mathcal{A}$, $\alpha_{k,i} \neq \alpha_{k,j}$ consists of all EVal such that $\ker(\mathcal{F}(\alpha_{k,i})) = Lin(\vec{v_k})$. Sets $\mathcal{A}_0, \mathcal{A}_1, ..., \mathcal{A}_s$ haven't any intersection and $\cup_{k=0}^{s} \mathcal{A}_k = \mathcal{A}$.

**Lemma 1.** *For any invertible matrices* $\mathbf{V} \in \mathsf{M}_2(\mathbb{F}_p)$, $\mathbf{V} = [\vec{v}_1\ \vec{v}_2]$ *and nonzero constants* $\beta, \gamma \in \mathbb{F}_p$ *we have* $\mathbf{V} \cdot \mathsf{diag}(\alpha_1, \alpha_2) \cdot \mathbf{V}^{-1} = \mathbf{V}' \cdot \mathsf{diag}(\alpha_1, \alpha_2) \cdot \mathbf{V}'^{-1}$ *where* $\mathbf{V}' = [\beta \cdot \vec{v}_1\ \gamma \cdot \vec{v}_2]$.

**Algorithm 2:** $\mathsf{ConstructEigenPairs}_2(\mathcal{F}(\lambda))$

**Input:** $\lambda$-matrix $\mathcal{F}(\lambda) \in \mathsf{M}_2(\mathbb{F}_p)[\lambda]$
**Output:** Set of pairs $\{(\cdot, \mathcal{A}_0), (\vec{v}_1, \mathcal{A}_1), ..., (\vec{v}_s, \mathcal{A}_s)\}$

1 $\mathcal{P} := \varnothing$; $\mathcal{A}_0 := \varnothing$;
2 $\{\alpha_1, ..., \alpha_r\} := \mathsf{ConstructEigenValues}(\mathcal{F}(\lambda))$;
3 **for** $i = 1$ *to* $r$ **do**
4     **if** $(\mathcal{F}(\alpha_i) = \mathbf{0})$ **then**
5         $\mathcal{A}_0 := \mathcal{A}_0 \cup \{\alpha_i\}$;
6     **end**
7     **else**
8         Compute $\vec{v}_*$, such that $\ker(\mathcal{F}(\alpha_i)) = \mathsf{Lin}(\vec{v}_*)$;
9         **foreach** $(\vec{v}, \Lambda) \in \mathcal{P}$ **do**
10             **if** $\mathsf{Lin}(\vec{v}, \vec{v}_*) = 1$ **then**
11                 $\Lambda := \Lambda \cup \{\alpha_i\}$;
12                 **goto** 3;
13             **end**
14         **end**
15         $\mathcal{P} := \mathcal{P} \cup \{(\vec{v}_*, \{\alpha_i\})\}$;
16     **end**
17 **end**
18 **if** $|\mathcal{A}_0| > 0$ **then**
19     $\mathcal{P} := \mathcal{P} \cup \{(\cdot, \mathcal{A}_0)\}$
20 **end**
21 **return** $\mathcal{P}$;

According to Lemma 1 one should not use all EP $(\alpha_{0,i}, \vec{v}), \vec{v} \in \mathbb{F}_p^2$ for constructing solvents (3). Only some subset is enough to exclude repetitions.

**Lemma 2.** *Let* $\vec{e}_1\ \vec{e}_2 \in \mathbb{F}_p^2$ *be a basis.* $\mathbb{F}_p^2$ *is a union of disjoint subspaces (excluding* $\vec{0}$*):*

$$\mathbb{F}_p^2 = Lin(\vec{e'}_1) \cup Lin(\vec{e'}_2) \cup Lin(\vec{e'}_3)\ \ldots\ \cup Lin(\vec{e'}_{p+1}),$$

$\vec{e'}_1 = \vec{e}_1$, $\vec{e'}_2 = \vec{e}_2$, $\vec{e'}_3 = \vec{e}_1 + \vec{e}_2$, $\vec{e'}_4 = \vec{e}_1 + 2 \cdot \vec{e}_2$, ..., $\vec{e'}_{p+1} = \vec{e}_1 + (p-1) \cdot \vec{e}_2$.

To exclude repetitions during computing by formula (3) we should use only EP $(\alpha_{0,k}, \vec{e'}_i)$, $1 \le i \le p+1$. Similarly we assume that in pair $(\vec{v}_i, \mathcal{A}_i)$ there is $\vec{v}_i \in \mathcal{E}$, where $\mathcal{E} = \{\vec{e'}_1, ..., \vec{e'}_{p+1}\}$. Finally we get algorithm $\mathsf{ConstructAllDiagonalizableSolvents}$, computing $\ker_{\mathcal{D}}(\mathcal{F}(X))$.

The complexity of $\mathsf{ConstructAllDiagonalizableSolvents}$ is $O(d \cdot (\log d + p)) + |\ker_{\mathcal{D}}(\mathcal{F}(X))|)$. ($O(d \cdot \log d)$ is complexity of computing $\det(\mathcal{F}(X))$, $O(d \cdot p)$ is complexity of finding EVal $\mathcal{F}(\lambda)$, using deterministic procedure. he number of diagonalizable matrices in $Eig_2(\mathbb{F}_p)$ is $\frac{p^2 \cdot (p^2-1)}{2} + p$. Then in the worst case the complexity is $O(d \cdot (\log d + p) + p^4)$. Also one can use probabilistic algorithm to find all EVal – then complexity in average is $\approx O(d \cdot (\log d + \log p) + |\ker_{\mathcal{D}}(\mathcal{F}(X))|)$.

IOP Conf. Series: Journal of Physics: Conf. Series **955** (2017) 012031 doi:10.1088/1742-6596/955/1/012031

**Algorithm 3:** ConstructAllDiagonalizableSolvents$_2(\mathcal{F}(X))$

**Input:** $\mathcal{F}(X)$
**Output:** $\ker_{\mathcal{D}}(\mathcal{F}(X))$

```
D := ∅;
(·, A_0), (v⃗_1, A_1), ..., (v⃗_s, A_s) := ConstructEigenPairs(F(λ));
foreach (i, j), such that 0 ≤ i < j ≤ s do
    foreach α ∈ A_i, β ∈ A_j do
        if α = β then
            D := D ∪ {diag(α, β)};
        end
        else if i = 0 then
            if j = 0 then
                foreach b⃗_1, b⃗_2 ∈ E, such that b⃗_1 ≠ b⃗_2 do
                    V = [b⃗_1 b⃗_2];
                    D := D ∪ {V · D · V^{-1}};
                end
            end
            else
                foreach b⃗ ∈ E do
                    V = [b⃗ v⃗_j];
                    D := D ∪ {V · D · V^{-1}};
                end
            end
        end
        else
            V = [v⃗_i v⃗_j];
            D := D ∪ {V · D · V^{-1}};
        end
    end
end
return D;
```

*4.2. Algorithm for non-diagonalizable solvents*

Now we consider the algorithm for finding solvents (4). The set of all these solvents we denote by $\ker_{\mathcal{N}}(\mathcal{F}(X))$ and let $\ker_{\alpha,\mathcal{N}}(\mathcal{F}(X))$ be a subset of such solvents for fixed EVal $\alpha$.

Computing of EVal is similar to Algorithm 1, but it takes into account the algebraic multiplicity of each EVal. The result is truncated set $\mathcal{A}' = \{\alpha'_1, ..., \alpha'_m\} \subseteq \mathcal{A}$ of EVal, having multiplicity $> 1$.

Let's fix some $\alpha \in \mathcal{A}'$ and consider how to solve (2). We have the following cases.

- If $\mathcal{F}(\alpha) = 0$, $\mathcal{F}'(\alpha) = 0$, then solvents are all matrices of the form $\mathbf{S} = \mathbf{N} + \mathsf{diag}(\alpha, \alpha)$, where $\mathbf{N} \in \mathsf{M}_2(\mathbb{F}_p)$ is arbitrary nilpotent matrix. $|\ker_{\alpha,\mathcal{N}}(\mathcal{F}(X))| = p^2$.
- If $\mathcal{F}(\alpha) \neq 0$, $\mathcal{F}'(\alpha) = 0$ then $|\ker_{\alpha,\mathcal{N}}(\mathcal{F}(X))| = 0$.
- If $\mathcal{F}(\alpha) = 0$, $\mathcal{F}'(\alpha) \neq 0$ then we have two cases:
  a) if $\mathsf{det}(\mathcal{F}'(\alpha)) \neq 0$ then $|\ker_{\alpha,\mathcal{N}}(\mathcal{F}(X))| = 0$;
  b) if $\mathsf{det}(\mathcal{F}'(\alpha)) = 0$ then for $\forall \vec{v}_1 \in \ker(\mathcal{F}'(\alpha))$ the second equation of (2) becomes an identity $\mathbf{0} \cdot \vec{v}_2 = \mathbf{0}$. Then $\mathbf{V}$ can be constructed using $\vec{v}_1 \in \ker(\mathcal{F}'(\alpha)) \setminus \{\vec{0}\}$, $\vec{v}_2 \in \mathbb{F}_p^2 \setminus \ker(\mathcal{F}'(\alpha))$

and we obtain $|\ker_{\alpha,\mathcal{N}}(\mathcal{F}(X))| < p^3$.

- If $\mathcal{F}(\alpha) \neq 0$, $\mathcal{F}'(\alpha) \neq 0$ then $\mathbf{V}$ consists of $\vec{v}_1 \in \ker(\mathcal{F}'(\alpha))$ and $\vec{v}_2$ that is a solution of a linear system $\mathcal{F}(\alpha) \cdot \vec{x} = -\mathcal{F}'(\alpha) \cdot \vec{v}_1$ and we obtain $|\ker_{\alpha,\mathcal{N}}(\mathcal{F}(X))| < p^2$.

The above is summarized into algorithm ConstructNonDiagonalizableSolvents for computing $\ker_{\mathcal{N}}(\mathcal{F}(X))$. Its complexity in the worst case is $O(d \cdot (\log d + \log p) + p^3 \cdot d)$.

**Algorithm 4:** ConstructNonDiagonalizableSolvents$_2(\mathcal{F}(X))$

**Input:** $\lambda$-matrix $\mathcal{F}(\lambda) \in \mathsf{M}_2(\mathbb{F}_p)[\lambda]$
**Output:** $\ker_{\mathcal{N}}(\mathcal{F}(X))$

$\mathcal{N} := \varnothing$;
$\mathcal{A}' =$ ConstructEigenValuesOfMultiplicityGreaterThanOne$(\mathcal{F}(\lambda))$;
**foreach** $\alpha \in \mathcal{A}'$ **do**
  $\mathbf{J} := \begin{pmatrix} \alpha & 1 \\ 0 & \alpha \end{pmatrix}$;
  **if** $\mathcal{F}(\alpha) = 0$ ***and*** $\mathcal{F}'(\alpha) = 0$ **then**
    **foreach** $\mathbf{N} \in \mathsf{M}_2(\mathbb{F}_p)$, *such that* $\mathbf{N}^2 = \mathbf{0}$ **do**
      $\mathcal{N} := \mathcal{N} \cup \{\mathbf{N} + \mathrm{diag}(\alpha, \alpha)\}$;
    **end**
  **end**
  **if** $\mathcal{F}(\alpha) = 0$ ***and*** $\mathcal{F}'(\alpha) \neq 0$ ***and*** $\det(\mathcal{F}'(\alpha)) = 0$ **then**
    **foreach** $(\vec{v}_1, \vec{v}_2)$, $\vec{v}_1 \in \ker(\mathcal{F}'(\alpha)) \setminus \{\vec{0}\}$, $\vec{v}_2 \in \mathbb{F}_p^2 \setminus \ker(\mathcal{F}'(\alpha_i))$ **do**
      $\mathbf{V} = [\vec{v}_1\ \vec{v}_2]$;
      $\mathcal{N} := \mathcal{N} \cup \{\mathbf{V} \cdot \mathbf{J} \cdot \mathbf{V}^{-1}\}$;
    **end**
  **end**
  **if** $\mathcal{F}(\alpha) \neq 0$ ***and*** $\mathcal{F}'(\alpha) \neq 0$ **then**
    **foreach** $\vec{v}_1 \in \ker(\mathcal{F}'(\alpha)) \setminus \{\vec{0}\}$, $\vec{v}_2$, *such that* $\mathcal{F}(\alpha) \cdot \vec{v}_2 = -\mathcal{F}'(\alpha) \cdot \vec{v}_1$ **do**
      **if** $\dim(\mathrm{Lin}(\vec{v}_1, \vec{v}_2)) = 2$ **then**
        $\mathbf{V} = [\vec{v}_1\ \vec{v}_2]$;
        $\mathcal{N} := \mathcal{N} \cup \{\mathbf{V} \cdot \mathbf{J} \cdot \mathbf{V}^{-1}\}$;
      **end**
    **end**
  **end**
**end**
**return** $\mathcal{N}$;

## 5. Conclusion

In this work the study was done of how existing techniques and methods for finding of UMP's solvents over $\mathbb{C}$ using $\lambda$-matrices can be adapted for the prime finite field $\mathbb{F}_p$. On the basis of this analysis we described algorithms for computing solvents of MUMP $\in (\mathsf{M}_2(\mathbb{F}_p))[X]$ and estimated their asymptotical complexity.

### Acknowledgments

The work was supported by the RFBR grant no. 16-37-00125 mol_a. The author is also very grateful to Professor V. P. Gerdt (LIT, JINR, Dubna), for participating in the fate of this work.

IOP Conf. Series: Journal of Physics: Conf. Series **955** (2017) 012031 doi:10.1088/1742-6596/955/1/012031

## References

[1] Dennis J E, Traub J F and Weber R P 1976 The algebraic theory of matrix polynomials. *SIAM Journal on Numerical Analysis* **13**(6) 831–845.

[2] Dennis J E, Traub J F and Weber R P 1978 Algorithms for solvents of matrix polynomials *SIAM Journal on Numerical Analysis* **15**(3) 523–533

[3] Gelfand S I 2004 O chisle reshenij kvadratnogo uravnenija [On the number of quadratic equation's solutions] *Obshhematematicheskij seminar "Globus". [All-mathematical workshop "Globus"].* **1** 124–133 (In Russian)

[4] Wilson R L 1999 Polynomial equations over matrices. *Rutgers University preprints.* `https://pdfs.semanticscholar.org/57bf/9e36f6bc8ed7648b70d1ff7b1fa80d024f23.pdf`

[5] Slusky M 2010 Zeros of $2 \times 2$ matrix polynomials.*Communications in Algebra.* **38**(11) 4212–4223.

[6] Shavarovskii B Z 2007 Finding a complete set of solutions or proving unsolvability for certain classes of matrix polynomial equations with commuting coefficients. *Computational Mathematics and Mathematical Physics.*, **47**(12) 1902–1911.

[7] Gantmakher F 1959 The theory of matrices – Volume 1. Chelsea, New York.

[8] Pereira E 2003 On solvents of matrix polynomials. *Applied numerical mathematics.* **47** 197–208.

[9] Burtyka F B 2014 Paketnoe simmetrichnoe polnost'ju gomomorfnoe shifrovanie na osnove matrichnyh polinomov [Batch symmetric fully homomorphic encryption using matrix polynomials]. *Trudy Instituta sistemnogo programmirovanija RAN [The Proceedings of the Institute for System Programming RAS]* **26**(5) 99–115. (In Russian)

[10] Burtyka F B and Makarevich O B 2014 Symmetric fully homomorphic encryption using decidable matrix equations. *Proceedings of the 7th International Conference on Security of Information and Networks.* ACM.

[11] Kaylor L and Offner D 2014 Counting matrices over a finite field with all eigenvalues in the field. *Involve, a Journal of Mathematics.* **7**(5) 627–645.

[12] Higham N J and Kim H M 2001 Solving a quadratic matrix equation by Newton's method with exact line searches. *SIAM Journal on Matrix Analysis and Applications.* **23**(2) 303-316.

[13] Peichang G 2014 The Newton-Shamanskii method for solving a quadratic matrix equation arising in quasi-birth-death problems arXiv preprint arXiv:1406.1075.

[14] Fernando M and Pereira E 2010 A fixed point method to compute solvents of matrix polynomials *Mathematica Bohemica* **135**(4) 355-362.

[15] Seo Jong-Hyeon and Hyun-Min Kim 2008 Solving matrix polynomials by Newton's method with exact line searches *Journal of the Korea Society for Industrial and Applied Mathematics* **12**(2) 55-68.

IOP Conf. Series: Journal of Physics: Conf. Series **955** (2017) 012032 doi:10.1088/1742-6596/955/1/012032

# On the possibility of using the Galerkin projection method to simulate the two-dimensional diffusion of excitons generated by an electron beam

**E V Seregina[1], A N Polyakov[2], M A Stepovich[2, 3]**

[1] Bauman Moscow State Technical University (National Research University), Kaluga Branch, 2 Bazhenov Street, Kaluga, 248000 Russia
[2] Tsiolkovsky Kaluga State University, 26 Stepan Razin Street, Kaluga, 248023 Russia
[3] Ivanovo State University, 39 Ermaka Street, Ivanovo, 153025 Russia

**Abstract**. This paper presents the results of using the Galerkin projection method for modelling a two-dimensional diffusion of excitons, excited by electron beam in a semiconductor material. The problem is solved in a cylindrical coordinate system. The required concentration of excitons found in the form of partial double sums of Fourier series for system of modified Laguerre functions. The ordinal error estimation residuals corresponding to the approximate solution of the time-dependent diffusion are obtained. The simulation was used for parameters characteristic for the gallium nitride.

## 1. Introduction

Mathematical models of the processes occurring from the interaction of an electron beam with a semiconductor target, as a rule, take into account quite a few factors affecting the formation of an informative signal (for example, a signal of current, induced by an electron beam, a cathodoluminescence signal, etc.), which leads to the necessity of using approximate methods for realization the mathematical model. Therefore, in this paper, we explore some of the possibilities of using the Galerkin method of projection to simulate the two-dimensional diffusion of excitons generated by an electron beam.

In [1] an error estimation of discrepancy corresponding to the approximate solution of the stationary diffusion equation by the Galerkin projection method was obtained. The present work continues such investigations and poses the problem of estimating the error of discrepancy corresponding to the approximate solution of the nonstationary diffusion equation.

## 2. Problem statement

In this paper, we used a model of two-dimensional diffusion of excitons, excited by electron beam in a semiconductor material. Such model can be used when using cathodoluminescent radiation excited by a low-energy electron beam [2].

In this case, the concentration of the excitons $c(x, y, t)$ can be found as a solution of unsteady diffusion equation

$$c_t = D\Delta c - \frac{c}{\tau} \tag{2.1}$$

with boundary conditions

$$c(x, y, 0) = n(x, y). \tag{2.2}$$

Here $\Delta = \partial^2/\partial x^2 + \partial^2/\partial y^2$ — the two-dimensional Laplace operator, $D$ — the diffusion coefficient and $\tau$ — lifetime of excitons, the function $n(x, y)$ can be found as solution of stationary differential equations describing diffusion of excitons in quasi-equilibrium:

$$D\Delta n - \frac{n}{\tau} = -\Phi(x, y), \tag{2.3}$$

$\Phi(x, y)$ — the source function generate excitons, which is determined by the nature and properties of the material. In the case of narrow electron beam, a volume of generated excitons one can describe by the Gaussian distribution function that gives for the right side (2.3) the product $G_0\varphi(x, y)$. Here $\varphi(x, y)$ — the density of a two-dimensional Gaussian normal distribution, and $G_0$ is the generation of excitons rate.

Then, in the cylindrical coordinate system, we obtain the following equation

$$r\frac{\partial^2 c(r,t)}{\partial r^2} + \frac{\partial c(r,t)}{\partial r} - \frac{1}{D} r\frac{\partial c(r,t)}{\partial t} - \frac{1}{\lambda^2} rc(r,t) = 0 \tag{2.4}$$

with boundary conditions

$$c(r, 0) = n(r), \quad c(\infty, \infty) = 0, \tag{2.5}$$

where $\lambda = \sqrt{D\tau}$ - diffusion length of excitons.

## 3. Projection approximation of the initial model based on using the Galerkin method

### *3.1. The algorithm of applying the projection Galerkin method for solving the problem under consideration*

In the Galerkin projection method, we choose the two-dimensional basis of modified Laguerre functions with parameters accelerating series convergence:

$$\varphi_{i,j}(r,t) = \varphi_i(r)\varphi_j(t) = \exp(-\gamma_1 r/2) L_i(\gamma_1 r) \exp(-\gamma_2 t/2) L_j(\gamma_2 t),$$

which are defined in terms of the Chebyshev-Laguerre polynomials on variable $r$ - $L_i(\gamma_1 r)$ and in terms of polynomials on variable $t$ - $L_j(\gamma_2 t)$ with the parameters $\alpha_1 = 0$ and $\alpha_2 = 0$ respectively, $i, j = 0, 1, 2, \ldots$. Here, $\gamma_1 > 0$ and $\gamma_2 > 0$ are used to optimize the computing scheme.

In the Galerkin method, it is assumed that the unknown function $c(r,t)$ can be represented by the approximate solution sufficiently exactly:

$$c_{m+1}(r,t) = c_m(r,t) + c_0(r,t), \tag{3.1}$$

where

$$c_m(r,t) = \sum_{i=0}^{m-1}\sum_{j=0}^{m-1} c_{ij}\varphi_i(r)\varphi_j(t)$$

is the rectangular partial sum of the double Fourier-Laguerre series on the order of $m \times m$ of the function $c(r,t)$,

$$c_0(r,t)=c_0(r)\varphi_m(t),$$

$$c_0(r)=\sum_{i=0}^{m-1}c_i^n\varphi_i(r)-\sum_{i=0}^{m-1}\sum_{j=0}^{m-1}c_{ij}\varphi_i(r)\varphi_j(0)\Big/\varphi_m(0).$$

The function $c_0(r,t)$ is introduced to satisfy the boundary conditions, and the unknown coefficients are determined from the solution for the following system of equations in accordance with the Galerkin method:

$$\left(R,\ \varphi_k(r)\varphi_l(t)\right)=0,\ k,l=\overline{0,m-1},\tag{3.2}$$

$$R=r\frac{\partial^2 c_{m+1}(r,t)}{\partial r^2}+\frac{\partial c_{m+1}(r,t)}{\partial r}-\frac{1}{D}r\frac{\partial c_{m+1}(r,t)}{\partial t}-\frac{1}{\lambda^2}rc_{m+1}(r,t)\tag{3.3}$$

is the residual of the initial equation

Let $C_{mm}$ and $C_m^n$ denote the columns (extended to matrix columns) of the coefficients of expansion of the unknown function $c(r,t)$ and the function $n(r)$ in terms of the chosen basis, respectively. Using the Kronecker product, we introduce the matrix of differentiation with respect to variable $r$ - $D_{mm}^r=D_m^{\gamma_1}\otimes E_m$ and with respect to variable $t$ - $D_{mm}^t=E_m\otimes D_m^{\gamma_2}$, where $E_m$ is the unit matrix and $D_m^{\gamma}$ is the differentiation matrix in the one-dimensional basis of modified Laguerre functions, whose elements are found using elementary algebraic operations [3]:

$$d_{ij}=\begin{cases}-\gamma/2, \text{ if } i=j,\\ -\gamma, \text{ if } i<j,\\ 0, \text{ if } i>j.\end{cases}.$$

Let $B_1$ denote the matrix which elements are found by the formula

$$b_{ij}^1=\begin{cases}-\gamma_1(j-1)/4, \text{ if } i+1=j,\\ \gamma_1\left(1-(i-1)/4\right), \text{ if } j+1=i,\\ \gamma_1(3/4-i/2), \text{ if } i=j,\\ 0, \text{ if } i+1<j,\\ \gamma_1, \text{ if } j+1<i,\end{cases}$$

and let $B_2$ denote the matrix with the elements

$$b_{ij}^2=\begin{cases}-i/\gamma_1, \text{ if } i+1=j,\\ -(i-1)/\gamma_1, \text{ if } j+1=i,\\ (2i-1)/\gamma_1, \text{ if } i=j,\\ 0, \text{ if } i+1<j,\\ 0, \text{ if } j+1<i,\end{cases}$$

which are determined with the following recurrent relations for the Laguerre functions:

$$r\varphi_i''(r) = -(\gamma_1/4)(i+1)\varphi_{i+1}(r) + ((\gamma_1/4)(2i+1) - \gamma_1 i)\varphi_i(r) +$$
$$+(i(-3\gamma_1/4) + (1/\gamma_1)(1-i))\varphi_{i-1}(r) + (1/\gamma_1)\sum_{k=0}^{i-2}\varphi_k(r),$$

$$r\varphi_i(r) = (1/\gamma_1)\left[-(i+1)\varphi_{i+1}(r) - i\varphi_{i-1}(r) + (2i+1)\varphi_i(r)\right].$$

The last relations are easily derived from the known recurrent relations for the Laguerre polynomials [4]:

$$(i+1)L_{i+1}(r) + (r-2i-1)L_i(r) + iL_{i-1}(r) = 0,$$

$$rL_i'(r) = iL_i(r) - iL_{i-1}(r) = 0$$

and from the formulas

$$rL_i''(r) + (1-r)L_i'(r) + iL_i(r) = 0,$$

$$\sum_{k=0}^{i} L_k(r) = L_i(r;1),$$

where $L_i(r;1)$ are the Chebyshev-Laguerre polynomials in variable $r$ with the parameter $\alpha_1 = 1$. Then we introduce the matrices: $D^{r2}_{mm} = B_1^T \otimes E_m$, $D^{t2}_{mm} = B_2^T \otimes D^{\gamma_2}_m$, $Ii_{mm} = B_2^T \otimes E_m$. We rewrite system (3.2) in the matrix form:

$$\left(D^{r2}_{mm} - \frac{1}{D}D^{t2}_{mm} + D^{r}_{mm} + \frac{1}{D}\tilde{D}^{t}_{mm} - \frac{1}{\lambda^2}Ii_{mm}\right)C_{mm} = \frac{1}{D}C^{\Phi_1}_{mm}, \qquad (3.4)$$

where $\tilde{D}^{t}_{mm} = B_2^T \otimes B_3$, and $B_3$ is the matrix with the elements: $b^3_{ij} = -\gamma_2$, and column $C^{\Phi_1}_{mm}$ is by stretching the matrix: $C^{\Phi_1} = B_2^T C^n_m \left(-\gamma_2 \varphi'''(0)\right)^T$. Substituting $c_{ij}$ (found from the system (3.2)) into (3.1) we get the sought approximate solution

*3.2. Investigation of the convergence of the residual R*

Following [5], we introduce the notation:

$$D = r\frac{d^2}{dx^2} + (1-r)\frac{d}{dr} + t\frac{d^2}{dt^2} + (1-t)\frac{d}{dt};$$

$L_2^n(D)$ $(n = 0,1,...)$ is the class of functions $f$ such that the functions $\tilde{f}(r,t) = f(r,t)\exp(\gamma_1 r/2)\exp(\gamma_2 t/2) \in L_2$ and have generalized partial derivatives in the Levy sense $\partial^k \tilde{f}(r,t)/\partial r^i \partial t^j$, $i + j = k$, $k = 0, 1, ...$, belonging to the space $L_2$, for which $D^n \tilde{f} \in L_2$, $n = 0, 1, ...$, where $D^0 \tilde{f} = \tilde{f}$, $D^n \tilde{f} = D\left(D^{n-1}\tilde{f}\right)$, $n = 1,2,...$, $L_2^0(D) = L_2$.

Let the function $c(r,t) \in L_2^{n+1}(D)$, $n > 2$ and has continuous partial derivatives up to the order $2n+2$ with respect to both spatial directions; then the following

$$\|R\|_{L_2} < \left[3\sqrt{2}/\gamma_1 + 2 + 2\sqrt{2}/(D\gamma_1) + \sqrt{2}/(\lambda^2\gamma_1)\right]\left[1-\left(1-m^{-1/2}\right)^m\right]^{-k} \times$$
$$\times(m)^{-n}\,\Omega\left(m^{-1/2}\right) + O\left(m^{-n+\frac{7}{4}}\right)\omega\left(m^{-1/2}\right),\ m \to \infty.$$

Here, $\Omega\left(m^{-1/2}\right)$ is the majorant of the generalized moduli of continuity for functions from the space $L_2$, and $\omega\left(m^{-1/2}\right)$ is the given majorant of the continuity moduli of differentiated functions.

## 4. Results of calculations

Mathematical modeling was carried out for the diffusion of excitons for semiconductor parameters specific for gallium nitride [2]. The frequency of exciton generation was assumed $G_0 = 10^{-13}\,\mathrm{s}^{-1}$, and the electron energy distribution profile in the beam was given by the Gauss function with zero mathematical expectation and the dispersion of 60 nm, which corresponds to the conditions of real experiments [6].

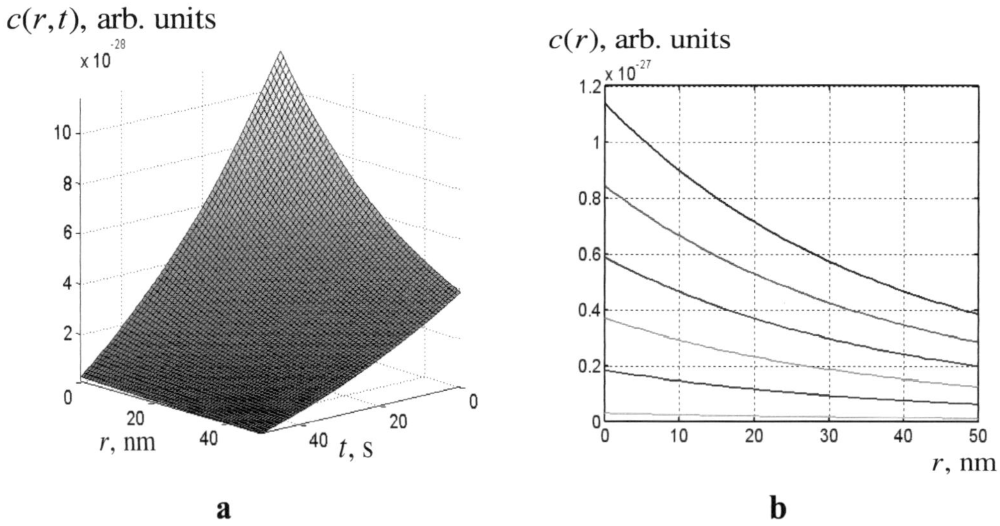


**Figure 1.** a – calculated concentration of excitons $c(r,t)$; b – surface cross-section $c(r,t)$ - dependencies $c(r)$ obtained for various values of time $t = 0$, 10, 20, 30, 40, 50 s (from top to bottom). Calculations carried out in 12th approximation for parameters characteristic for gallium nitride. Electron energy of the beam 5 keV.

Some results of solution of non-stationary equation of diffusion (2.4), (2.5) for excitons are shown in the Figure 1a. Figure 1b presents the results of mathematical modeling for various time: 0, 10, 20, 30, 40, 50 s.

## 5. Conclusions

Considered the possibility of using Galerkin method for mathematical modelling of processes of heat and mass transfer, used in a thin layer near the surface of the target and have obtained the order estimate of the residual error corresponding to approximate solution to the unsteady diffusion equation.

**References**

[1] Makarenkov A M, Seregina E V, and Stepovich M A 2017 *Comput. Math. Math. Phys.* **57** 802
[2] Polyakov A N, Noltemeyer M, Hempel T, Christen J and Stepovich M A 2012 *Journal of Surface Investigation. X-ray, Synchrotron and Neutron Techniques* **6** 901
[3] Lapin S V and Egupov N D 1997 *Theory of Matrix Operators and its Application to Problems on Automatic Control* (Bauman Moscow State Technical Univ., Moscow) 496 pp. [in Russian]
[4] Suetin P K 2005 *Classical Orthogonal Polynomials* (Fizmatlit, Moscow) 480 pp. [in Russian]
[5] Abilov V A, Abilov M V and Kerimov M K 2015 *Comput. Math. Math. Phys.* **55** 1094
[6] Polyakov A N, Noltemeyer M, Hempel T, Christen J, Stepovich M A 2012 *Bulletin of the Russian Academy of Sciences. Physics* **76** 970

# "Escargot Effect" and the Chandler Wobble Excitation

**Leonid Zotov**[1], **Christian Bizouard**[2]

[1] National Research University Higher School of Economics, Moscow Institute of Electronics and Mathematics, Russia
[2] SYRTE, Observatoire de Paris, PSL Research University, CNRS, Sorbonne Universites, UPMC Univ. Paris 06

E-mail: wolftempus@gmail.com

**Abstract.** We study the Chandler wobble (CW) of the pole from 1846 to 2017 extracted by the Panteleev filtering. The CW has period of 433 days, average amplitude of 0.13 milliarcseconds (mas) which is changing, and phase jump by $\pi$ in 1930-th. The CW amplitude strongly (almost to zero) decreases in 1930-th and 2010-th with the phase jump in 1930th. The envelope model contains 83- and 42-years quasi-periodicities. We think the first one can be represented by the 166-years changes of the envelope, crossing zero in 1930th. We reconstruct Chandler input excitation based on the Euler-Liouville equation. Its amplitude has $\sim$ 20-years variations. We explain this based on simple model and prove, that they appear in consequence of 42-years modulation of CW. The excitation amplifies the amplitude of CW for $\sim$ 20 years then damps it for another $\sim$ 20 years.

The analysis of the modulated CW signal in a sliding window demonstrates the specific effect, we called the "escargot effect", when instantaneous "virtual" retrograde component appears in the purely prograde (at long-time interval) signal. Chandler excitation envelope shape is similar to this instantaneous retrograde component, which reflects the changes of ellipticity of the approximation ellipse.

## 1. Introduction

Chandler wobble is one of the most enigmatic component of the polar motion (PM) of the Earth [6, 9]. Discovered at the end of XIX century it is observed for almost 160 years, but the origin of its amplitude changes remains unexplained. Being the resonant oscillation with the damping time of about $\sim$ 50-years, the Chandler wobble maintenance requires small excitation, which is supposed to be provided by the ocean and atmosphere [1, 2, 4, 7]. Under this assumption the random phase jumps and amplitude changes of the CW are caused by the integration of noise, coming from the atmosphere and ocean (winds, currents, and pressure changes) [3]. In our previous paper [13] we developed several methods of Chandler wobble filtering and its excitation reconstruction based on singular spectrum analysis and/or the Panteleev filtering [12]. They gave similar results. Figure 1 shows CW and its excitation, whose amplitude variations have some regularities. We believe, they are causal.

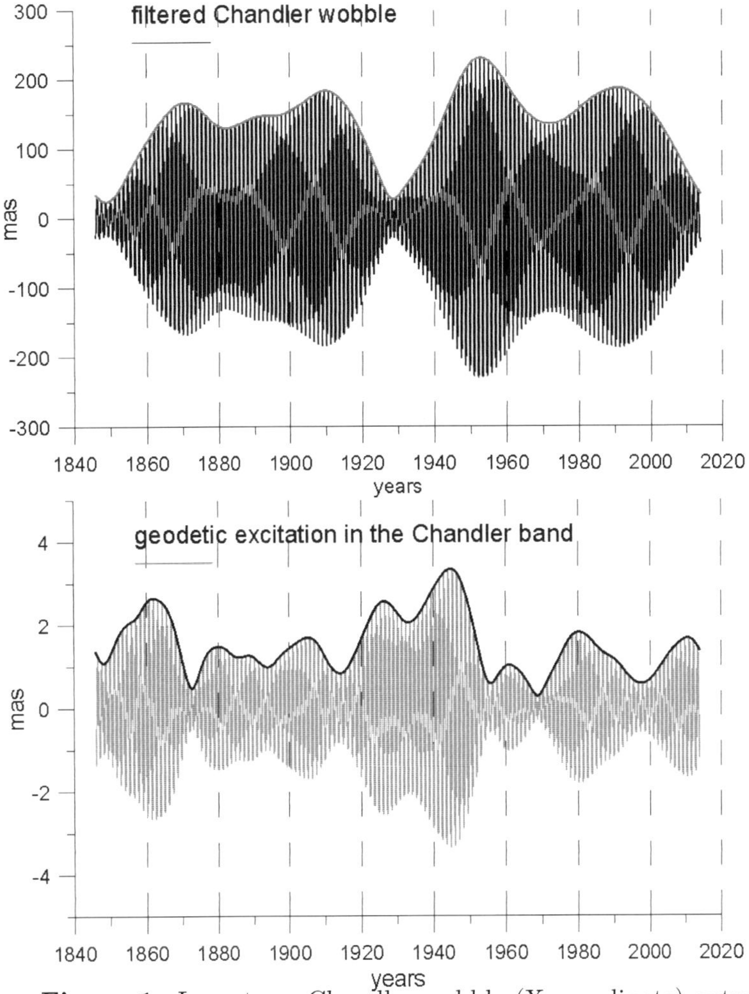


**Figure 1.** Long-term Chandler wobble (X-coordinate) extracted by the Panteleev filter (top) and reconstructed excitation for it (bottom).

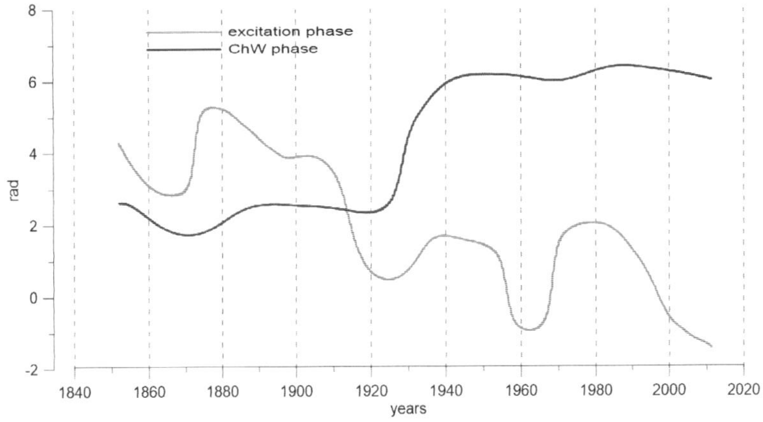


**Figure 2.** Phase changes of the Chandler wobble and its excitation.

IOP Conf. Series: Journal of Physics: Conf. Series **955** (2017) 012033 doi:10.1088/1742-6596/955/1/012033

**Table 1.** Components (for cos) of the Chandler wobble amplitude model.

| Period | Amplitude | Phase (1880) |
|---|---|---|
| 83.4 yrs | 42.6 mas | $40.8^0$ |
| 42.0 yrs | 54.6 mas | $-101.5^0$ |
| mean | 134.8 mas | |

## 2. Chandler wobble modelling

Non-linear least squares (LS) fit of the CW amplitude reveals (Table 1) 42-years and 83-years quasi-periodicities, which can be used for the CW amplitude prediction [11]. The phase changes represented in Fig. 2 are much more difficult to predict. If the strong decrease of the CW amplitude in 2010th has the same origin as its decrease in 1930th [10, 8] we can expect the phase jump by $\pi$ in CW phase now (in 2010th).

If the CW envelope in recent century can be modelled through simple combination of harmonics, what about its excitation? From Fig. 1 it is seen, that the excitation envelope has quasi-20 years periodicity. To explain its origin let's generate the signal with the Chandler carrying frequency $f_c = 0.843$ cpy (cycles per year), and the envelope of 42-years period $f_m = 1/42$ cpy (83-year component is skipped)

$$m(t) = [a_0 + a_1 \cos(2\pi f_m t)] \exp(i2\pi f_c t + \phi_\pi). \tag{1}$$

We used the amplitudes $a_0 = 135$ mas, $a_1 = 55$ mas, and zero-epoch for $t$ argument $t_0 = 1869.3$ yr. Through $\phi_\pi$ we introduced the phase jump in 1930th, constant value $\phi_\pi = 0$ before 1930 is abruptly changed to $\phi_\pi = \pi$ in 1930.0. We introduced constant phase on both intervals of time and in calculations we assume that its jump-like input into derivative can be ignored.

The excitation for the model (1) was reconstructed by the same method of Panteleev corrective filtering in frequency domain [13]. As seen from Fig. 3, the result is very similar to the real excitation, obtained from the real CW, except for the border effects we can skip. The pikes behaviour are quite in phase thanks to $\phi_\pi$ introduction and $t_0$ choice.

Let's look at the Euler-Liouville equation, governing Earth rotation [9]:

$$i\frac{dm(t)}{dt} + \sigma_c m(t) = \sigma_c \Psi(t). \tag{2}$$

Here $m(t) = m_1 + im_2$ is the 2D pole trajectory, $\Psi(t)$ is the excitation; the complex frequency $\sigma_c = \omega_c(1 + i/2Q)$ depends on the real Chandler frequency $\omega_c = 2\pi f_c$ with $f_c = 0.843$ cpy and the quality factor $Q$ (we used $Q = 100$); $i = \sqrt{-1}$. It is easy to obtain the derivative of the model (1)

$$\frac{dm(t)}{dt} = i\omega_c m(t) + \frac{d|m(t)|}{dt} \exp(i\omega_c t + \phi_\pi).$$

Substitutution of model (1) into (3) gives

$$\sigma_c \Psi(t) = -\omega_c m(t) + \sigma_c m(t) + i\frac{d|m(t)|}{dt} \exp(i\omega_c t + \phi_\pi). \tag{3}$$

The terms $-\omega_c m$ and $+\sigma_c m$ almost neutralize each other, but the term multiplied by $1/2Q$ factor remains

$$\Psi = \frac{i}{\sigma_c}\left(\frac{\omega_c}{2Q}[a_0 + a_1 \cos(\omega_m t)] - \omega_m a_1 \sin(\omega_m t)\right) \exp(i\omega_c t + \phi_\pi). \tag{4}$$

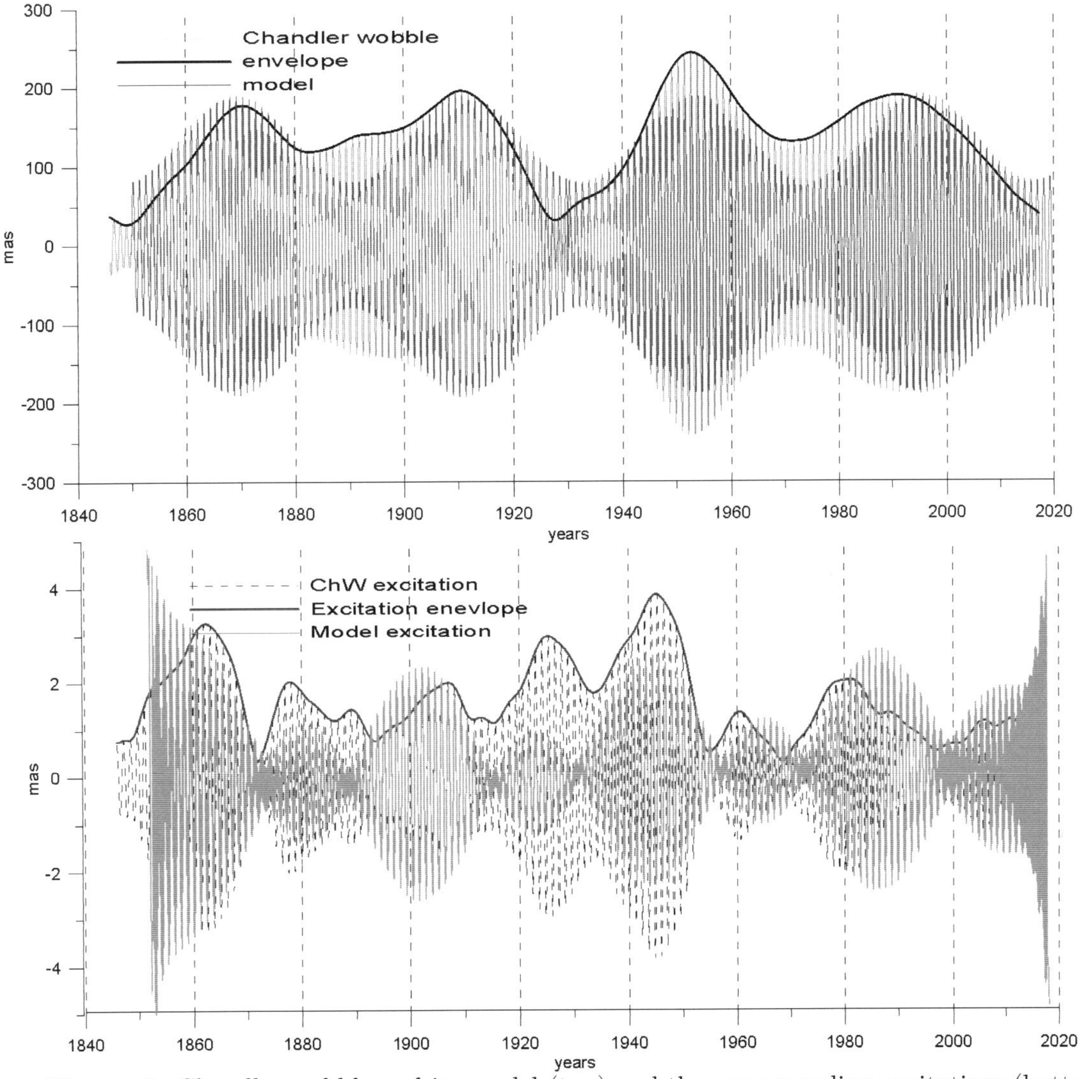


**Figure 3.** Chandler wobble and its model (top) and the corresponding excitations (bottom).

The sum in the large brackets defines the amplitude modulation of $|\Psi|$. Though $\omega_c = 2\pi f_c > \omega_m = 2\pi f_m$, the first term would dominate the second, related to the CW envelope, if not divided by $2Q$. The ratio $\omega_c/2Q$ relative to the modulation frequency $\omega_m$ plays important role, defining the phase and amplitude of $\sin(\cdot)$ and $\cos(\cdot)$ terms combination. The resulting excitation and its envelope are zoomed in Fig. 4, left, over the background of the CW model (1). The asymmetry of this envelope with respect to zero level is responsible for different amplitudes of excitation in maxima. If $a_0$ is zero, the amplifying and damping counterparts are equal in size and last exactly 20-years each. Other combinations of $a_0$, $a_1$, $f_m$ and $Q$ (for example for 82-years modulation) would result in different amplitude ratio, but in reality we clearly observe $\sim$ 20-years wave packages in excitation, corresponding to the epochs of CW amplification and deceleration.

Thus, in simple words, the $\sim$ 20-years modulation of excitation is required to get $\sim$ 40-years

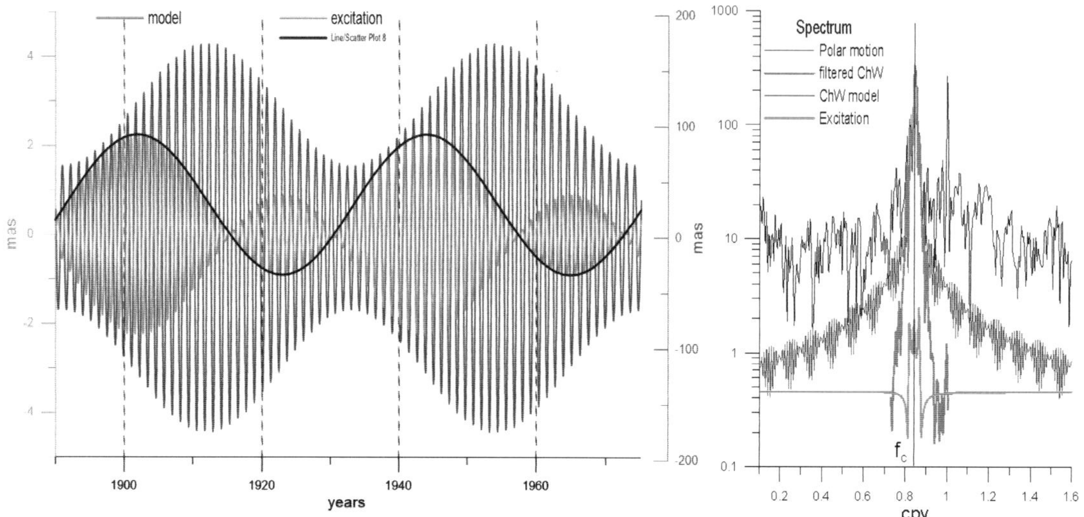


**Figure 4.** Chandler wobble model and corresponding excitation (left), the amplitude spectra of CW, its model and excitation (right).

amplitude changes of CW. At the beginning the excitation amplifies the wobble for $\sim$ 20-years, firstly slowly, then quicker $(d|m|/dt = max)$, then slowly again $(d|m|/dt = 0)$. Then for $\sim$ 20-years the excitation damps CW in the same manner, but with the opposite sign of derivative. Due to opposite sign of the envelope, the phase delay between excitation and CW is changing by $\pi$ every $\sim$ 20 years (see Figs. 2 and 4). The excitation pikes are ahead of CW for $\sim$ 20 years, then the phase changes by $\pi$ and the excitation pikes go behind the CW pikes. Now, lets pay attention to the $\phi_\pi = \pi$ jump, occurred in 1930th. It reverses the pikes, and, according to the conclusions made above, it can be represented by long-term $83 \cdot 2 = 166$ years CW envelope, crossing zero-level in 1930th. Under this model we should expect the phase jump in 2010th.

## 3. "Escargot effect"

The Panteleev filter designed to extract Chandler wobble has purely prograde transfer function [13]. Thus, the CW component studied here has purely prograde spectral pike, Fig. 4. The CW pike is split, having the side lobes representing the amplitude modulations. But when treating the exact frequency $f_c$ at every moment, we should use instantaneous amplitude and phase. Being purely prograde at the 150-years time interval, the extracted CW being approximated at the short time interval shows ellipticity. This happens because the CW pole trajectory slightly twists and untwists every $\sim$ 20-years, which produces the changes of the approximation ellipse semiaxes ratio and brings the "virtual" retrograde component to life. We called it "escargot effect" after the nice dish of French cousin (the snails with filling). In result, the LS-estimation of the prograde and retrograde complex amplitudes $C_m^+$, $C_m^-$ of the filtered CW from Fig. 1 in the 12-years sliding window has shown the presence of non-zero $C_m^-$, with module $|C_m^-|$ oscillating with $\sim$ 20-years period. Our attention was attracted to this effect, because the change of $|C_m^-|$ reminds the shape of the CW excitation envelope. Further studies are needed to explain how "escargot effect" and the changes of ellipticity of the approximation ellipse are related to the difference between $\sin(\cdot)$ and $\cos(\cdot)$ in eq. (4). Such study can be done through the consideration of the sliding least-squares filter's transfer function [5].

## 4. Conclusion

In this short paper we explain why, if the Chandler wobble envelope has 40-years modulations, 20-years amplitude changes appear in its excitation. We do not discuss physical causes hers, our goal is to prove this effect mathematically and to confirm it through simple modelling. We think that the Chandler wobble amplitude transition from positive to negative amplitudes can represent the 1930-th phase jump, which could repeat in 2010th epoch of CW amplitude minima.

The mathematical "escargot effect" observed while studying the CW amplitude in the sliding window is reported. The purely prograde long-term signal, if modulated, produce non-zero retrograde amplitude at instantaneous consideration. In case of CW, the behaviour of this "virtual" retrograde amplitude repeats the behaviour of the excitation envelope.

*Acknowledgments*

This work is supported by the Chinese "111 Plan", NRU HSE support, RFBR grant No 16-05-00753, and monthly position at Paris observatory for the first author.

## References

[1] Bizouard C., F. Remus, S.B. Lambert, L. Seoane and D. Gambis, 2011, *The Earth's variable Chandler wobble,* (Astronomy & Astrophysics, vol. 526, A106 )
[2] Brzezinski A., Dobslaw H., Dill R. et al., 2012, *Geophysical Excitation of the Chandler Wobble Revisited,* (Geod. for Planet Earth Int. Ass. of Geodesy Symposia, 136(3), 499-505)
[3] Chao B.F. and Wei-Yung Chung, 2012, *Amplitude and phase variations of Earth's Chandler wobble under continual excitation,* ( J. of Geodynamics, Vol. 62, p. 35-39)
[4] Gross R., 2000, *The excitation of the Chandler wobble,* (Geophys. Res. Lett., 27(15), 2329-2332).
[5] Gubanov V.S., 2009, *Dynamics of the Earth core from VLBI observations,* (Astronomy letters. Vol. 35, Iss. 4, pp 270277).
[6] Lambeck K., 1980, *The Earth's Variable Rotation; Geophysical Causes and Consequences,* Cambridge University Press
[7] Liao D.C., Zhou Y.H., Liao X.H., 2007, *Comparison of Wind Contributions to Chandler Wobble Excitation,* (Chinese Astronomy and Astrophysics, 31, p. 57-65)
[8] Wang G, Liu L., Su X., Liang X., Yan H., Tu Y., Li Z., Li W, 2016, *Variable Chandler and Annual Wobbles in Earths Polar Motion During 1900 2015,* Surveys in Geophysics, Vol 37, DOI: 10.1007/s10712-016-9384-0
[9] Munk W., G. MacDonald, 1960, *The rotation of the Earth,* (Cambridge Univ. Press)
[10] Miller N.O., 2011, *Chandler Wobble in Variations of the Pulkovo Latitude for 170 Years* (Solar System Research, Vol. 45, No. 4, p. 342-353)
[11] Nastula J., A. Korsun, B. Kolaczek, W Kosek, W Hozakowski, 1993, *Variations of the Chandler and annual wobbles of polar motion in 1846-1988 and their prediction* (Manuscripta geodaetica, No 18 p. 131-135)
[12] Panteleev V.L., Chesnokova T.S., 2011, *Problem deconvolution in inertial gravimetry,* (Moscow University Physics Bulletin, Vol. 66, N 1)
[13] Zotov L., C. Bizourad, 2012, *On modulations of the Chandler wobble excitation* (Journal of Geodynamics, special issue "Earth rotation", Vol. 62, p. 30-34), DOI: 10.1016/j.jog.2012.03.010

# Science and ethics meet: a mathematical view on one kind of violation of publication ethics

**Taisiya S Shinyaeva, Yuri Yu Tarasevich**

Astrakhan State University, 20A Tatishchev Street, Astrakhan, 414056, Russia

E-mail: tarasevich@asu.edu.ru

**Abstract.** When a person who did not make a significant intellectual contribution to a published research is included into the co-author list, the person is called gift or guest author depending on the reason why the person has been added to the co-authors. Essential deviation of properties of a particular co-author network from typical values may evidenced that the network is artificial. Using network analysis, we have performed an attempt to characterize a typical co-author network. We performed analysis of the co-author networks using references in the thesis on Physics and Mathematics, Economics defended from 2012 to 2017 and planned to be defended in 2017 and 2018 in Russia. Properties of the co-author networks are expected to be a reference sample in future research.

## 1. Introduction

During last decade, a problem of malpractice and misconduct in scholarly publications attracts more and more attention from the academic community [1, 2]. The problem has many manifestations, including, but not limited to, duplicate publications and plagiarism [1], predatory journals [3], and closely related to them authors, diploma, and degree mills [4, 5]. Incorrect list of co-authors (i.e., gift, guest, ghost authors) is one of the possible kinds of malpractice [6]. Gift and guest authors are persons who did not make a significant intellectual contribution to a published research but included into the co-author list because of different reasons. During the last couple of years, "author contribution statement" became a mandatory section of articles in many journals. This fact evidenced that improper co-author lists are often suspected. Unfortunately, explicit "author contribution statement" is hardly able to stop Mr. Fraud. Academic community badly needs a tool to detect and prevent malpractice. Scientific methods are successfully used for plagiarism detection and analysis of publications in predatory journals [7], nevertheless, possibilities to use scientific methods for "gift author" detection are not clear yet [8]. Application of network analysis [9, 10, 11, 12, 13, 14, 15] to co-author networks looks very promising as a potential tool for improper author lists detection. Social network analysis based techniques are used for anomaly detection in different types of networks [16, 17, 18]. Method of finding transitive triads was used to identify the degree mill — the dissertational councils involved in the production of plagiarism, which allow successful protection of substandard works [19].

In this conference paper, we present some new attempts applying network analysis to indicate improper co-author lists. Our research is based on an assumption that there may exist some stable groups or persons which permanently include so-called "gift authors" into co-author lists. If this assumption is correct then such the artificially enlarged co-author networks presumptively have some anomalies in their properties. https://www.dissernet.org/ evidenced that malpractice and misconduct in Russia in Physics and Mathematics is minimal in compare with other sciences, humanities, and arts. The greatest number of violations falls on the economics, jurisprudence and pedagogical sciences. This fact allows considering co-author networks in physical science as a potential reference sample.

## 2. Materials

In our study, we investigated the theses defended from 2012 to 2017 and planned to be defended in 2017 and 2018 in Russia relating to the two branches of science: physics and mathematics, economics. Their open access full texts can be found at the official portal of the Higher Attestation Commission of the Ministry of Education and Science of Russian Federation http://vak.ed.gov.ru/dis-list. Information about 11976 doctoral and candidate dissertations on examined fields of sciences is presented on this site.

Each record of the thesis at the portal contains the following information

- thesis' ID,
- the type of dissertation (candidate/PhD or doctoral/Habilitation),
- full name of the applicant,
- title of the dissertation,
- scientific specialty,
- branch of science,
- ID of the dissertational council,
- name of the organization in which the defense took place,
- thesis (link to file),
- date of defense of the thesis.

The records are uniform that allows automatic processing with very small amount of improper identification. Each record has to be accompanied with full text of the thesis as a file in pdf format. We were able to process only 7093 files out of 11976 records, because in some cases the files were improper encoded, absent or secured (Table 1).

Table 1: Result of the automatic search of theses. $N$ — number of theses on the http://vak.ed.gov.ru/dis-list from 2012 to 2018

| Subject area | $N$ | Improper encoded files | Absent files | Secured files |
|---|---|---|---|---|
| Physics and mathematics | 6199 | 1500 (24.2%) | 823 (13.3%) | 105 (1.7%) |
| Economics | 5777 | 232 (4%) | 2200 (38.1%) | 23 (0.4%) |

An analysis of obtained files was carried out to obtain the necessary additional information

- the organization in which the thesis was fulfilled,
- information about supervisors and opponents:
  - full name,
  - scientific degree and title,
  - affiliation,
- list of journal articles published on the topic of the dissertation.

Automatic extraction of metadata from these files are obstructed by lack of uniformity in presentation of information about supervisors and opponents, as well as in the preparation of lists of publications on the topic of the dissertation. Thus, we were extracted necessary information from slightly less than 90% of all files. Information about found defense participants (persons who were involved in the defense of the thesis: the applicant, the supervisor or the opponent), main publications and authors is presented in Table 2.

Table 2: Result of files processing. $N$ — number of files, $P$ — number of entries about defenses participants, $Auth$ — total number of entries about publications authors, $Pub/D$ — average number of publications per thesis, $Auth/Pub$ — average number of authors per publication

| Subject area | $N$ | $P$ | $Auth$ | $Pub/D$ | $Auth/Pub$ |
|---|---|---|---|---|---|
| Physics and math. | 3771 | 13600 | 88338 | 8.3 | 3.47 |
| Economics | 3322 | 11730 | 38362 | 9.4 | 1.45 |

It was necessary to determine the specific person for each such record because one person can participate in several defenses and to act both as an applicant and as an opponent or supervisor. Due to the fact that 98.4% of all participant records contain full names, and 89.7% of records contain information on the degree, rank or position of the defense participant, only 0.4% of the participants could not be identified (linked to the specific person). Searching for people in the list of authors of publications is a non-trivial task. The complexity of identification is primarily due to incomplete names (names and initials are indicated), with various versions of transliteration, misprints. Automatical linking assumes errors. Therefore, when there were ambiguous situations, no binding was performed. Thus, about 3% of the authors could not be identified.

The list of persons who took part in defenses with incorrect borrowings is available on the "Disseropedy" website (http://rosvuz.dissernet.org/person) and include 2647 records. 434 matches with people from our database were found. All coincidences refer to the economics.

## 3. Methods

We built network of the defenses participants and network of co-authors of the main publications on the topic of the dissertation. Depth-first search algorithm [20] has been applied to the networks. We built egonets (the 1-step neighborhood of a node) and studied the following characteristics of obtained ego networks.

- $N$ — number of nodes,
- $E$ — number of edges,
- $AvD$ — average degree,
- $d$ — diameter,
- $AvPath$ — average path length,
- $D$ — density,
- $C$ — average clustering coefficient [21],
- $T$ — total triangles,
- $N_T$ — number of triangles related to node,
- $r$ — degree assortativity of graph [22]. Assortativity measures the similarity of connections in the graph with respect to the node degree.

## 4. Results

### *4.1. Network of defense participants*

Network of defense participants contains 15434 nodes and 29853 edges, including 1299 weakly connected components. Two giant connected components contains 10231 nodes and 23067 edges (figure 1), their characteristics are presented in table 3.

Table 3: $N$ — number of nodes, $E$ — number of edges, $AvD$ — average degree, $d$ — diameter, $AvPath$ — average path length, $D$ — density, $C$ — average clustering coefficient, $T$ — total triangles, $r$ — degree assortativity of graph.

| Sciences (Subject) area | $N$ | $E$ | $AvD$ | $d$ | $AvPath$ | $D$ | $C$ | $T$ | $r$ |
|---|---|---|---|---|---|---|---|---|---|
| Physics and mathematics | 3710 | 7734 | 4.169 | 38 | 12.033 | 0.001 | 0.804 | 5571 | -0.05 |
| Economics | 6521 | 15333 | 4.703 | 21 | 7.314 | 0.001 | 0.808 | 11091 | -0.03 |

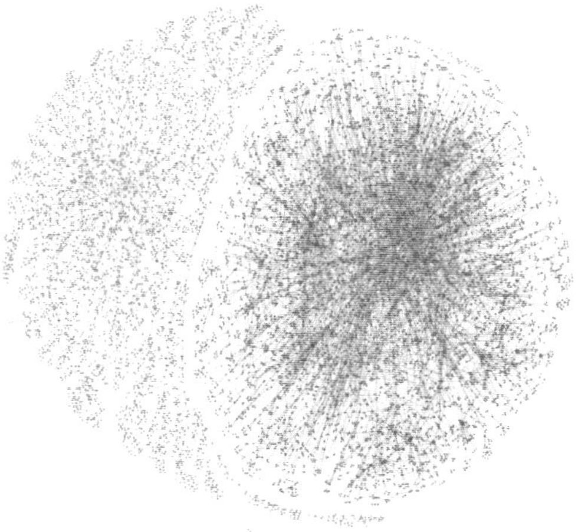

Figure 1: Two giant connected components of defense participants network. On the left side, there is component related to physics and mathematics (green color), on the right side — to economics (blue color). The participants of the defenses included in the "Disseropedy" (http://rosvuz.dissernet.org/person) are marked in red.

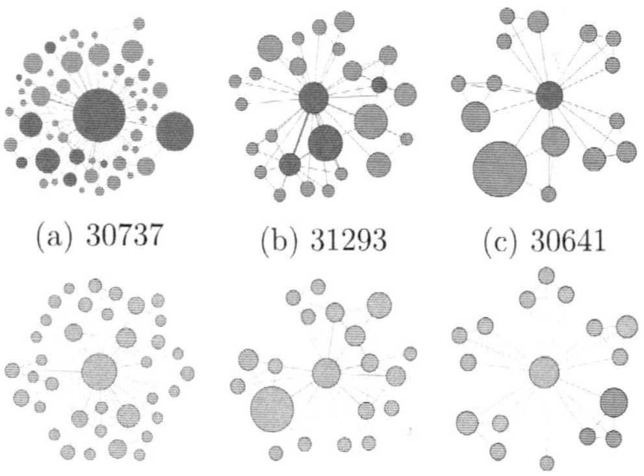

(a) 30737 (b) 31293 (c) 30641
(d) 530 (e) 3164 (f) 3169

Figure 2: Some ego networks in defense participants network. Green color indicates participants related to the physics and mathematics, blue — to the economics, red color indicates participants of the defenses included in the "Disseropedy".

Some ego networks in defense participants network are shown in the figure 2, their characteristics are presented in table 4.

### *4.2. Co-author networks*

Co-author network contains 31195 nodes and 109333 edges, including 2341 weakly connected components. 58.3% of nodes form giant connected component of co-author network (figure 3). Some ego networks are shown in the figure 4, their characteristics are presented in the table 5.

Table 4: Charateristics of ego networks. $N_T$ — number of triangles related to node, $N$ — number of nodes, $E$ — number of edges, $AvD$ — average degree, $D$ — density, $C$ — average clustering coefficient, $T$ — total triangles, $r$ — degree assortativity of graph.

| Node Label | $N_T$ | $N$ | $E$ | $AvD$ | $D$ | $C$ | $T$ | $r$ |
|---|---|---|---|---|---|---|---|---|
| 30737 | 70 | 54 | 123 | 4.556 | 0.086 | 0.875 | 97 | -0.297 |
| 31293 | 35 | 26 | 60 | 4.615 | 0.185 | 0.877 | 50 | -0.324 |
| 30641 | 15 | 16 | 30 | 3.75 | 0.25 | 0.946 | 20 | -0.333 |
| 530 | 41 | 37 | 78 | 4.216 | 0.117 | 0.932 | 58 | -0.313 |
| 3164 | 31 | 20 | 50 | 5 | 0.263 | 0.861 | 49 | -0.296 |
| 3169 | 15 | 17 | 31 | 3.647 | 0.228 | 0.945 | 20 | -0.364 |

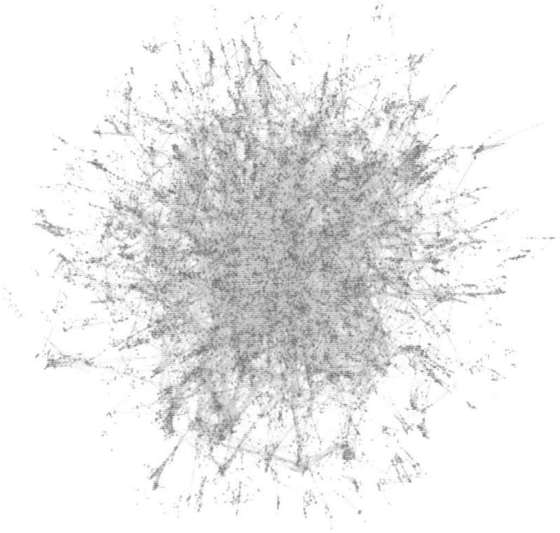

Figure 3: The giant connected component of co-author network contains 18173 nodes and 88247 edges.

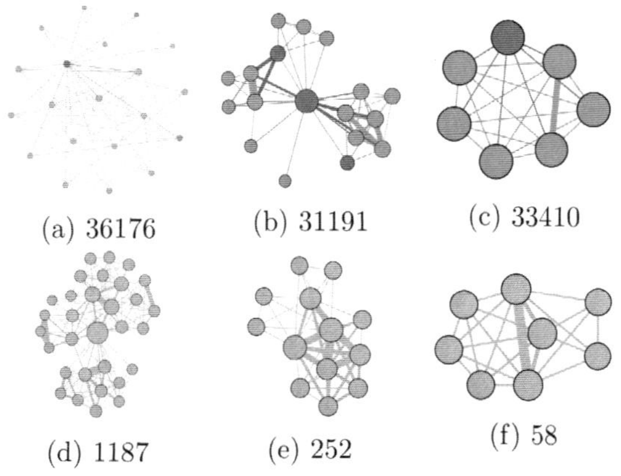


Figure 4: Ego networks. Green color indicates participants related to the physics and mathematics, blue — to the economics, red color indicates participants of the defenses included in the "Disseropedy".

Table 5: Charateristics of ego networks. $N_T$ — number of triangles related to node, $N$ — number of nodes, $E$ — number of edges, $AvD$ — average degree, $D$ — density, $C$ — average clustering coefficient, $T$ — total triangles, $r$ — degree assortativity of graph.

| Node Label | $N_T$ | $N$ | $E$ | $AvD$ | $D$ | $C$ | $T$ | $r$ |
|---|---|---|---|---|---|---|---|---|
| 36176 | 114 | 20 | 133 | 13.3 | 0.7 | 0.921 | 521 | 0.005 |
| 31191 | 32 | 18 | 49 | 5.444 | 0.32 | 0.861 | 62 | -0.307 |
| 33410 | 15 | 7 | 21 | 6 | 1 | 1 | 35 | 0 |
| 1187 | 114 | 31 | 144 | 9.29 | 0.31 | 0.847 | 316 | -0.198 |
| 252 | 32 | 13 | 44 | 6.769 | 0.564 | 0.854 | 69 | -0.396 |
| 58 | 15 | 8 | 22 | 5.5 | 0.786 | 0.893 | 29 | -0.369 |

## 5. Conclusion

In our research, we have analysed some open access information presented at the official portal of the Higher Attestation Commission of the Ministry of Education and Science of Russian Federation http://vak.ed.gov.ru/dis-list. The information allows building co-author networks and networks as applicant–supervisor–opponent. Average clustering coefficient, graph density, and average degree have been calculated for particular clusters. We suppose that such the quantities may be used for identification of suspicious co-author networks which presumably are artificially built and may contain gift and/or guest authors.

Research is in progress.

## Acknowledgments

The reported study was funded by RFBR according to the research project No. 6-33-00006.

Author contribution statement: YYT proposed the principal idea of the research, TSS performed data processing, network and statistical analysis, both co-authors took a part in discussion and interpretation of the results as well in the manuscript preparation.

## References

[1] Fang F C, Steen R G and Casadevall A 2012 *PNAS* **109** 17028–17033 ISSN 1091-6490
[2] Gasparyan A Y, Yessirkepov M, Diyanova S N and Kitas G D 2015 *J. Korean Med. Sci.* **30** 1010 ISSN 1011-8934
[3] Beall J 2012 *Nature* **489** 179–179 ISSN 0028-0836
[4] Sorokowski P, Kulczycki E, Sorokowska A and Pisanski K 2017 *Nature* **543** 481–483 ISSN 0028-0836
[5] Angulo A 2016 *Diploma mills: How for-profit colleges stiff students, taxpayers, and the American dream* (Baltimore: Johns Hopkins University Press) ISBN 978-142142008-0
[6] Wislar J S, Flanagin A, Fontanarosa P B and DeAngelis C D 2011 *BMJ* **343** ISSN 0959-8138
[7] Xia J, Harmon J L, Connolly K G, Donnelly R M, Anderson M R and Howard H A 2015 *J. Assoc. Inf. Sci. Tech.* **66** 1406–1417 ISSN 2330-1643
[8] Shinyaeva T S and Tarasevich Y Y 2017 *Procedia Comput. Sci.* **106** 3–10 ISSN 1877-0509
[9] Yose J, Kenna R, MacCarron P, Platini T and Tonra J 2016 *Advs. Complex Syst.* **19** 1650008 ISSN 0219-5259
[10] Kenna R and Mac Carron P 2016 *Phys. World* **29** 22 ISSN 2058-7058
[11] Kenna R and Mac Carron P 2016 *J. Phys.: Conf. Ser.* **681** 012002 ISSN 1742-6596
[12] Sarkanych P, Holovatch Y, Kenna R and Mac Carron P 2016 *J. Phys. Studies* **20** 4801 ISSN 2310-0052
[13] Mac Carron P and Kenna R 2013 *Significance* **10** 12–17 ISSN 1740-9713
[14] Mac Carron P and Kenna R 2013 *Eur. Phys. J. B* **86** 407 ISSN 1434-6036
[15] Mac Carron P and Kenna R 2012 *EPL (Europhysics Letters)* **99** 28002 ISSN 1286-4854
[16] Akoglu L, Tong H and Koutra D 2015 *Data Mining and Knowledge Discovery* **29** 626–688 ISSN 1573-756X
[17] Vengertsev D and Thakkar H 2015 Anomaly detection in graph: Unsupervised learning, graph-based features and deep architecture Tech. rep. Department of Computer Science, Stanford University
[18] Kaur R and Singh S 2016 *Egyptian Informatics Journal* **17** 199 – 216 ISSN 1110-8665
[19] Safonova M A and Sokolov M M 2016 *Partija Vernadskogo. VVIPartija* In russian URL `http://vviparti.esrae.ru/2-23`
[20] Tarjan R 1972 *SIAM J. Comput.* **1** 146–160 ISSN 0097-5397
[21] Watts D J and Strogatz S H 1998 *Nature* **393** 440–442 ISSN 0028-0836
[22] Newman M E J 2003 *Phys. Rev. E* **67**(2) 026126 ISSN 2470-0053 (online), 2470-0045 (print)

IOP Conf. Series: Journal of Physics: Conf. Series **955** (2017) 012035 doi:10.1088/1742-6596/955/1/012035

# How the methods of natural sciences can help in the studies of ethnically mixed families?

**E Soroko**
Institute of Demography at the National Research University Higher School of Economics
20 Myasnitskaya str. 101000 Moscow Russia
Email: esoroko@hse.ru

**Abstract**. Statistical physics is the branch that uses different mathematical methods in solving not only physical problems. The field of application may be the interdisciplinary studies of many social phenomena. The reason is that they have a stochastic nature. The aim of the paper is to display the opportunities of using the methods of natural sciences in the social sciences. The example is suggested of the research of ethnically mixed families. These are the marital couples where a husband and a wife consider themselves as belonging to different ethnicities. It was demonstrated that application of the reasons used in the kinetic theory helps us to introduce new measure that describes mutual attitudes for a specific combination of ethnicities. The idea of this measure calculation is quite simple. We directly relate the number of marriages established from the reasons of full randomness of collisions of "particles" (persons) and their connection irrespective to their type, and the phenomenology – the actual number of families for a given combination of husband's and wife's ethnicity observed form the population censuses. What we mean by "collision" is any form of personal or social interaction. This measure may be called inter-ethnic propensity, or its inverse value as a mutual inter-ethnic distance.
It was shown that in such multiethnic country like Russia both measures cannot be estimated as the good ones. However this does not mean that the measures introduced are the wrong ones in principle. Simply before their calculation we require to perform co-called "geographical" decomposition that explicitly takes into account the fact and the extent of territorial distribution of population of all the ethnicities in this country by regions. In terms of kinetic approach for gases it may have the analogy of various density of different particles by the volume they are placed in, that is required at consideration of their physical properties.

## 1. Setting up a problem

Topic of the paper is an interesting and promising interdisciplinary research in demography, sociology, statistics, and ethnography on ethnically mixed families. These are the marital couples where a husband and a wife consider themselves as belonging to different ethnicities. Unfortunately, the standard methods of demographic analysis such as age-sex-specific profiling or life table calculation are not applicable here. Thus it is a rather difficult and curious question "Is 75060 a great deal or a negligible number of Bashkir-Tatar and Tatar-Bashkir families?" (according to the 2010 population census). We cannot answer it simply using some set of appropriate indicators. What are the main reasons? The first one is peculiarity of marriage market formation. On the one hand it is appended by unmarried men and women depending on their age structure and intensity, on the other hand the numbers of husbands and wives who contract their marriage must correspond to each other. This marriage balance cannot be obtained via usual age-sex indicators. The second reason is that mixed marriage formation is influenced by multiple factors of inter-ethnic interaction. We do not have yet any available indicators for measuring it.

## 2. Aim of the paper and analogies used

The aim of the paper is to display the opportunities of using the methods of natural sciences in the social sciences. The following analogies may be suggested for better understanding of models for marriage market of mixed marriages.

Table 1. Some analogies in the natural and social sciences used

| Natural sciences | Social sciences |
|---|---|
| Gas kinetic theory | Market of mixed marriages |
| Particles (molecules) | Bridegrooms, brides (husbands, wives) |
| Collision | Meeting, conversation, acquaintance, joint work, recreation, education, tourism, journey, sports, etc. In sociology: participation in different small groups |
| Quantum emission, chemical reaction | Contracting a marriage |
| Partial pressure of gas A | Proportion of ethnicity A |

These analogies give us the opportunity to introduce a set of indicators for estimating the frequency and proportion of the events on the marriage market.

## 3. How the first approximation may be obtained?

At this step all the absolute numbers and the indicators refer to the country as a whole. Let us introduce the following variables.

$M_i$ – total number of men of ethnicity *I* in a mixed marriage,

$F_j$ – total number of women of ethnicity *j* in a mixed marriage,

$F$ – total number of women in a mixed marriage (in the country, equals to sum of $F_j$),

$N_{ij}$ – number of marital ouples with husband of ethnicity *I* and wife of ethnicity *j* (actual, according to the census),

$N_{ij}^R$ – number of marital couples with husband of ethnicity *I* and wife of ethnicity *j* (estimated)

Assuming that formation of marital couple (*I,j*) does not depend on the combination of ethnicities this number may be estimated as

$$N_{ij}^R = M_i * \frac{F_j}{F}, \tag{1}$$

That is formed according to the principle of random equiprobable combination of given ethnicities. From the point of view of kinetic equations of demographic dynamics this assumption was called by Orlov and Suslin "uniform assimilation when there are no preferences at entering a mixed marriage" [1]. It is a result of applying the analogy of random collision of particles in a mixture of several gases. In this case formula (1) may be also interpreted in this approximation as a result of influence of partial pressure of female particles – future spouses of type *j* on male particles of type *i*.

Apparently it is not the fact in the general case. That is the number of marital couples depends on this ethnic combination (*I,j*). For some combinations we observe more mixed marriages and for the other ones – less compared with "uniform assimilation". Let us denote this dependence by multiplier $E_{ij}$. Then the real number of mixed families (*I,j*) may be represented in the form

$$N_{ij} = E_{ij} * N_{ij}^{R} \qquad (2)$$

This relation is used for estimating the inter-ethnic preferences $E_{ij}$ :

$$E_{ij} = \frac{N_{ij}}{N_{ij}^{R}} \qquad (3)$$

The approach in this approximation was suggested for the first time in the paper [2] (see in more details).

## 4. How the matrix of inter-ethnic couples looks like?

Calculations of inter-ethnic indicators are based on the matrix of marriages for all combinations of ethnicities $N_{ij}$ obtained as a demographic table from the census micro-data. New information technology of such census data processing became possible due to the special system developed by Rosstat [3] and described in [4]. From the reasons of visibility a rather restricted set of 11 ethnicities is considered here. In order to provide the demographic balance an additional ethnic group is required. This 12-th conditional ethnic group is suggested to call "the rest". It includes the set of ethnicities not included in the list of selected ones.

Table 2. Matrix of ethnic combinations in marital couples. Russia, 2010, 11 selected ethnicities, number of couples.

| Ethnicity of husband | Ethnicity of wife | | | | | | | | | | | |
|---|---|---|---|---|---|---|---|---|---|---|---|---|
| | Russians | Tatars | Ukrai-nians | Bashkirs | Chu-vash | Che-chens | Arme-nians | Mordo-vians | Azer-baijani | Kazakhs | Udmurts | The rest |
| **Russians** | 12255162 | 158277 | 172145 | 27963 | 49315 | 725 | 14879 | 39267 | 3608 | 11491 | 28291 | 262802 |
| **Tatars** | 170031 | 521326 | 4402 | 40381 | 5477 | 109 | 294 | 1646 | 292 | 3178 | 3026 | 12404 |
| **Ukrainians** | 212218 | 5408 | 58808 | 1205 | 1716 | 25 | 548 | 1154 | 118 | 537 | 597 | 14018 |
| **Bashkirs** | 26484 | 34679 | 839 | 156002 | 973 | 7 | 35 | 198 | 49 | 946 | 440 | 2544 |
| **Chuvash** | 47401 | 5126 | 1354 | 1277 | 148189 | 3 | 56 | 1632 | 20 | 170 | 315 | 3014 |
| **Chechens** | 4443 | 245 | 164 | 45 | 16 | 210093 | 42 | 14 | 37 | 130 | 13 | 2303 |
| **Armenians** | 52546 | 1042 | 1557 | 201 | 196 | 22 | 148271 | 147 | 427 | 103 | 106 | 5035 |
| **Mordovians** | 46975 | 2026 | 1111 | 286 | 2130 | 11 | 68 | 55521 | 16 | 123 | 130 | 1276 |
| **Azerbaijani** | 29395 | 2134 | 1003 | 394 | 256 | 53 | 662 | 135 | 80870 | 371 | 185 | 6032 |
| **Kazakhs** | 16055 | 3770 | 552 | 1296 | 245 | 39 | 41 | 111 | 40 | 77389 | 72 | 2531 |
| **Udmurts** | 25391 | 2379 | 400 | 428 | 242 | 3 | 18 | 66 | 11 | 47 | 51477 | 1336 |
| **The rest** | 408629 | 18422 | 15787 | 4728 | 4174 | 3284 | 4132 | 1596 | 3549 | 3824 | 2113 | 1315065 |

In this case the matrix of all ethnic combinations of spouses represents a following 12*12 matrix. It contains 15.08 million mono-ethnic marital couples on the diagonal and 2.06 million ethnically mixed couples outside it. Thus, the last ones comprise 12 per cent of all couples and the diagonal – 88 per cent in the Russian Federation in 2010.

Following to the model (1-3) for calculation inter-ethnic preferences we may estimate the distances to the Russians. The results are placed in the following list (ethnicity – distance): Ukrainian

– 0.66, Mordovians – 0.68, Udmurts – 0.71, Armenian – 0.74, Chuvash – 0.75, Tatars – 0.84, Kazakh – 0.97, Azerbaijani – 1.03, Bashkirs – 1.55, Chechens – 2.03 (ethnicities are listed in ascending order of distance).

This matrix may be used for one more new measure in order to describe a propensity to form ethnically mixed marriage with a spouse of any different ethnicity. Numerically it is calculated as a share of ethnically mixed families of a given ethnicity among all the families of this ethnicity. Similar to chemistry, it may be called "valency".

## 5. The second approximation

More detailed analysis of the results obtained in the first approximation for such multiethnic country like Russia shows that measures introduced cannot be estimated for all the peoples as the good and adequate ones. The reason is a significant inhomogeneity of ethnicity distribution by territory of the country. Some of such peoples have their own national republics, some do not have such administrative-territorial organization but reside in a few number of regions.

Two extreme examples: 98.7 per cent of the Tuvunians reside in their national republic, and 41.2 per cent of the Mordovians. Accordingly, the strong variation for the share of mixed marriages among them is observed. It is 1.9 per cent for the Tuvunians and 49.4 for the Mordovians.

At the second step we require to take into account that ethnic composition of Russia has several significant features that influence its regional peculiarities and the level of inter-ethnic interaction. In addition to the indicators introduced above, some new variables are used for regional population distribution:

$M_{iq}$ – number of men of ethnicity *i*, in a mixed marriage in a region *q*,

$F_q$ – total number of women in a mixed marriage in region *q*,

$F_{jq}$ – total number of women of ethnicity *j* in a mixed marriage in region *q*,

$N_{ijq}$ – actual number of marital couples with husband of ethnicity *i* and wife of ethnicity *j* in region *q*,

$N^R_{ijq}$ – number of marital couples with husband of ethnicity *i* and wife of ethnicity *j* in region *q* (estimate),

$N^{RQ}_{ij}$ – number of marital couples with husband of ethnicity *i* and wife of ethnicity *j* in the country (estimate based on the regional estimates)

Similarly to (1) in case of full absence of dependence from the specific combination (*i,j*) a hypothetic number of mixed marriages in region *q* may be estimated as:

$$N^R_{ijq} = M_{iq} * \frac{F_{jq}}{F_q}, \quad (4)$$

This may be interpreted according to the analogy from the gas kinetics as a result of the effect of local partial pressure of *j* type females on *i* type males. In this case the total number of these marriages in Russia may be obtained as a sum of these numbers by regions:

$$N^{RQ}_{ij} = \sum_q N^R_{ijq}, \quad (5)$$

In the general case the estimates calculated from (1) and (4-5) will differ due to significant differences in displacement of the ethnicities by regions of the country.

In order to describe qualitatively the "geographical" factor of displacement of marital couples of ethnicities (*i,j*) by regions is introduced:

$$G_{ij} = \frac{N_{ij}^{RQ}}{N_{ij}^{R}}, \tag{6}$$

This variable has the following properties: the value «1» takes place if grooms are uniformly distributed by regions; similarly if brides are uniformly distributed by regions; the values at the diagonal ($i=j$) for monoethnic couples are much more compared with 1 for ethnicities the major part of which resides in the national republic. For instance, according to the 2010 population census this value exceeds 100 for Kabardians, Ossetians, Yakuts, Ingush, Komi, Tuvinians, Carachai.

Similarly to the first approximation we introduce the "new" inter-ethnic preferences $E_{ij}^{N}$, free from the factor of population displacement by regions, taken into account as a multiplicative factor:

$$N_{ij} = E_{ij}^{N} * G_{ij} * N_{ij}^{R} \tag{7}$$

Interpretation of this formula is the following. In the absence of influence of inter-ethnic preferences and uniform settlement of ethnicities by territory of the country the number of mixed marriages coincides with the estimate calculated (the first two multipliers are equal to 1). In the absence of influence of inter-ethnic preferences the number of mixed marriages will be distorted by 'geographical' factor of ethnicities settlement compared with the estimate (the first multiplier is equal to 1). If there is no influence of settlement nonuniformity for ethnicities involved (the second multiplier equals to 1) the actual number of mixed marriages for a given combination of ethnicities will differ from the estimated one only due to the influence of inter-ethnic preferences upon these mixed marriages. In the general case "ethnical" and "geographical" factors act combined multiplicatively. Following the relation (7) the estimates of enter-ethnic preferences may be calculated basing on the number of mixed marriages from the population census and the calculated estimates of "geographical" factor from the formula:

$$E_{ij}^{N} = \frac{N_{ij}}{G_{ij} * N_{ij}^{R}} \tag{8}$$

## 6. "Geographical" decomposition: illustration

As an example for application of the geographical decomposition method for data from the population census the estimated inter-ethnic distances between the Russians and the most common peoples in the Russian Federation may be suggested as illustration in the following table.

Table 3. 32 ethnicities in ascending order of inter-ethnic distance to the Russians. Russia, 2010.

| | | | | | | | |
|---|---|---|---|---|---|---|---|
| 1 | Ukrainian | 9 | Mari | 17 | Bashkir | 25 | Avar |
| 2 | Mordovian | 10 | Moldavians | 18 | Chechen | 26 | Uzbek |
| 3 | Chuvash | 11 | Armenian | 19 | Ingush | 27 | Kabardian |
| 4 | German | 12 | Lezgian | 20 | Azerbaijanian | 28 | Tadjik |
| 5 | Belorussian | 13 | Georgian | 21 | Kumyk | 29 | Kazakh |
| 6 | Udmurt | 14 | Ossetian | 22 | Tabasaran | 30 | Buryat |
| 7 | Komi | 15 | Laks | 23 | Dargin | 31 | Karachai |
| 8 | Tatar | 16 | Gipsy | 24 | Yakut | 32 | Tuvinian |

## 7. Problems not solved yet

As a new problem statement in ethnography not solved yet an analogy with thermodynamics is suggested for analysis of ethnical population structure and its evolution. Some questions in this field are: Is the entropy actually growing over time as applied to the composition of population by ethnicities? May the dynamics of the population of the USA considered as the well-known "melting pot" for ethnicities be interpreted in the way similar to the second law of thermodynamics? Why this law is not valid in the general case for population ethnic structure at the level of city or country?

## 8. Conclusion

Using of methods from natural sciences lets us produce much more clear explanation, more simple understanding, modeling, interpretation of the processes under consideration. Models for estimating the inter-ethnic distances give a set of new quantitative measures in ethnography and demography for important processes at the borders of ethnicities – mixed marriages.

## 9. References

[1] Orlov Yu N and Suslin V M 2003 Kinetic equations for some demographic models *Math. Model.* **15 No 3**, pp 43-54 (in Russian)
[2] Soroko E 2014 Ethnically mixed married couples in the Russian Federation *Dem Rev* **4** pp 96-123, http://demreview.hse.ru (in Russian)
[3] Rosstat 2012 The 2002 and 2010 All-Russian population censuses. SuperWEB2. URL: http://std.gmcrosstata.ru/webapi/jsf/tableView/customiseTable.xhtml (in Russian)
[4] Soroko E and Voronin V 2014 New information technology in data processing of population census results *Innovative Information Technologies: Materials of Int sci-prac conf* **Part 3** ed S U Uvaysov pp 430-434

# Simulation System for Making Political and Macroeconomical Decisions and Its Development

**A A Vnukov**[1], **A.E. Blinov**[1]

[1]National research University Higher school of Economics, 20 Myasnitskaya ulitsa, Moscow 101000 Russia

**Abstract.** Object of this research are macroeconomic indicators, which are important to descript economic situation in a country. Purpose of this work is to identify these indicators and to analyze how the state can affect these figures with available instruments. Here was constructed a model where the targets can be calculated from raw data – tools in the field of economic policy. Software code that implements all relations among the indicators and allows to analyze with high accuracy, sufficiently successful economic policies and with the help of some tools, you can achieve better results. This model can be used to forecast macroeconomic scenarios. The corresponding values of the objective (outcome) variables are set as a consequence of the configuration data of the previous period, subject to external influences and depend on the instrumental variables. The results may be useful in economical predictions. The results were successfully checked on real scenarios of Russian, European and Chinese economics. Moreover, the results can be applied in the field of education. Program is available to use as "economical game" the educational process of the University, in which you can virtually implement various macroeconomic scenarios, draw conclusions about their success.

## 1. Introduction

What are the benefits of business games as tools for modeling and analysis of economic processes? It is possible to recreate the actual events in the framework of the model very accurately, while simulation modeling is highly appropriate to conduct in cases where real experiments are impossible, because of serious costs or impossibility of backsliding and repeating the process once more. Of course, one of cases is when decisions are made on macroeconomic level. Creating dynamic models incrementally and with the ability to analyze the changes due to each specific conversion, allow to find the causes of economic problems with high accuracy. and to make preconditions for economic stability and prosperity.

As in the business economic game, you can simulate different conditions and scenarios without severe time and economic and political risks, that's why such models are very useful for application and analysis. Good scientific tool for economical game helps to make simulation in realistic conditions. The traditional mathematical modelling is complemented by modern computer technologies, and also get wide economic experience and the external environment that allows the model to reproduce reality. Moreover, economic-mathematical models are useful for understanding the principles of governance, and can help students to improve their knowledge in this field and to make researches through a variety of scenarios themselves. [2] [3] [5]

## 2. Task

Accomplish a model of the general economic development of a country, using modern programming language. Describe possible scenarios of state economy' transformation.

Analyze the model' success with specific examples. To make changes to the original mathematical General economic model of the country to the current economic situation with the aim of the study scenarios. To ensure the possibility of changing the model, the initial conditions in the software.

## 3. Data

We will use existing macroeconomic model, which was implemented in the environment of "Transform" as an economic game. Unfortunately, the aforementioned software is not compatible with modern operating systems. The objective of this work is to write code to apply the model on modern computing machines. Furthermore, the goal of this work is to determine which indicators have the most important impact on changes in the economic situation, and secondly, to analyze how it is related to possible actions by the state. The theme of affecting macroeconomic stability and the methods of achieving it, is very crucial, because each state at different stages of development (e.g., industrial society, transition to a post-industrial, developing country) is facing a key issue of economic policy – how to regulate economic life or by basic solutions, such as nationalization of enterprises or decisions, expressed numerically – for example, the change in the key interest rate. In the best case, the state respects a middle ground between the commitment to a planned economy, that is the ideology of absolute control of economic processes and production by state structures and market economy, in which the current market conditions creates everything, without exception, macro - and microeconomic situations. [1] [4] [6]

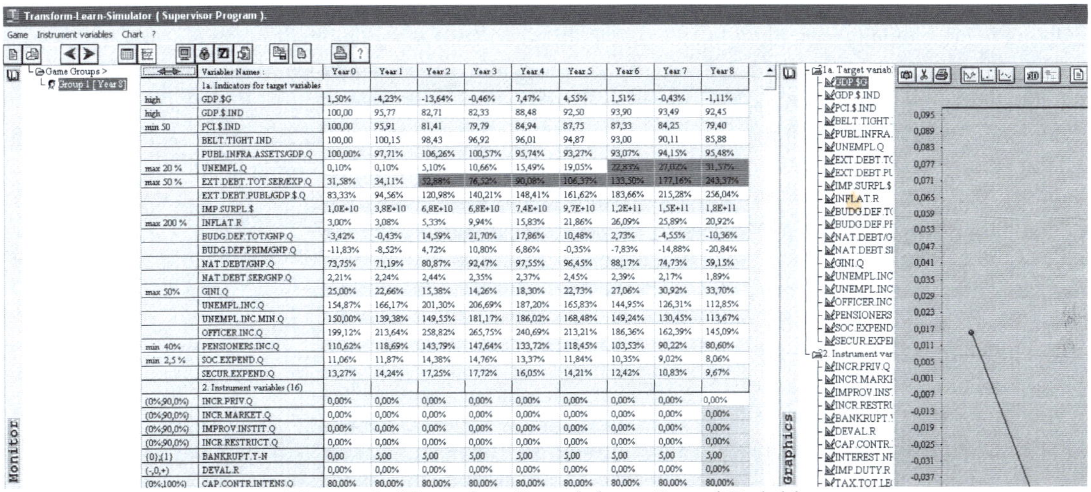


**Figure 1.** "Transform" module on Pascal Delphi.

The "Transform" module on Pascal Delphi contains 3 large opening Windows: monitor, decision-making, and graphs. In the monitor window contains a variety of group decision-making, carrying out different governance and modelling with the forecast. In the window of decision-making in the column on the left displays the restrictions and conditions in the form of upper and lower limits of recommended values for output constraints computed target variables and fields allowed values of the input conditions of instrumental variables. In the middle column for all variables visible, their short names, and at the bottom of their list is a window to the full variable pointed to by the cursor. In the right-hand group of columns displays the initial condition (year 0) and all the years of the modelling forecast, which introduced instrumental variables and the calculated target variable values. The chart window is organized as follows. In the left part of the window there is a tree with two branches to select a target and separately for choice of instrumental variables. The right most part of the window is used for graphical representation of the selected target or instrument variable.

These three Windows can be open and close their boundaries across the width of the corresponding window and display data and graphs can be dynamically changed by the movement of the cursor.

When you run the software possible to select the operating language of data: English, German and Russian that allowed you to use the program in research and international education programs

| Условия | Расшифровка переменных | Имена переменных: | Год 0 | Год1 | Год2 | Год3 |
|---|---|---|---|---|---|---|
| | | 1. Целевые переменные(21) | | | | |
| high | Темп прироста реального ВНП (Валового Национального продукта) | GDP.$.G | 1,50% | 2,14% | 0,45% | -4,69% |
| high | Индекс реального ВНП | GDP.$.IND | 100,00 | 102,14 | 102,59 | 97,78 |
| min 50 | Индекс реального дохода на душу населения | PCI.$.IND | 100,00 | 103,14 | 103,15 | 97,24 |
| max 20% | Число безработных | UNEMPL.Q | 0,10% | 0,10% | 0,09% | 0,09% |
| max 200% | Уровень инфляции | INFLAT.R | 3,00% | 15,31% | 31,15% | 161,16% |
| max 50% | Коэффициент Gini | GINI.Q | 25,00% | 24,93% | 27,53% | 31,74% |
| min 40% | Отношение дохода пенсионеров к среднему доходу | PENSIONERS.INC.Q | 110,62% | 110,87% | 101,90% | 87,37% |
| min 2,5% | Отношение социальных расходов бюджета к общим расходам бюджета | SOC.EXPEND.Q | 11,06% | 11,09% | 10,19% | 8,74% |
| | | 2.Инструментальные переменные(16) | | | | |
| (0%;90,0%) | Увеличение доли приватизированных предприятий | INCR.PRIV.Q | 0,00% | 0,00% | 0,00% | 0,00% |
| (0%;90,0%) | Увеличение доли рыночной экономики | INCR.MARKET.Q | 0,00% | 0,00% | 0,00% | 0,00% |
| (0%;90,0%) | Улучшение уровня образованности | IMPROV.INSTIT.Q | 0,00% | 0,00% | 0,00% | 0,00% |
| {0};{1} | Применение закона о банкротстве (да - нет) | BANKRUPTCY-Y-N | 0,00 | 0,00 | 0,00 | 0,00 |
| (-,0,+) | Уровень девальвации | DEVAL.R | 0,00% | 0,00% | 0,00% | 0,00% |
| (0%;100%) | Интенсивность управления выводом капитала | CAP.CONTR.INTENS.Q | 80,00% | 80,00% | 80,00% | 80,00% |
| (-,0,+) | Номинальная процентная ставка | INTEREST.NR | 3,00% | 3,00% | 3,00% | 3,00% |
| >0 | Уровень импортных пошлин | IMP.DUTY.R | 0,70 | 0,70 | 0,70 | 0,70 |
| >0 | Общий уровень налогов | TAX.TOT.R | 0,55 | 0,55 | 0,55 | 0,55 |
| (-,0,+) | Темп прироста общественных вложений | PUBL.INV.G | 6,00% | 6,00% | 6,00% | 6,00% |
| (-,0,+) | Темп прироста доходов безработных на душу населения | UNEMPL.INC.PC.G | 6,00% | 6,00% | 6,00% | 6,00% |
| (-,0,+) | Темп прироста доходов служащих на душу населения | OFFICER.INC.PC.G | 6,00% | 6,00% | 6,00% | 6,00% |
| (-,0,+) | Темп прироста доходов пенсионеров на душу населения | PENSIONER.INC.PC.G | 6,00% | 6,00% | 6,00% | 6,00% |
| (-,0,+) | Темп прироста социальных расходов | SOC.EXPEND.G | 6,00% | 6,00% | 6,00% | 6,00% |
| (-,0,+) | Темп прироста расходов на безопасность | SECUR.EXPEND.G | 6,00% | 6,00% | 6,00% | 6,00% |

**Figure 2.** “Transform” module for VBA.

The “Transform” module for VBA after running visually reflected on the screen from left to right the following information. The left column contains the restrictions and conditions respectively for the target and instrumental variables. In the next column to the right contains the decoding variables in Russian full name of each variable. In the third column are the short names of all groups of variables and constants used in this version of the software. The remaining right part of the screen is gradually filled with speakers, which are the starting initial conditions (year 0) and successive years of decision making on simulation of the development of economic scenarios for the country (year 1, year 2, etc.).

In Figure 1 and 2 shows the modules of “Transform” module on Pascal Delphi and our module for VBA. Our model consists of more than 200 variables, which we will divide into several groups:

1. Target variables. These are the indicators based on which we draw conclusions about the financial and economic welfare of the country; the values of which for the successful development of the state should remain within certain limits. Target variables help evaluate the success of modeling.

2. The instrumental variables. These are variables that we can change. In fact, it is the input data for the formulas to calculate the target variables.

3. Intermediate variables. On the basis of their calculated target parameters, in fact, is intermediate between the instrumental and target values. Using intermediate variables the user can evaluate more precisely, what is the reason for the unsuccessful simulation results, where the errors.

4. Constant development of the world market. Constants indicating the level of development of the world market. They are defined for the whole simulation period. They are used to calculate the target and instrumental parameters.

5. Constants used in various iterative algorithms for computing the intermediate and target variables. A number of variables that the user entering the instrumental variables cannot change this - constants involved in the calculation of intermediate variables and thus affect the objectives of the modelling forecast, which evaluate, for example, the elasticity, i.e. the effect of some factors on others.

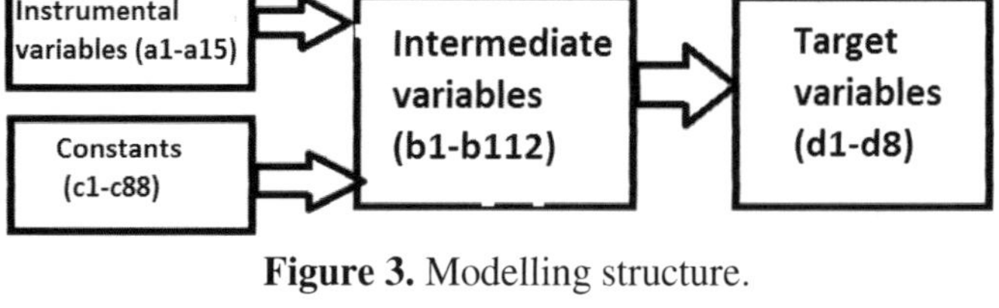


**Figure 3.** Modelling structure.

IOP Conf. Series: Journal of Physics: Conf. Series **955** (2017) 012036 doi:10.1088/1742-6596/955/1/012036

Also shown in figure 3 variables, there are variables through which the external influence on the macroeconomic system, e.g. a change in interest rates in the markets provide external loans.

## 4. Model

It is possible to consider the General and simplified mathematical model. Both models are step-by-step models, in which for step N mathematical formulas and expressions tie the values of a set of target variables with values sets target, tool and intermediate variables of step N-1 step N. At each step can be performed any number of calculations to determine values of a specific set of targeted, instrumental variables and intermediate step N, then the selected values are captured and used for calculations in step N+1.

For the management and monitoring plans are used to target and instrumental variables, and intermediate variables only for calculations. The model has the ability to monitor individual processes within the planning target and instrument variables of these processes and to monitor processes on a joint action plan to achieve the strategic goals defined by the integrated target of the calculated and measured variables and parameters. On the value of the target and instrumental variables can be specified restrictions ("corridors" change of variables), the boundaries of which are either invalid or undesirable.

The system model management and monitoring to design and predict the operational and strategic development plans involves the use of linked systems of equations and mathematical relationships that describe a variety of macroeconomic processes.

To visualize the model applies a simplified computational scheme that contains the most important relationships between variables. It uses recursive relationships, functional chains, input and output variables related by mathematical formulae.

Variables are connected with each other with lots of formulas. For example, real Gross National Product is expressed with its previous value and with its raise. So real income does.

VNPReal(n)=VNPReal(n-1)*(1+VNPRaise)

IncomeReal(n)=IncomeReal(n-1)*(1+IncomeRealRaise)

The model includes not only the formula connecting the variables, constants and functions, but also the initial conditions of the simulation, various constraints and conditions for a targeted, instrumental, exogenous and intermediate variables, and constants. Our model can be modified to change the formulas, their number, terms and conditions, if necessary.

## 5. Real scenarios

For example, as for model, with interest rate raise, unemployment will also raise, Gini coefficient will also grow up, but rate of state social expenses should descend. We could see this situation in China, where government elevated interest rate during 2008-2011.

The following are the equations used for plotting. Here are some of them.

SOC.EXPEND.1 := SOC.EXPEND.0*(1+SOC.EXPEND.G.1) ;

SOC.EXPEND.Q.1 := SOC.EXPEND.1/GNP.1 ;

SOC.EXPEND/EXPEND.Q.1 := SOC.EXPEND.1/BUDG.EXPEND.TOT.1 ;

SECUR.EXPEND.1 := SECUR.EXPEND.0*(1+SECUR.EXPEND.G.1) ;

SECUR.EXPEND.Q.1 := SECUR.EXPEND.1/GNP.1 ;

SECUR.EXPEND/EXPEND.Q.1 := SECUR.EXPEND.1/BUDG.EXPEND.TOT.1 ;

Below are the equations used for plotting. Here are some of them.

The three first and three second expression describing the growth rate of expenditures for social needs and security.

Variables SOC.EXPEND.0 and SECUR.EXPEND.0 are the monetary expression of the corresponding social spending and security in the previous period. Variables SOC.EXPEND.G.1 and SECUR.EXPEND.G.1 - the tools the nominal growth rate of social spending and security in the current period. Spending on social services represent expenditures for social sectors, except pensions to the elderly and the unemployed. Expenditure on internal and external security does not include the payment of salaries of staff. Variables SOC.EXPEND.1 and SECUR.EXPEND.1 are the monetary expression of the corresponding social spending and security in the current period. They affect the target variables SOC.EXPEND.Q.1 and SECUR.EXPEND.Q.1 , the corresponding relations of social spending and safety in the current period to the GDP (gross domestic product). Variables SOC.EXPEND/EXPEND.Q.1 and SECUR.EXPEND/EXPEND.Q.1 Express the ratio of expenditure on social services and safety in the current period to budget all expenses.

To build other charts used by a significant number of equations that calculates the corresponding target variables, and expressing: the GINI coefficients of the bundle.Q.1 , the unemployment rate is UNEMPLOYED.1 , all taxes TAX.TOT.FACT.Q.1 , gross national product GNP.$.1 , revenue officials OFFICER.INC.PC.1 , the payments to the retired PENSIONER.INC.PC.1 , the income of unemployed UNEMPL.INC.MIN.Q.1 .

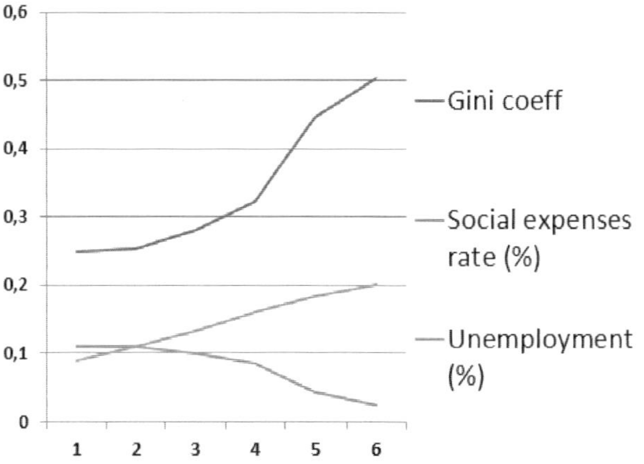


**Figure 4.** China example.

Other example is from French economics. New president E. Macron is going to increase security costs and to decrease level of taxes. Experts think that social expenses will not raise as fast as industrial ones. Moreover, unemployment will grow up. On the other hand, raise of real income and GNP is expected. We can see these changes from our model too in figures 5 and 6.

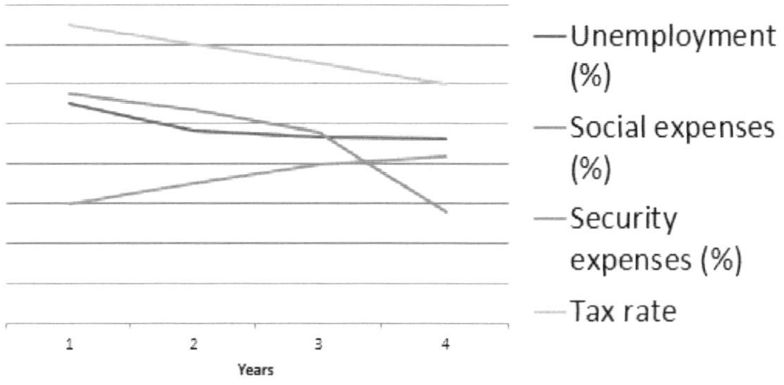


**Figure 5.** French example.

The model predicts the growth of real incomes and GDP. But the real growth may be less and in General will not happen, for example, due to changes in the external economic situation and of the restrictions or sanctions.

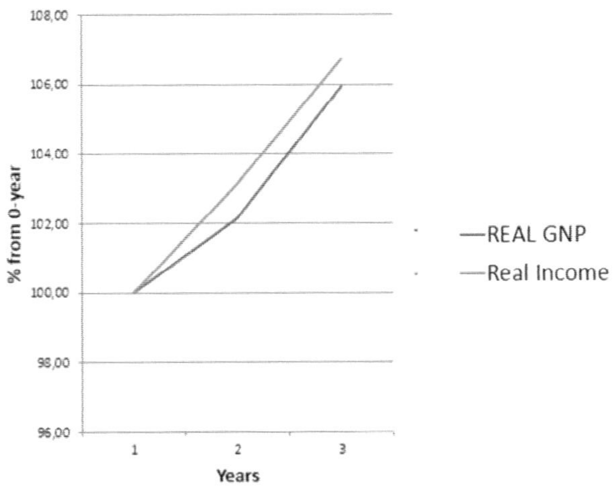


**Figure 6.** French example.

In Russia in April 2017 was already the second time this year the key interest rate was cut. Military expenses are growing currently. We will simulate this situation using our model:

As we can see from this table, GNP and welfare will start decreasing during the third year. As for Gini coefficient, property inequality will grow permanently.

**Table 1.** Russian example.

| | Year 1 | Year 2 | Year 3 |
|---|---|---|---|
| Percent rate (%) | 9,75% | 9,25% | 9% |
| Gini | 25% | 25,13% | 28,53% |
| Real GNP (% from 0-year) | 100 | 102,09 | 99,61 |
| Real income (% from 0-year) | 100 | 103,08 | 99,97 |
| Social expenses (%) | 11,5% | 11,09% | 9,92% |
| Security expenses (%) | 31% | 32% | 33% |

## 6. Conclusion

In this work the model of the economic game Transform is realised. Written code implements all the relationships between indicators and allows to analyze with high accuracy successful economic policies. This model can be used for forecasting macroeconomic scenarios and in the process of learning mathematical disciplines, as it offers the opportunity to simulate the real economic scenarios and to compare them with each other.

## 7. References

[1] Vnukov A. A., Raevsky S. N., Program for creating and editing business game parameters // *Computer programs. Databases*. 2011. № 4(77)(IIp.) . P. 487-487.

[2] Agapova T. A., Seregina S. F., *«Macroeconomics»*. Moscow, 2004. – 448 p.

[3] Tarasevich L. S., Grebennikov P. I., Leussky A. I., *«Macroeconomics»*. Moscow, 2006. – 328 p.

[4] Rudaya I. L., *«Strategic Business game Nixdorf Delta»*. Moscow, 2000. – 115 p.

[5] Pavlovsky Y. N., *«Simulation models and systems»*. Moscow, Phasis. - 131p.

[6] Vnukov A. A., Zhazhin A. A., *«Business game Delta. Enterprise management simulation system»* Moscow, MIEM, 2006.

IOP Conf. Series: Journal of Physics: Conf. Series **955** (2017) 012037 doi:10.1088/1742-6596/955/1/012037

# A framework to monitor activities of satellite data processing in real-time

**M D Nguyen**[1], **A P Kryukov**[1]

[1] Skobeltsyn Institute of Nuclear Physics, Lomonosov Moscow State University, Moscow, Russia

E-mail: nguyendmitri@gmail.com

**Abstract.** Space Monitoring Data Center (SMDC) of SINP MSU is one of the several centers in the world that collects data on the radiational conditions in near-Earth orbit from various Russian (Lomonosov, Electro-L1, Electro-L2, Meteor-M1, Meteor-M2, etc.) and foreign (GOES 13, GOES 15, ACE, SDO, etc.) satellites. The primary purposes of SMDC are: aggregating heterogeneous data from different sources; providing a unified interface for data retrieval, visualization, analysis, as well as development and testing new space weather models; and controlling the correctness and completeness of data. Space weather models rely on data provided by SMDC to produce forecasts. Therefore, monitoring the whole data processing cycle is crucial for further success in the modeling of physical processes in near-Earth orbit based on the collected data. To solve the problem described above, we have developed a framework called Live Monitor at SMDC. Live Monitor allows watching all stages and program components involved in each data processing cycle. All activities of each stage are logged by Live Monitor and shown in real-time on a web interface. When an error occurs, a notification message will be sent to satellite operators via email and the Telegram messenger service so that they could take measures in time. The Live Monitor's API can be used to create a customized monitoring service with minimum coding.

## 1. Introduction

One of the most critical tasks of a space monitoring data centre is providing correct data collected from satellites which are suitable to be used in research. The data collected from satellites are usually called the raw data. These raw data need to be processed and converted into a proper format (CSV, TXT, JSON, CDF, HDF, etc.) so that they can be read by analytical programs. An automatic data processing, storage and distribution system called SDDS has been created in 2016 at Skobeltsyn Institute of Nuclear Physics for this purpose. SDDS automates the whole cycle of satellite data processing which consists of the following steps: 1) connecting to data sources of each satellite; 2) checking for new data; 3) downloading them to a temporary storage; 4) decoding data if they were encoded; 5) extracting raw instrumental data from decoded data; 6) producing scientific data suitable for further analysis from the raw ones; 7) inserting scientific data into a unified database; 8) moving both raw and scientific data to long-term data storage with compression on demand. In each step, a set of programs are involved. Processed data are considered to be correct if every program involved in processing completed with no error. So to be sure of the data correctness, we need a mechanism that can monitor all activities of each program during a processing cycle and produce a report at the end. If an error occurs during a processing cycle, the mechanism must detect the component that caused the error and inform satellite operators and developers immediately so that they could take measure in time.

To solve the task we have developed a monitoring subsystem called Live Monitor. Later we decided to transform Live Monitor into an independent framework with RESTful API so that other satellite developers could build their own monitoring system based on Live Monitor.

This paper is organized as follows. In the second section, we consider several existing solutions to the problem compared to our. In section 3, we give a more detailed view of the overall architecture of the Live Monitor framework and show how it works. In section 4, we demonstrate an use case where the framework was used as the monitoring system in processing data of the Meteor-M2 satellite. In conclusion, we give a short resume of our completed work and describe our vision of the future perspective.

## 2. Related works

Our framework has been developed for use mostly in Linux operating systems. In practice, there are two ways to monitor activities of a running program: the passive way and the active way. The idea of the passive way is as follows. There is a master server running permanently on a machine and waiting for incoming requests. This server also serves as a web server for displaying statistics and a notification server to broadcast messages to subscribed users. On the machine where the target program is running there is another slave server running. The slave server triggers a checking script to get the current condition of the target program and send it to the master server on a regular basis. The idea of the active way, so-called lightweight event-driven or push notification, is that whenever an event arises the target program sends a short message (or a signal) to a master server, and the master server, in turn, broadcasts the message to all subscribed users.

Popular IT infrastructure monitoring solutions, such as Zabbix [1], Nagios [2], MMonit [3], and collectd [4], use the passive way to gather metrics of target programs. In our case, to monitor a program involved in data processing using the passive way we need to write a wrapper which runs a number of tests and returns certain metrics as a result. The client monitoring server then triggers the wrapper in a regular basis, for example, a 5-minute interval, to check the target program. This approach is not suitable for us because of the delay time between checks. Writing a wrapper for each target program would lead to a big amount of source code to be maintained. Furthermore, additional checks on a regular basis will produce overhead and affect the overall performance of the operating system.

On the other hand, push notification would inform us about the current condition of the target program immediately when an event arises. Currently, there are many available push notification solutions with API ready in the market such as Amazon Simple Notification Service [5], Urban Airship [6], Appsfire [7], One Signal [8] and so on. The primary problem is that most of these solutions are commercial. Free solutions often imply different constraints on how it can be used: a limited number of message queues or a limited amount of messages generated by one user. One Signal, for example, requires that one must have a website with a public domain and messages must be generated only by the application bound to the website. Appsfire provides SDK only for mobile platforms. Thus, we decided to create our own solution - a flexible framework with an open API and no constraint.

## 3. Live Monitor's architecture

Live Monitor framework consists of the following components: a backend logging library, RabbitMQ message broker, a RESTful API backend, a frontend UI library. An illustration of the architecture is shown below in figure 1. In the figure, the target programs to be monitored are components of the SDDS system such as satellite controllers, instrument data decoders, DB loader, and others.

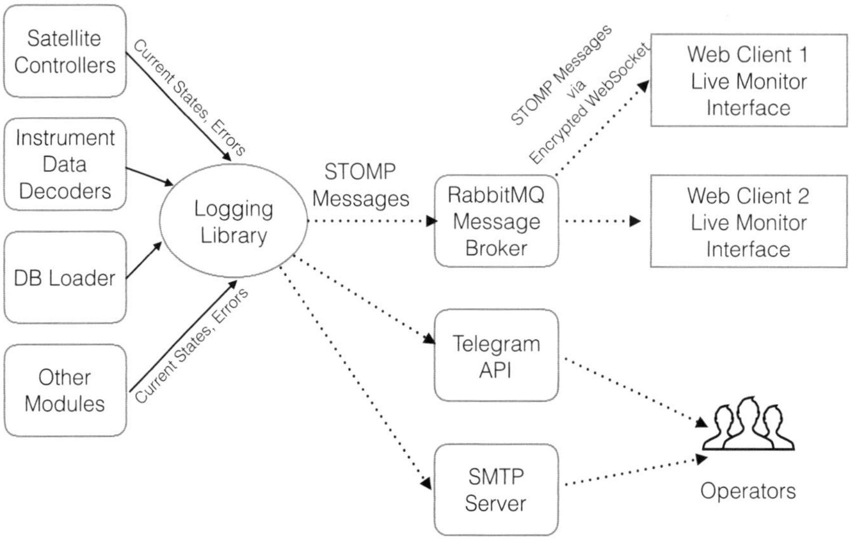


**Figure 1.** Live Monitor's architecture

The backend logging library is based on the standard Python Logging module. The logging library supports four levels of logging messages: debug, info, warning, and error. Error messages are logged when an error occurs during the data processing cycle which could lead to incorrectly processed data or cause a component failure. Error messages are also sent directly to the operators via the Telegram messenger service and/or email. Warning messages are logged for minor errors that do not affect the data correctness and normal functioning. Info messages are just normal text descriptions of events during the data processing cycle. Debug messages include diagnostic information that is helpful in failure investigation. Besides the standard behaviour which is writing short text messages in different log levels to a log file, the logging library sends these text messages to the RabbitMQ message broker [9] via a TCP socket. Messages are formatted using the STOMP protocol [10]. RabbitMQ, in turn, broadcasts received text messages to all subscribed frontend clients that use the frontend UI library to display messages on web pages. Notification features can be turned on and off by editing proper configuration files or using the RESTful API. Currently, the following operations are supported by the RESTful API:

- create/delete a customised monitoring service;
- change the logging level;
- switch the whole monitoring service on/off;
- switch a component of the monitoring service on/off;
- switch Telegram message delivery for a service on/off;
- switch e-mail delivery on/off.

The main goal of the frontend library is to control how each step of a data processing cycle should be shown on the web interface. The frontend library sends a GET request to the backend to retrieve necessary information of what should be shown. The answer from the backend is a JSON object that consists of a number of steps. After that, the frontend library establishes a WebSocket [12] connection with the RabbitMQ message broker. When a text message of a step arrives, the frontend library parses its content and changes the visual appearance of the step. In figure 2 and figure 3 all steps in the processing cycle of the Meteor-M2 satellite and possible states of the `connect` step are illustrated.

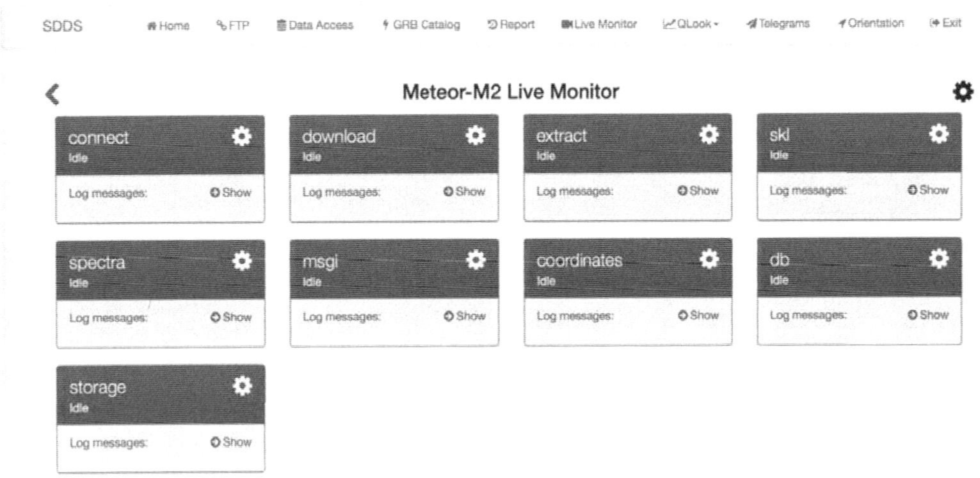


**Figure 2.** The data processing cycle of Meteor-M2

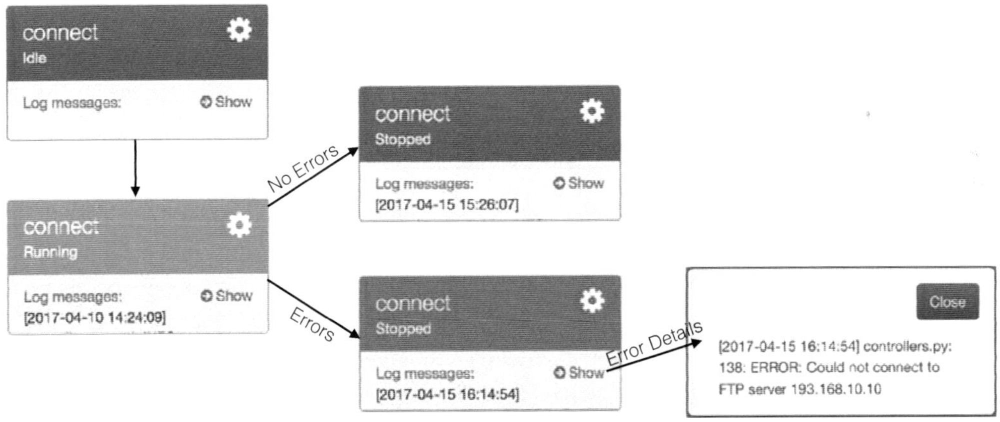


**Figure 3.** Different states of the connect step of Meteor-M2

## 4. Monitoring Meteor-M2 using Live Monitor

In this section, we demonstrate how we used Live Monitor to track activities in data processing of the Meteor-M2 satellite. To register the monitoring service, we sent a POST request to the backend. We described the monitoring service as an object using the JSON format and passed the object as the data parameter of the request. The backend generated a JSON configuration file according to input parameters and created a new message queue.

```
# POST request parameters
```

```
{
  "satellite": "meteor_m2",
  "instruments": ["msgi", "skl"],
  "norad_id": 40069,
  "sources": {...}
}

# a part of the generated configuration file
...
"notification": {
  "email_enabled": true,
  "email": "***",
  "smtp_server": "***",
  "smtp_protocol": "tls",
  "smtp_port": 587,
  "smtp_user": "***",
  "smtp_pass": "***",
  "push_enabled": true,
  "push_host": "localhost",
  "push_port": 61613,
  "push_user": "***",
  "push_pass": "***"
  "push_queue": "/topic/meteor_m2",
  "telegram_enabled": true
},
...
```

In each target program, firstly, we created a logger object using the get_logger function provided by the backend library. Then whenever we need to inform about an event, we called the logger object with either debug, info, warning, or error functions. It is possible to add a customised text as a parameter to the function. The logging library will format the text and send it further.

```
# importing the logging library
import log
...
# creating the logger
self.logger = log.get_logger( module_name = module_name,
                              stomp_cfg = cfg['notification'],
                              telegram = True,
                              stdout = True,
                              log_file = log_file,
                              log_level = log_level )
...
# using the logger to log event
self.logger.error('Error running decoder, return code = %r' % ret_code)
```

On the frontend side, we created a web page and included the Javascript library provided by Live Monitor with authentication credentials taken from the generated JSON configuration file as well as WebSocket and STOMP libraries. When we open the web page, the Live Monitor library makes a request to get the configuration that describes the appearance of notification messages, initialises a WebSocket connection to the RabbitMQ STOMP server, and subscribes to the message queue. Each time after receiving a new event from the message queue the Live Monitor library will trigger a proper function to handle the event and to show it on the web page.

## 5. Conclusion

The Live Monitoring framework is now actively used in developing new components of the satellite data processing system (SDDS) at SINP MSU. For 18 months of operation, Live Monitor has been helping us identify and localise the scope of any occurred problem in data processing in time and hence prevent or fix the problem quickly. We conducted a stress test which showed that Live Monitor was able to deliver up to 10000 messages per second on the following hardware: E5-2650 2.60GHz / 8GB memory / 1Gb Ethernet. In future, we plan to support more operations to control the behaviour of monitoring services such as changing components of a monitoring service dynamically and controlling the service via Telegram Bot.

## Acknowledgments

We would like to thank Dr Vladimir Kalegaev for helpful discussions on satellite data processing and clear problem statements. This project was supported by RSF grant #16-17-00098.

## References

[1] Zabbix SIA, `Zabbix Documentation` (`https://www.zabbix.com/documentation/3.0/manual`, Version 3.0)
[2] Ethan Galstad, Nagios Core Development Team, and Community Contributors `Nagios Core Documentation` (`https://assets.nagios.com/downloads/nagioscore/docs/nagioscore/4/en/index.html`, October 4th, 2016)
[3] Tildeslash Ltd, `M/Monit User Manual` (`https://mmonit.com/documentation/mmonit_manual.pdf`, Version 3.7.1)
[4] Collectd, (`https://mmonit.com/documentation/mmonit_manual.pdf`, Version 3.7.1)
[5] `Amazon Simple Notification Service` (`https://aws.amazon.com/ru/sns/`)
[6] Urban Airship, `User Guides` (`https://docs.urbanairship.com`)
[7] Appsfire, `Appsfire iOS SDK` (`http://docs.appsfire.com/sdk/ios/integration-reference/Introduction`)
[8] One Signal, `One Signal Documentation` (`https://documentation.onesignal.com/docs`, Version 3.0)
[9] Pivotal Software, Inc. `RabbitMQ Server Documentation` (`https://www.rabbitmq.com/admin-guide.html`, Version 3.6.9)
[10] `STOMP Protocol Specification` (`https://stomp.github.io/stomp-specification-1.2.html`, Version 1.2)
[11] Hiram Chirino, `Stomp Benchmark` (`https://github.com/chirino/stomp-benchmark`)
[12] Internet Engineering Task Force, `The WebSocket Protocol` (`https://tools.ietf.org/html/rfc6455`, December 2011)

# Distributed Time Synchronization Algorithms and Opinion Dynamics

**Anatoly Manita**[1,2] **and Larisa Manita**[2]

[1] Lomonosov Moscow State University, Faculty of Mechanics and Mathematics, 119991, GSP-1, Moscow, Russia

[2] National Research University Higher School of Economics, Moscow Institute of Electronics and Mathematics, Russia

E-mail: manita@mech.math.msu.su, lmanita@hse.ru, amanita@hse.ru

**Abstract.** We propose new deterministic and stochastic models for synchronization of clocks in nodes of distributed networks. An external accurate time server is used to ensure convergence of the node clocks to the exact time. These systems have much in common with mathematical models of opinion formation in multiagent systems. There is a direct analogy between the *time server/node clocks* pair in asynchronous networks and the *leader/follower* pair in the context of social network models.

## 1. Agreement algorithms for distributed networks

Distributed algorithms have a variety of applications in computer science [1], mobile multiagent systems [2, 3], biology [4, 5], social network dynamics [6, 7] and are also important for physicists [8, 9] interested in modeling of coordinated behavior. The present study is devoted to a special class of distributed algorithms adapted to the local clocks synchronization in asynchronous networks and to convergence to a leader opinion in large social networks. From the mathematical point of view, as we will soon show, the both problems lead to the same type of linear iterative algorithms

$$\boldsymbol{x}(s) = W(s)\,\boldsymbol{x}(s-1) + D(s)\boldsymbol{\delta}(s), \qquad s = 1, 2, \ldots, \tag{1}$$

where $\boldsymbol{x} \in \mathbb{R}^N$ is a state of an $N$-node network, $\boldsymbol{\delta}(s) \in \mathbb{R}^N$ is a random noise vector added on step $s$ of the algorithm, $W(s)$ and $D(s)$ are nonnegative $N \times N$-matrices to be precised later. Proposed models belong to (or, it would be better to say, are modifications of) famous agreement algorithms [10] known also as network consensus (NC) algorithms. One says that the network $\boldsymbol{x}(s)$ reaches a consensus if for any initial state $\boldsymbol{x}(0)$ there is $c \in \mathbb{R}^1$ such that $\boldsymbol{x}(s) \to c\mathbf{1}$ as $s \to \infty$. Here and henceforth $\mathbf{1}$ denote a colunm vector of all ones. Let $M_N$ be the set of all real $N \times N$-matrices. We write $A = (a_{jk})_{j,k=1}^N \in M_N^{(+)}$ and call $A$ *nonnegative* if $A \in M_N$ and all $a_{jk} \geq 0$. We call $A$ a stochastic matrix and write $A \in M_N^{\text{st}}$ if $A \in M_N^{(+)}$ and $A\mathbf{1} = \mathbf{1}$.

DeGroot [6] proposed the algorithm $\boldsymbol{x}(s) = A\,\boldsymbol{x}(s-1)$ where $A \in M_N^{\text{st}}$. It can be considered as a deterministic time invariant version of (1) and thus this is the simplest agreement algorithm. DeGroot interpreted it as a model of reaching a common decision in a group of experts by pooling their individual opinions. Technically the analysis of the consensus problem can be based on the discrete Markov chains theory (as in [6]) or on the direct study of powers $A^n$ in the framework of the Perron-Frobenius theory. A condition for a consensus to be reached was given in [6] in terms the matrix powers $A^n$.

While in [6] there was no word *network*, it is straightforward to introduce a digraph $\mathcal{G} = (\mathcal{N}, \mathcal{E})$ with the set of vertices $\mathcal{N} = \{1, \ldots, N\}$ and the set $\mathcal{E}$ of directed edges $(m_1, m_2)$ such that $a_{m_1 m_2} > 0$, $m_1 \neq m_2$. Denote by $\mathcal{O}_j := \{k : a_{jk} > 0\}$ the set of neighbours of a node $j$ in the graph $\mathcal{G}$. Hence the DeGroot algorithm takes the form $x_j(s) = \sum_{k \in \mathcal{O}_j} a_{jk}\, x_k(s-1)$ and has the meaning of *averaging* of the values $x_k(s-1)$. We may say that the entry $a_{jk}$ quantifies the importance of the value of $x_k(s-1)$ for evaluation of $x_j(s)$. For this reason $A$ is called the interconnection matrix. If $|\mathcal{O}_j| < |\mathcal{N}|$ for all $j$ then the network is called distributed [11].

Cybenko [12] considered a multiprocessor network and studied a *dynamics load balancing* model $\boldsymbol{x}(s) = A\,\boldsymbol{x}(s-1) + \boldsymbol{\delta}(s)$ with *symmetric* $A \in M_N^{\mathrm{st}}$ and independent identically distributed (i.i.d.) noise vectors $\boldsymbol{\delta}(s)$. Some optimization problem for a similar model was considered in [13].

Vicsek *et al.* [8] introduced and *numerically* studied a model of moving particles where the velocity of the particles is determined by a simple rule and random fluctuations. The simple rule is that at each time step a given particle driven with a constant absolute velocity assumes the *average direction* of motion of the particles in its neighborhood of radius $r$. Depending on the density and a noise strength parameter the model exhibits such rich behavior as clustering, transport and phase transition. The Vicsek model (VM) immediately became very popular. In [2] the authors *analytically* studied a *linearized* deterministic version of the VM and found conditions when all particles reach an agreement on direction of motion. An agreement algorithms of [2] has the form $\boldsymbol{\theta}(s) = A(s)\,\boldsymbol{\theta}(s-1)$ where $\theta_j$ is the direction of the particle $j$ and the sequence of matrices $\{A(s)\}_{s\geq 1}$ reflects a changing topology of the network. The spirit of the VM suggests

$$a_{jk}(s) = \frac{\mathbf{1}\left(|q_k(s) - q_j(s)| \leq r\right)}{\sum_{m=1}^{N} \mathbf{1}\left(|q_m(s) - q_j(s)| \leq r\right)}$$

where $q_j$ is the position of the particle $j$. However, the paper [2] deals with abstract sequences $\{A(s)\}_{s\geq 1}$ taken from a finite set of *symmetric* stochastic matrices $\{A_1, \ldots, A_M\}$. The above mentioned sufficient conditions were given purely in terms of this time-varying topology, i.e., in terms of connectivity properties of the sequence of graphs $\{\mathcal{G}(s)\}$. This approach was developped by other authors [14, 15]. Another result of the paper [2] is related to the "leader-followers" assumption and is important for the opinion dynamics modeling. According to this assumption the particle 1 is a leader and does not change its direction: $\theta_1(s) \equiv \theta_1$. It was proved that under suitable conditions the direction of motion of any other particle $j \geq 2$ tends to the direction of the leader: $\theta_j(s) \to \theta_1$ as $s \to \infty$.

Hegselmann and Krause [7] introduced an opinion dynamics model $\boldsymbol{x}(s) = A(s)\,\boldsymbol{x}(s-1)$ with

$$a_{jk}(s) = \frac{\mathbf{1}\left(|x_k(s) - x_j(s)| \leq \varepsilon_j\right)}{\sum_{m=1}^{N} \mathbf{1}\left(|x_m(s) - x_j(s)| \leq \varepsilon_j\right)} \tag{2}$$

where $\varepsilon_j > 0$ are *confidence* levels. So, generally speaking, the matrices $A(s)$ are not symmetric. It was shown in [7] that sometimes, depending on the choice of $\{\varepsilon_j\}$ and an initial configuration $\boldsymbol{x}(0)$, the model can exhibit an opinion fragmentation phenomenon.

Other opinion dynamics models (Abelson, Friedkin and Johnsen etc.) are out of the present discussion because they are not so closely related to the network consensus problem as (2).

## 2. Clock synchronization problem for WSNs

Wireless sensor networks (WSNs) are typical examples of distributed networks. The sensor nodes $\mathcal{N} = \{1, \ldots, N\}$ communicate via radio waves. Their interconnection matrices are naturally determined by range and power characteristics of the radio communication (see [16]).

From now and onwards, the possibility of direct communications between pairs of nodes will be described by an $N \times N$ stochastic matrix $W = (w_{jk})_{j,k \in \mathcal{N}}$. Elements of $\mathcal{N}$ will be called client nodes. There is also a stand-alone time server formally not belonging to the network but being reffered in the sequel as the node 0. The system is observed at discrete times. Any client node $j \in \mathcal{N}$ is equipped with an unperfect clock which current value is $\tau_j \in \mathbb{R}$. The time server (the node 0) has a perfect clock providing exact time $\tau_0 \in \mathbb{R}$. Let $v > 0$ be the rate of the clocks with respect to $t$. Being isolated these clocks evolve as

$$\tau_0(t) = \tau_0(t-1) + v, \qquad \tau_j^\circ(t) = \tau_j^\circ(t-1) + v + \delta_j(t), \quad j \in \mathcal{N}, \tag{3}$$

where $t \in \mathbb{T} := \{1, 2, \ldots\}$ and $\delta(t) = (\delta_j(t),\, j \in \mathcal{N}) \in \mathbb{R}^N$ are i.i.d. random vectors representing random noise related to unperfect clocks of client nodes. We assume that $\mathsf{E}\,\delta_j(t) = 0$.

The evolutions (3) are discrete time versions of the clock models used in [17–20]. Since rates of clients are equal to the server rate $v$ the above system of clocks is drift-free [21].

To perform a common task the WSN needs a common time. However, the client clocks $\tau_j^\circ(t)$ are desynchronized for two reasons: 1) the presense of random noise $\delta(t) \neq 0$; 2) initial values $\tau_j^\circ(0)$, $j \in \mathcal{N}$, are different and unknown. The second reason is present even in the deterministic model ($\delta(t) = 0$). A natural idea is to run the following agreement algorithm: for any $j \in \mathcal{N}$

$$\tau_j(t) = \tau_j(t-1) + v + \delta_j(t) + \sum_{k \neq j} w_{jk} \left( \tau_k(t-1) - \tau_j(t-1) \right). \tag{4}$$

The sum in (4) is in fact taken over $k \in \mathcal{O}_j = \{k : w_{jk} > 0\}$ and so (4) is a distributed algorithm. The last summand in (4) is a correction made by the node $j$ by using information on local time values $\tau_k(t-1)$ obtained from the neighbourhood $\mathcal{O}_j$. Equally, one can say that the sum in (4) introduces a special interaction between client clocks. Since $w_{jj} = 1 - \sum_{k \neq j} w_{jk}$ the matrix form of (4) is $\boldsymbol{\tau}(t) = W\boldsymbol{\tau}(t-1) + v\mathbf{1} + \boldsymbol{\delta}(t)$ where $\boldsymbol{\tau}$ and $\boldsymbol{\delta}$ are column vectors of length $N$.

To eliminate the influence of a constant drift we introduce a new vector $\boldsymbol{\tau}'(t) = \boldsymbol{\tau}(t) - vt\mathbf{1}$. The iterative scheme (4) turns into $\boldsymbol{\tau}'(t) = W\boldsymbol{\tau}'(t-1) + \boldsymbol{\delta}(t)$ which is a stochastic version of the network consensus (NC) algorithm. Consider first the *deterministic case* $\delta(t) = 0$. If the matrix $W$ satisfies any known conditions ( [1,6]) sufficient for reaching concensus then $\boldsymbol{\tau}'(t) \to c\mathbf{1}$ as $t \to \infty$. This means that the algorithm (4) provides an *internal synchronization* of the clients clocks, i.e., $\tau_{j_0}(t) - \tau_{k_0}(t) \to 0$ for any nodes $j_0$ and $k_0$. For the *stochastic model* with $\boldsymbol{\delta}(t) \neq 0$ the local times $\tau_j(t)$ are random variables. So the best we can expect is that under suitable conditions (see [12, 13]) the differences $\tau_{j_0}(t) - \tau_{k_0}(t)$ converge in distribution to some limit probability laws. Nevertheless, the algorithm (4) keeps the client clocks far from the *accurate time* of the time server 0 and does not provide a *global synchronization.* To see this consider deviations $x_j = \tau_j - \tau_0$. The vector $\boldsymbol{x}(t) = (x_1(t), \ldots, x_N(t))$ evolves as $\boldsymbol{x}(t) = W\boldsymbol{x}(t-1) + \boldsymbol{\delta}(t)$ so all above arguments are applicable. For example, in the deterministic case $\boldsymbol{\delta} = 0$ under suitable assumptions one has $\tau_j(t) - \tau_0(t) \to c_0\mathbf{1}$ as $t \to \infty$ but $c_0 \neq 0$ depends on initial values $\boldsymbol{\tau}(0)$ of the clients clocks which are unknown. More details can be found in [22].

## 3. Interaction with the time server

To synchronize client clocks $\tau_j(t)$ with the accurate time $\tau_0(t)$ we *modify* the model (4) by adding the time server (TS) to the system. The time server 0 can send messages to some (but not to all) client nodes. A message $m^{0\to j'}$ sent on some step $t = t'$ from 0 to $j'$ contains the value $\tau_0(t')$. The message $m^{0\to j'}$ *instantly* reaches the destination node $j'$ which *immediately* adjusts its clock to the value recorded in $m^{0\to j'}$: $\tau_{j'}(t') = \tau_0(t')$. This is the usual message passing mechanism with zero delays. If on step $t$ there is no message from 0 to $j$ then the clock value $\tau_j(t)$ is adjusted according to the $j$th row of (4). A node just received a message from the time server *is aware* that its newly adjusted clock value is more precious than ones of its neigbours. If on step $t_1$ a client clock $\tau_{j_1}$ was set to the value $\tau_0(t_1)$ then during the time interval $t_1+1,\ldots,t_1+\Delta_{j_1}$ the node $j_1$ decides to ignore opinions of its neighbours. Let $T^{0,j} = \left\{t_n^{(j)}\right\}_{n\in\mathbb{N}}$ denote the sequence of steps $t$ when $j$ receives messages from 0. Hence to define the modified model one needs to specify $T := (T^{0,j},\ j\in\mathcal{N})$ and $\Delta := (\Delta_j,\ j\in\mathcal{N})$. Now the variables $x_j = \tau_j - \tau_0$ evolve as in (1) where $W(s) = W(s;T,\Delta)$ is some $N\times N$-matrix with entries $w_{jk}(s)\in\{0, w_{jk}, 1\}$ uniquely determined by the above algorithm. Notation $D(s) = D(s;T)$ stands for the diagonal matrix of 0s and 1s, $D(s) := \mathrm{diag}(\mathbb{I}_{\{s\notin T^{0,j}\}},\ j\in\mathcal{N})$, indicating that random noise terms $\delta_j(s)$ are not added to $\tau_j(s)$ on steps $s = t_n^{(j)}$.

Define $\mathcal{R}(s) := \left\{j \mid T^{0,j}\ni s\right\}\subset\mathcal{N}$, the subset of client nodes receiving messages from the time server on the step $s$. From the viewpoint of the node 0 the set $\mathcal{R}$ defines a prescribed sequence of recipients $\mathcal{R}(s)$ to whom it should consequently send messages. We assume henceforth that the *scheduling sequence* $\mathcal{R}$ is not empty and *periodic* with period $d$, i.e., $\mathcal{R}(s+d) = \mathcal{R}(s)$, $s\in\mathbb{T}$. It is easy to see that $\{D(s)\}_{s\in\mathbb{T}}$ and $\{W(s)\}_{s\geq s_0}$ are $d$-periodic too where $s_0 = s_0(\mathcal{R},\Delta)$ is sufficiently large.

It is convenient to consider separately *two variants* of the above defined model: TS+DN (deterministic model with $\boldsymbol{\delta}(t) = 0$) and TS+SN (stochastic model with $\boldsymbol{\delta}(t)\neq 0$). It is readily seen that the mean $\boldsymbol{m}(s) = \mathsf{E}\,\boldsymbol{x}(s)$ of the TS+SN behaves as the state $\boldsymbol{x}(s)$ of the TS+DN.

An asymptotic behavior of the both TS+DN and TS+SN models was studied in [22]. Under some *additional* conditions on $W$ and $\boldsymbol{\delta}(t)$ we have the following results as $s\to\infty$.

1) In TS+DN model all clients synchronize with the time server, i.e., $x_j(s) = \tau_j(s) - \tau_0(s)\to 0$.

2) $\mathsf{Var}\,\boldsymbol{x}(s)$, the covariance matrix of synchronization errors of the TS+SN, is bounded in $s$.

3) Any subsequence $\{\boldsymbol{x}(nd+i)\}_{n\in\mathbb{N}}$ has a limit in distribution as $n\to\infty$.

The first item is similar to results on the NC problem in the presence of leaders [2,23]. *Additional* sufficient conditions assumed in [22] for proving 1–3 are easy to verify. We omit details.

## 4. Performance of the modified algorithm and network design problem

*4.1. Spectral radius.* Let $\mathcal{U}(s)$ be the set of clients *which* are receiving messages from the server node 0 on the step $s$ *or* are ignoring opinions of neighbors on the step $s$. As it follows from [22] the rate of convergence of the modified algorithms depends of the *spectral radius* $\rho$ of the matrix $\Pi = \overset{\circ}{W}(s_0+d-1)\overset{\circ}{W}(md+i-1)\cdots\overset{\circ}{W}(s_0)$ where $\overset{\circ}{W}(s) = Z(s)W$ and $Z(s) = \mathrm{diag}(\mathbb{I}_{\{j\notin\mathcal{U}(s)\}})$. Note that $\rho = \rho(W,\mathcal{R})$. It is clear that $\Pi$ is substochastic and $0\leq\rho<1$. Below we describe some important cases when effective evaluation of $\rho$ is possible.

*4.2. Boundary and internal nodes.* Denote by $\mathcal{B} = \left\{j : T^{0,j}\neq\emptyset\right\}$ the set of nodes receiving messages from the time server node 0. Evidently, $\mathcal{B} = \bigcup_s\mathcal{U}(s)$. Nodes of $\mathcal{B}$ will be called a *boundary* and the set of nodes $\mathcal{I} = \mathcal{N}\backslash\mathcal{B}$ will be called an *internal network*. The most interesting situation is that where the set $\mathcal{B}$ is only a small part the whole distributed network. For WSNs it means that the internal network is large but the time server communicates only with relatively small number of boundary nodes. Let the nodes of $\mathcal{N}$ be renumbered in such a way that $\mathcal{B} = \{1,2,\ldots,k\}$ and $\mathcal{I} = \{k+1,\ldots,k+n\}$. Thus $N = k+n$.

*4.3. Design of the network.* Assume that the internal network $\mathcal{I} = \{k+1,\ldots,k+n\}$ and its interconnection (stochastic) matrix $W^{(\mathcal{I})} = (q_{j_1,j_2})_{j_1,j_2=k+1}^{k+n}$ are given. The task is to organize communication of $\mathcal{I}$ with the time server. For *WSNs* it is easy imagine situations when the direct radio connection between 0 and $\mathcal{I}$ is not possible due to natural obstacles (hills, walls etc.). An *opinion dynamics* interpretation of this task may be the following one. The opinion leader (node 0) and the client nodes speak different languages. It is necessary to add "translators" $\mathcal{B}$ to organize transmission of information from the node 0 to client nodes $\mathcal{I}$.

To build an accurate time synchronization algorithm the *network designer* needs
- to introduce a set of "boundary" nodes $\mathcal{B} = \{1, 2, \ldots, k\}$,
- to fix some periodic scheduling sequence $\mathcal{R} = \{\mathcal{R}(s),\, s \in \mathbb{T}\}$,
- to choose topology of connection links between $\mathcal{I}$ and $\mathcal{B}$.

The choice of topology here is the choice of $W = (w_{i_1,i_2})_{i_1,i_2=1}^{k+n} \in M_{k+n}^{\mathrm{st}}$ subject to the condition $w_{j_1,j_2} = r\, q_{j_1,j_2}$, $j_1, j_2 = k+1, \ldots, k+n$, where $r \in (0,1)$ is some coefficient.

*4.4. A special class of synchronization algorithms.* We say that a matrix $W \in M_{k+n}$ belongs to the class $\mathrm{HCRS}_{k,n}$ if
1) $w_{i_1 j} = w_{i_2 j} \quad \forall i_1, i_2 \in \{k+1, \ldots, k+n\}, \quad j \in \{1, \ldots, k\}$.
2) $\sum_{j=k+1}^{k+n} w_{ij} = \mathrm{const} \quad \forall i \in \{k+1, \ldots, k+n\}$.
In terms of the *WSNs* the item 1 means that all communications in the direction $\mathcal{B} \to \mathcal{I}$ (transmission of clock values information from nodes of the boundary set to nodes of the internal network) pass through a common gate. Note that there is no restriction on communications in direction $\mathcal{I} \to \mathcal{B}$ and communications between nodes of $\mathcal{B}$ are also arbitrary. The item 2 evidently holds for the network designs discussed above. For *opinion dynamics* models the item 1 means that while different nodes of $\mathcal{B}$ may have different information on opinion of the leader node 0, any boundary node distributes such information between nodes of $\mathcal{I}$ uniformly.

**Lemma 1.** *The class* $\mathrm{HCRS}_{k,n}$ *is closed with respect to multiplication.*

**Definition 1.** We write $W \in \mathrm{SD}(\mathcal{B}, \mathcal{I}, W^{(\mathcal{I})})$ and call $W$ a *specially designed* matrix if $W$ is designed with $\mathcal{B}$, $\mathcal{I}$ and $W^{(\mathcal{I})}$ as specified above and $W \in \mathrm{HCRS}_{k,n}$.

**Lemma 2.** *If* $W \in \mathrm{SD}(\mathcal{B}, \mathcal{I}, W^{(\mathcal{I})})$ *then any* $\overset{\circ}{W}(s) = Z(s)W \in \mathrm{HCRS}_{k,n}$.

**Definition 2.** Given $A \in M_{k+n}$ define $B^{(A)} = (b_{ij})_{i,j=1}^{k+1} \in M_{k+1}$ as follows

$$b_{ij} = a_{ij}, \quad (i,j) \in \{1, \ldots, k+1\} \times \{1, \ldots, k\},$$

$$b_{i,k+1} = \sum_{m=k+1}^{k+n} a_{im}, \quad i \in \{1, \ldots, k+1\}.$$

**Lemma 3.** *If* $A^{(1)}, \ldots, A^{(m)} \in \mathrm{HCRS}_{k,n}$ *then* $B^{\left(A^{(1)} \cdots A^{(m)}\right)} = B^{\left(A^{(1)}\right)} \cdots B^{\left(A^{(m)}\right)}$.

*4.5. Spectral problem in the case of specially designed W.* Let $W \in \mathrm{SD}(\mathcal{B}, \mathcal{I}, W^{(\mathcal{I})})$. Then from Lemmas 1 and 2 we see that $\Pi \in \mathrm{HCRS}_{k,n}$. Consider the $(k+1) \times (k+1)$ matrix $B^{(\Pi)}$. By Lemma 3 $B^{(\Pi)} = B^{\left(\overset{\circ}{W}(s_0+d-1)\right)} B^{\left(\overset{\circ}{W}(s_0+d-2)\right)} \cdots B^{\left(\overset{\circ}{W}(s_0)\right)}$. Hence while $\Pi$ is a large-size matrix, $B^{(\Pi)}$ can be obtained as a product of small-size matrices.

Denote by $\Lambda_B$ and $\Lambda_Q$ complete lists of eigenvalues of $B^{(\Pi)}$ and $W^{(\mathcal{I})}$, respectively, including all repeated eigenvalues. Clearly, $|\Lambda_B| = k+1$. Let $\Lambda'_Q$ be obtained from $\Lambda_Q$ by excluding only one element 1. Denote by $\left(\Lambda'_Q\right)^d$ the list of $d$-powers of elements of $\Lambda'_Q$. Note that $\left|\Lambda'_Q\right| = \left|(\Lambda'_Q)^d\right| = n-1$.

**Theorem 1.** *The complete list of eigenvalues of the matrix* $\Pi$ *is* $\Lambda_B \bigcup \left(r\,\Lambda'_Q\right)^d$.

This theorem is useful when some information on the spectrum of $W^{(\mathcal{I})}$ is *a priori* known and $\mathcal{B}$ is relatively small. Proofs of the above statements are purely algebraic and are omitted.

## 5. Conclusions

We have analytically studied new algorithms for synchronization of local clocks in nodes of distributed networks. The proposed algorithms can also be applied to studying social networks dynamics in the presence of opinion leaders. We show that they fit to the general scheme (1) of distributed iterative algorithms with time-dependent topologies and a time-nonhomogeneous random noise. However, structures of matrices $W(s)$ and $D(s)$ in our algorithms are quite original. Our approach is to combine agreement algorithms with the message passing mechanism. The latter is widely used in various synchronization models [24–27]. We have discussed some network design problem naturally issued from our approach. We have described a class of algorithms that admit an evaluation of their spectrum, and, hence, a quantitative estimation of their performance.

## References

[1] Olshevsky A and Tsitsiklis J N 2011 Convergence speed in distributed consensus and averaging *SIAM Review* **53**(4) 747–772
[2] Jadbabaie A, Lin J and Morse A S 2003 Coordination of groups of mobile autonomous agents using nearest neighbor rules *IEEE Transactions on automatic control* **48**(6) 988–1001
[3] Ren W and Beard R W 2008 *Distributed consensus in multi-vehicle cooperative control* (Springer)
[4] Ben-Jacob E, Schochet O, Tenenbaum A, Cohen I, Czirók A and Vicsek T 1994 Generic modelling of cooperative growth patterns in bacterial colonies *NATURE* **368**
[5] Cucker F and Smale S 2007 Emergent behavior in flocks *IEEE Transactions on automatic control* **52**(5) 852–862
[6] DeGroot M H 1974 Reaching a consensus *Journal of the American Statistical Association* **69**(345) 118–121
[7] Hegselmann R, Krause U et al. 2002 Opinion dynamics and bounded confidence models, analysis and simulation *Journal of Artificial Societies and Social Simulation* **5**(3) 1–33
[8] Vicsek T, Czirók A, Ben-Jacob E, Cohen I and Shochet O 1995 Novel type of phase transition in a system of self-driven particles *Phys. Rev. Lett.* **75** 1226–1229
[9] Vicsek T and Zafeiris A 2012 Collective motion *Physics Reports* **517**(3) 71–140
[10] Kshemkalyani A D and Singhal M 2008 *Distributed computing: principles, algorithms, and systems* (Cambridge University Press)
[11] Olfati-Saber R and Murray R M 2004 Consensus problems in networks of agents with switching topology and time-delays *IEEE Transactions on automatic control* **49**(9) 1520–1533
[12] Cybenko G 1989 Dynamic load balancing for distributed memory multiprocessors *J. Parallel Distrib. Comput.* **7**(2) 279–301
[13] Xiao L, Boyd S P and Kim S J 2007 Distributed average consensus with least-mean-square deviation *J. Parallel Distrib. Comput.* **67**(1) 33–46
[14] Hendrickx J M and Blondel V D 2006 Convergence of linear and non-linear versions of Vicsek's model *Proceedings of the 17th Symposium on Mathematical Theory of Networks* (Kyoto) pp. 1229–1240
[15] Shi G and Johansson K H 2012 Persistent graphs and consensus convergence *Proceedings of the 51st Annual Conference on Decision and Control* (IEEE) pp. 2046–2051
[16] Simeone O, Spagnolini U, Bar-Ness Y and Strogatz S H 2008 Distributed synchronization in wireless networks *IEEE Signal Processing Magazine* **25**(5) 81–97
[17] Manita A 2014 Clock synchronization in symmetric stochastic networks *Queueing Systems* **76**(2) 149–180
[18] Manita L 2014 Controlling of clock synchronization in WSNs: structure of optimal solutions `arXiv:1408.1434 [math.OC]`
[19] Manita L 2015 Singular extremals in control problems for wireless sensor networks *Stochastic Modeling, Data Analysis and Statistical Applications* (ISAST) pp. 461–474
[20] Manita L 2016 Optimization problems for wsns: trade-off between synchronization errors and energy consumption *Journal of Physics: Conference Series* **681**(1) 012009
[21] Patt-Shamir B and Rajsbaum S 1994 A theory of clock synchronization *Proceedings of the 26th Annual ACM Symposium on Theory of Computing* (ACM) pp. 810–819
[22] Manita L and Manita A 2017 Stochastic time synchronization models based on agreement algorithms *Lecture notes in computer science* **10684** 361–375
[23] Lin F, Fardad M and Jovanovic M R 2014 Algorithms for leader selection in stochastically forced consensus networks *IEEE Transactions on Automatic Control* **59**(7) 1789–1802
[24] Fujimoto R 2000 *Parallel and Distributed Simulation Systems* (Wiley)
[25] Shchur L N and Shchur L 2015 Parallel discrete event simulation as a paradigm for large scale modeling experiments. *Data Analytics and Management in Data Intensive Domains* (Obninsk, Russia) pp. 107–113
[26] Manita A 2017 Probabilistic issues in the node synchronization problem for large distributed systems *Lobachevskii Journal of Mathematics* **38**(5) 948–953
[27] Manita A 2016 On behavior of stochastic synchronization models *Journal of Physics: Conference Series* **681**(1) 012024

# Docker Container Manager: A Simple Toolkit for Isolated Work with Shared Computational, Storage, and Network Resources

**S P Polyakov[1], A P Kryukov and A P Demichev**

Skobeltsyn Institute of Nuclear Physics, M.V.Lomonosov Moscow State University (SINP MSU), 1(2), Leninskie gory, GSP-1, Moscow 119991, Russian Federation

[1] s.p.polyakov@gmail.com

**Abstract.** We present a simple set of command line interface tools called Docker Container Manager (DCM) that allow users to create and manage Docker containers with preconfigured SSH access while keeping the users isolated from each other and restricting their access to the Docker features that could potentially disrupt the work of the server. Users can access DCM server via SSH and are automatically redirected to DCM interface tool. From there, they can create new containers, stop, restart, pause, unpause, and remove containers and view the status of the existing containers. By default, the containers are also accessible via SSH using the same private key(s) but through different server ports. Additional publicly available ports can be mapped to the respective ports of a container, allowing for some network services to be run within it. The containers are started from read-only filesystem images. Some initial images must be provided by the DCM server administrators, and after containers are configured to meet one's needs, the changes can be saved as new images. Users can see the available images and remove their own images. DCM server administrators are provided with commands to create and delete users. All commands were implemented as Python scripts. The tools allow to deploy and debug medium-sized distributed systems for simulation in different fields on one or several local computers.

## 1. Introduction

If we want to let several users share access to computational, storage, and network resources of a server, we can use several approaches. The most simple one is to let them log in to the server directly. It imposes no overhead, but the server administrators have to configure the server and install any software the users need but cannot install without privileges. This becomes particularly problematic if two or more users need different versions of the same software. Another approach is to give users privileged access to virtual machines. Virtual machines can be configured independently and can even use different operating systems (OS), but the overhead may be significant. Container virtualization [1] provides a middle ground between these two approaches and combines many of their benefits.

We present a set of tools for container virtualization platform Docker that allows users to create and manage their own Docker containers (analogs of virtual machines) with preconfigured SSH access. The users can configure containers, install any software compatible with the server OS kernel, and run network services and computational programs, including microservices. This allows to deploy and debug medium-sized distributed systems that can be used for simulation in different fields.

The rest of the paper is organized as follows: Section 2 outlines some possibilities of container virtualization and features of Docker. Section 3 presents the Docker Container Manager (DCM) toolkit we developed and describes its functionality. Section 4 discusses the security limitations of the tools. Section 5 concludes the paper and outlines some possibilities for further development of the tools.

## 2. Container virtualization and Docker

Container virtualization is an OS feature that allows to use the same OS kernel to run multiple isolated user-space instances called containers. It was developed from Linux chroot mechanism [2] that allowed to change apparent root directory for a process and its children. Unlike full virtualization, container virtualization does not allow containers to use different operating systems. But it does provide isolation between the processes, allowing each to use its own set of libraries and installed software without dependencies conflict, and imposes relatively small overhead [3]. Many container virtualization methods use copy-on-write technique [4]: resources used by two or more containers are not copied until one of the containers needs to modify it. Thanks to it, a server can run or store large number of containers.

Docker is a rapidly developing and very popular container virtualization platform for Linux. Docker uses copy-on-write technique: file system of a Docker container is based on a read-only image; changes made in a running container can be saved to a new image, resulting in an additional layer. Docker provides a number of tools for creating, monitoring, and managing containers, as well as building images, sharing them via repositories and looking for the images created by other users.

A server administrator can give users access to containers started for them, but if we want users to be able to leverage more of the Docker features, we could let them manage containers on their own, save changes to new images, start new containers from their own images, discard the containers they no longer need and so on. However, giving users full access to Docker capabilities on a server amounts to giving them unrestricted access to the server. In particular, volumes mechanism allows to map any host directory to a container directory (so the users will be able to modify anything from the container they mapped a host root directory to); containers may also be given privileges to access devices of the host. Therefore, to let users create their own containers we need to restrict their access to Docker commands to prevent them from accidentally disrupting the work of the server. Imposing such restrictions is one of the purposes of the toolkit presented in this paper, Docker Container Manager.

## 3. Docker Container Manager

Docker Container Manager (DCM) is a set of tools which has three primary purposes:

· to allow users to create Docker containers on a server, configure them and save the changes;

· to let users work with the containers remotely, using the computational, storage, and network resources of the server;

· to isolate users from each other and restrict their access to the Docker features that could potentially disrupt the work of the server.

DCM has a simple command-line interface accessible by SSH: when new DCM users are created, their access to the server is configured to redirect them to DCM interface as they log in. The interface allows users to execute a simplified set of Docker commands, with DCM keeping track of their containers and images to make sure only owners can change or see them.

After a container is created, DCM interface informs the user about the server port that can be used to connect to the container by SSH. The interface is not needed to connect to the container and work with it.

The following subsections give the overview of DCM commands in more detail.

IOP Conf. Series: Journal of Physics: Conf. Series **955** (2017) 012039 doi:10.1088/1742-6596/955/1/012039

### *3.1. Creating containers*

DCM users can create a container using create command. In its most basic form the command takes one argument, an image to create the container from. DCM then invokes a Docker command run with the options that ensure the following:
· port 22 (SSH) of the container is mapped to a random port of the server (the port number is also used as the container ID);
· two storage directories assigned to the user on the server are mapped to home directories /root and /home/<user> of the container.
As a result, the user can access the container by SSH using their own private key both as regular user and as root (the copies of the public key are placed to their storage directories when the DCM user account is created), and any changes made to /root and /home/<user> directories in any container are accessible in all other containers of the user.

Additionally, the user can select one or several container ports to be published on the server. These ports are mapped to the server ports without changing their numbers. More than one container can use the same port or ports if the server has IP aliases: DCM will choose an alias where none of the requested additional ports are currently in use by other containers.

Any other options of the Docker command run that are not available to DCM users by default can be specifically permitted to individual users by the DCM server administrators. Since many of the additional options create security weaknesses, an administrator is supposed to be careful when allowing users to use these options. A user can select one of the permitted options or combinations of options with preset arguments, but not combine them or change arguments at will.

### *3.2. Saving changes of containers to new images*

Administrators of a DCM server must provide users with at least one image with preconfigured SSH access. When creating a container for the first time, a user can select one of these images. After making any changes outside of /root and /home/<user> directories (e.g. installing new software), the user can save these changes to a new image using the commit command.

### *3.3. Managing containers and images*

The following DCM commands to manage containers are available to users: stop – stops a running container (without removing it), start – starts a stopped container, pause – freezes all processes within a container, unpause – unfreezes frozen processes within a container, show – shows all containers belonging to the user and their status, rm – removes a container.

There are also two commands to manage images:
images – shows all images available to the user (this includes basic images provided by the DCM server administrators and the images created by the user), rmi – removes an image.

### *3.4. Creating and removing DCM user accounts*

Administrators of DCM servers are provided with tools to create new DCM user accounts and remove them. Creating new user accounts includes setting up their access to DCM interface, creating storage directories and copying their public keys to these directories. Removing user accounts includes removing their containers, storage directories, and possibly their images.

## 4. Security

In the earlier versions of Docker, it was considered possible to break out of a container if users were allowed to execute arbitrary commands («Containers do not contain», [5]), even more so if they had root access to the container. Since Docker Engine version 1.10 released in 2016, two key security features are supported: seccomp filtering and user namespaces. These features make Docker considerably more secure (see [6] for analysis and recommendations).

However, user namespaces is not used by default, Docker has some other potential security weaknesses, and a mistake by a DCM server administrator may introduce further weaknesses. Therefore we recommend to allow only users who can be trusted to make no attempts to break out of their containers intentionally. The users should also be instructed to give no access to their accounts or privileged access to their containers to outside parties.

If necessary, DCM can be easily modified to use additional options when creating new containers. With some modifications it is possible to make containers more secure than Docker default, e.g. by further restricting root capabilities.

## 5. Conclusions and further work

We have presented a set of tools that give users indirect access to some of the Docker commands without full access to the host, allowing them to create Docker containers that can be configured by the user, save and access the changes made to the containers, and manage the containers. The users are isolated from each other. The tools can be accessed via a simple command line interface.

The ideas for further development of the toolkit include creating a Web interface and giving users the option to migrate their images to outside repositories.

## References

[1] Soltesz S, Pötzl H, Fiuczynski M E, Bavier A and Peterson L 2007 *Proc. 2nd ACM SIGOPS/EuroSys European Conf. on Computer Systems 2007* Container-based operating system virtualization: a scalable, high-performance alternative to hypervisors (New York: ACM New York) 275–87

[2] McFearin L D 2011 Chroot jail *Encyclopedia of Cryptography and Security* Springer US 206–7

[3] Felter W, Ferreira A, Rajamony R and Rubio J 2015 An updated performance comparison of virtual machines and linux containers *Performance Analysis of Systems and Software (ISPASS), 2015 IEEE Int. Symp. On* IEEE 171–2

[4] Fábrega F J T, Javier F and Guttman J D 1995 Copy on write. URL: http://citeseerx.ist.psu.edu/viewdoc/download?doi=10.1.1.33.3144&rep=rep1&type=pdf

[5] Walsh D 2014 Are Docker containers really secure? [Electronic resource]. URL: https://opensource.com/business/14/7/docker-security-selinux (accessed 10.09.2017)

[6] Grattafiori A 2016 Understanding and hardening Linux containers *NCC Group whitepaper* URL: https://www.nccgroup.trust/uk/our-research/understanding-and-hardening-linux-containers/

IOP Conf. Series: Journal of Physics: Conf. Series **955** (2017) 012040 doi:10.1088/1742-6596/955/1/012040

# On one peculiarity of the model describing the interaction of the electron beam with the semiconductor surface

**M A Stepovich[1, 2], A N Amrastanov[1], E V Seregina[3], M N Filippov[4]**

[1] Tsiolkovsky Kaluga State University, 26 Stepan Razin Street, Kaluga, 248023 Russia
[2] Ivanovo State University, 39 Ermaka Street, Ivanovo, 153025 Russia
[3] Bauman Moscow State Technical University (National Research University), Kaluga Branch, 2 Bazhenov Street, Kaluga, 248000 Russia
[4] Kurnakov Institute of General and Inorganic Chemistry RAS, 31 Leninsky Prospect Street, Moscow, 119991 Russia

**Abstract**. The problem of heat distribution in semiconductor materials irradiated with sharply focused electron beams in the absence of heat exchange between the target and the external medium is considered by mathematical modeling methods. For a quantitative description of energy losses by probe electrons a model based on a separate description of the contributions of absorbed in the target and backscattered electrons and applicable to a wide class of solids and a range of primary electron energies is used. Using the features of this approach, the nonmonotonic dependence of the temperature of the maximum heating in the target on the energy of the primary electrons is explained. Some modeling results are illustrated for semiconductor materials of electronic engineering.

## 1. Introduction

Some methods for local analysis of semiconductors are based on the excitation of an informative signal by a beam of accelerated electrons (electron microscopy, spectroscopy of characteristic losses of electron energy, X-ray spectral microanalysis, etc.). However, when the electrons of low (1-8 keV) and medium (8-50 keV) energies decelerate in the semiconductor, only a small part of energy goes to the formation of informative signals, the most part of the energy goes to heating the sample [1, 2]. When the target is irradiated by a sharply focused electron beam, the heating of the sample can be significant, which can lead to an increase in the local temperature and, as a consequence, to a change in the characteristics of the semiconductor – this may lead to the necessity to take this into account on carrying out the quantitative measurements. Modeling the heating of semiconductor targets through an electron beam is the subject of this paper.

## 2. Simulation of electron beam energy losses

Many studies have been devoted to description of the distribution of energy losses in the interaction of an electron probe with a substance – see, for example, [3-7], while most models give only a qualitative agreement with the experimental results, and the application of each of them is usually limited by a comparatively narrow range of materials.

Among the available models, the most general approach to the description of energy losses by electron beams of medium energies was deduced in [6, 7], in which a model based on the possibility of a separate quantitative description of the contribution of the energy of the electrons absorbed in the target and backscattered electrons is described. This model can be successfully used to perform quantitative calculations for a wide class of materials in a wide range of primary-electron energies. According to this model, the energy loss density of the beam electrons looks like as:

$$P(M)=\frac{1.085(1-\eta)P_0}{\pi^{\frac{3}{2}}a_1^2 z_{ms}\left(1-\eta+\eta\frac{z_{ss}}{z_{ms}}\right)}\left(\exp\left(-\left[\frac{x^2+y^2}{a_1^2}+\left(\frac{z-z_{ms}}{z_{ms}}\right)^2\right]\right)+\right.$$
$$\left.+\frac{\eta a_1^2}{(1-\eta)a_2^2}\exp\left(-\left[\frac{x^2+y^2}{a_2^2}+\left(\frac{z-z_{ss}}{z_{ss}}\right)^2\right]\right)\right). \quad (1)$$

Here the origin coincides with the point of incidence of the electron probe on the sample. $P_0 = IE_0/e$ – the primary beam power; $z_{ms}$ – the depth of maximum energy loss by primary electrons, that have experienced small-angle scattering and absorbed in the target, and $z_{ss}$ – the depth of the maximum energy loss by backscattered electrons [4-7], $z_{ss}=Z^{-1/3}z_{ms}$, $Z$ – ordinal number of an element in the periodic table; $\eta$ – backscattering coefficient of electrons, $\eta=1.085ae\pi^{-\frac{1}{2}}z_{ms}^{-1}\int_0^z \exp\left(-(z-z_{ss})^2 z_{ss}^{-2}\right)$, $a=0.024Z^2A^{-1}$, e – base of natural logarithms; $A$ – atomic weight of the element. Parameters $a_1$ and $a_2$ are determined from the relevant $a_1^2 = z_{ms}^2 + 0.72d_b^2$, $a_2^2 = 0.25z_{ss}^2 + 0.72d_b^2$, where $d_b$ – diameter of the probe [7].

The distributions of energy loss densities we calculated by (1) for kilovolt electrons in Si, GaAs, and CdTe normalized by the amount of energy released per target per unit time are illustrated in Fig. 1.

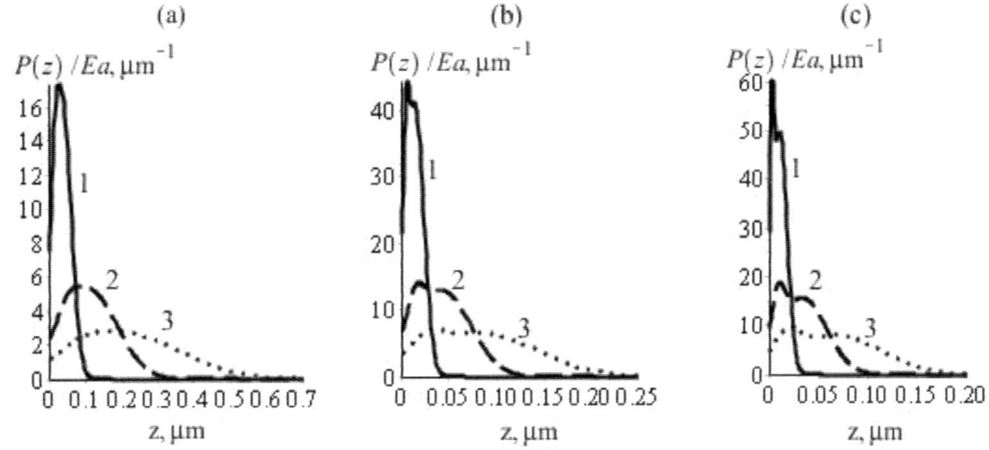


**Figure 1.** Distributions of the energy loss densities of beam of electrons in Si (a), GaAs (b) and CdTe (c) normalized by the amount of power $E_a$ released in these targets and calculated for the electron beam energy $E_0$ = 2 (curves 1), 4 (2 ) and 6 (3) keV.

The possibilities of using this model for estimating the heat distribution in various semiconductor targets are the subject of this work. In this case, unlike [8], in the present work, the simulation was carried out for the case of the absence of heat transfer of the target with the external medium, which is usually realized in electron-probe technology.

## 3. Modeling of heat distribution

The time to establish a steady-state temperature regime of the analyzed microvolume is usually much shorter than the data set time during which the electron probe is positioned at a given point on the surface of the sample [6]. This makes it possible to look for a temperature distribution in the region of interaction of the probe electrons with the sample and in the adjacent regions of the sample on the basis of the solution of the stationary heat equation

$$\left(\frac{\partial^2 \Delta T(x,y,z)}{\partial x^2}+\frac{\partial^2 \Delta T(x,y,z)}{\partial y^2}+\frac{\partial^2 \Delta T(x,y,z)}{\partial z^2}\right)=-\frac{P(x,y,z)}{k}\,. \tag{2}$$

Here $\Delta T(x,y,z)=T(x,y,z)-T_0$, where $T$ – sample temperature at point $(x,y,z)$, $T_0$ – temperature after the establishment of a stationary regime under the influence of an electron beam, and *P(x, y, z)* – function describing energy losses by an electron probe in a target; $\Delta T(x,y,z)$ must satisfy the following boundary conditions:

$$\lim_{x\to-\infty}\Delta T=0,\ \lim_{x\to+\infty}\Delta T=0,\ \lim_{y\to-\infty}\Delta T=0,\ \lim_{y\to+\infty}\Delta T=0,\ \lim_{z\to+\infty}\Delta T=0\,.$$

Since in electron-probe technology the target irradiation by an electron beam is mainly carried out under conditions close to vacuum, the following boundary condition, meaning the absence of heat exchange with the external medium:

$$\left.\frac{\partial \Delta T}{\partial z}\right|_{z=0}=0\,.$$

Equation (2) can be solved in a standard way using the Green's function. The corresponding equation for the Green's function will have the form:

$$\operatorname{div}\operatorname{grad} G=-\delta(x-x_0,y-y_0,z-z_0)$$

with boundary conditions:

$$\lim_{x\to-\infty}G=0,\ \lim_{x\to+\infty}G=0,\ \lim_{y\to-\infty}G=0,\ \lim_{y\to+\infty}G=0,\ \lim_{z\to+\infty}G=0,\ \left.k\frac{\partial G}{\partial z}\right|_{z=0}=0\,.$$

This equation as a Green's function of the Neumann problem in the upper half-space is solved by the method of mirror mappings and has a solution

$$G=\frac{1}{4\pi}\left(\frac{1}{\sqrt{(x-x_0)^2+(y-y_0)^2+(z-z_0)^2}}+\frac{1}{\sqrt{(x-x_0)^2+(y-y_0)^2+(z+z_0)^2}}\right).$$

Here $(x_0,y_0,z_0)$ – point source coordinate. Heat distribution $\Delta T(x,y,z)$ is defined as follows:

$$\Delta T=\int_D G(x_0,y_0,z_0,x,y,z)\frac{P(x_0,y_0,z_0)}{k}dV_Q\,, \tag{3}$$

where $D$ – heat distribution area ($-\infty<x_0<+\infty$, $-\infty<y_0<+\infty$, $0\le z_0<+\infty$), $dV_Q=dx_0dy_0dz_0$. Formula (3) was used to estimate the heating of semiconductor targets by an electron beam.

## 4. Results of calculations

The simulation results showed that the highest heating of all samples by medium energy electrons is observed at the lowest (from the considered range) energies, which is explained by the increasing of heating region with increasing primary electron beam energy. Therefore, the range of low electron energies of the probe is the most interesting for estimating target heating. At fixed beam energies, the smallest heating is observed for an easy target (Si), and the heaviest sample (CdTe) is heated most of all (see Fig. 2. This is due to the fact that as the ordinal number of the semiconductor increases, the interaction region of the electron probe with the target decreases.

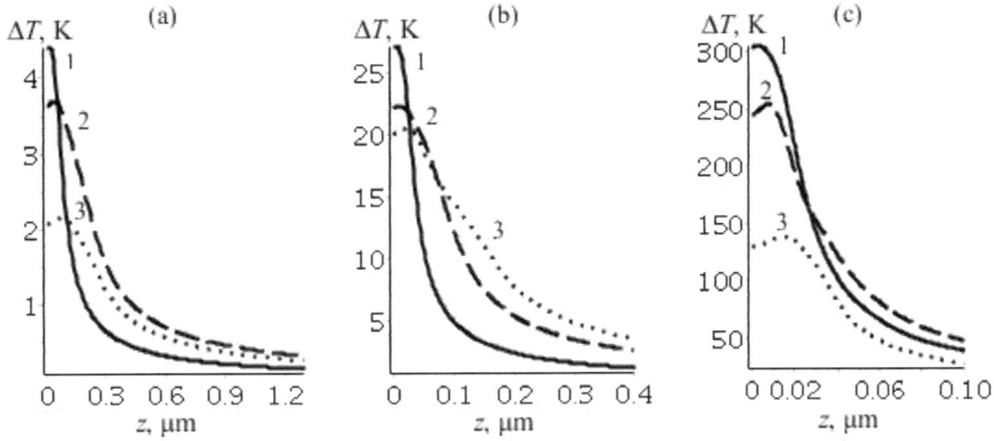


**Figure 2.** Temperature distribution in Si (a), GaAs (b) and CdTe (c) by coordinate $z$. The calculations were carried out at $x = 0$, $y = 0$, for primary electron energies $E_0 = 2$ (curves 1), 4 (2), 6 (3) keV, at probe current $10^{-7}$ A

Figures 3, 4 and 5 show the calculated distributions $\Delta T(x, y, z)$ in Si, GaAs, and CdTe semiconductor materials and the dependence of the maximum temperature in the target on the energy of the primary electrons $\Delta T(E_0)$; the contribution of the energy scattered in the target by the absorbed and backscattered (reflected) electrons is separately shown.

The contribution of the absorbed in the target electrons to the total energy losses of the electron probe in the target and, correspondingly, to its heating, is determinant at a probe energy of less than 4...5 keV. For heavy semiconductors, the energy losses by these electrons become practically zero at an energy of about 8 keV. For light samples and samples with average ordinal numbers, the energy losses absorbed in the target and backscattered electrons become commensurate at an electron energy of the probe of about 8 keV. As for backscattered electrons, they are characterized by the presence of a maximum of about 5 keV, and then the curves decrease monotonically.

The presence of a rather sharp increase in the contribution of backscattered electrons to the total curve $\Delta T(E_0)$ leads to the formation of a "step" on the dependences of the maximum temperature on the energy of the primary electrons. We also note that the backscattered electrons have the least effect on the value of the maximum heating for light samples; this effect is observed most in heavy semiconductor targets.

IOP Conf. Series: Journal of Physics: Conf. Series **955** (2017) 012040 doi:10.1088/1742-6596/955/1/012040

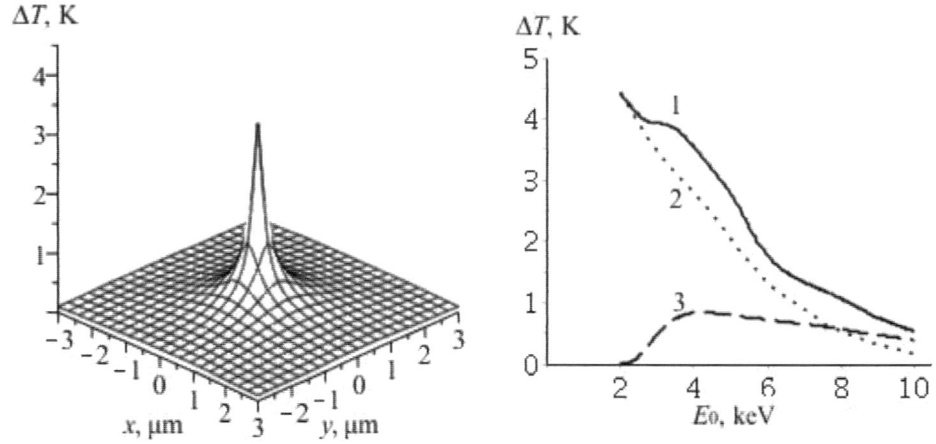


**Figure 3.** The results of modeling the heating of a semiconductor target for parameters characteristic of single-crystal silicon: in the left figure, modeling the temperature distribution for electron energy 4 keV; in the right figure, the simulation of the distribution of the dependence of the maximum temperature on the energy of the primary electrons (curve 1) also the contribution to the curve 1 of the energy scattered by the electrons absorbed in the target (curve 2) and reflected (curve 3) is shown. Calculations were carries out at the probe current $10^{-7}$ A

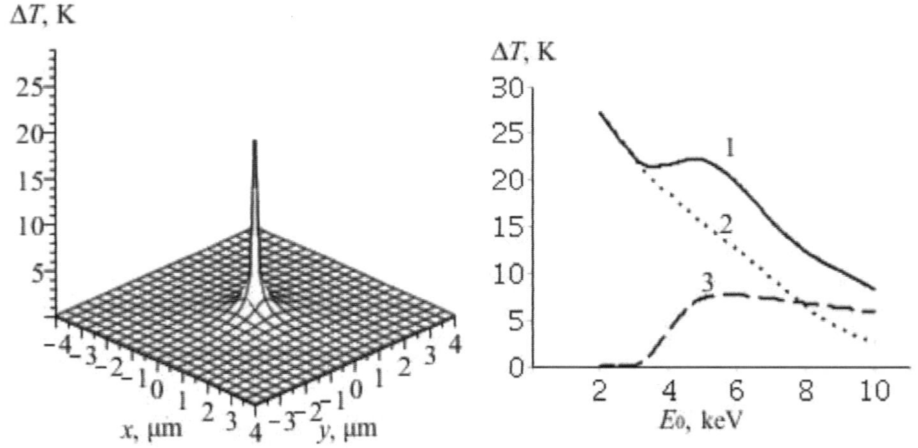


**Figure 4.** The results of modeling the heating of a semiconductor target for parameters characteristic of single-crystal gallium arsenide: in the left figure, modeling the temperature distribution for electron energy 4 keV; in the right figure, the simulation of the distribution of the dependence of the maximum temperature on the energy of the primary electrons (curve 1) also the contribution to the curve 1 of the energy scattered by the electrons absorbed in the target (curve 2) and reflected (curve 3) is shown. Calculations were carries out at the probe current $10^{-7}$ A

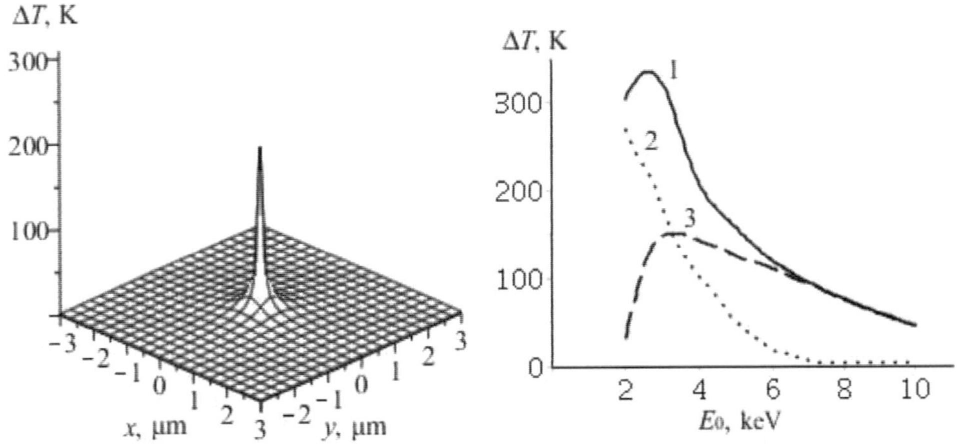


**Figure 5.** The results of modeling the heating of a semiconductor target for parameters characteristic of single-crystal cadmium telluride: in the left figure, modeling the temperature distribution for electron energy 4 keV; in the right figure, the simulation of the distribution of the dependence of the maximum temperature on the energy of the primary electrons (curve 1) also the contribution to the curve 1 of the energy scattered by the electrons absorbed in the target (curve 2) and reflected (curve 3) is shown. Calculations were carries out at the probe current $10^{-7}$ A

## 5. Conclusions

The methods of mathematical modeling were used to study the heating of semiconductor targets using an electron beam of medium and low energies. The greatest heating of all samples by medium-energy electrons is observed at the lowest (from the considered range) energies. At fixed beam energies, the smallest heating is observed for light targets. The obtained heating estimates make it possible to select the optimal mode of action of the beam electrons on a semiconductor target when planning the experiment.

## References

[1] Rau E I, Ditsman S A, Zaitsev S V, Lermontov N V, Lukyanov A E and Kupreenko S Yu 2013 *Bulletin of the Russian Academy of Sciences: Physics* **77** 951
[2] Amrastanov A N, Kuzin A Yu, Mityukhlyaev V B, Seregina E V, Stepovich M A, Todua P A and Filippov M N 2017 *Measurement Techniques* **60** 534
[3] Donolato C 1981 *Phys. Stat. Sol. (a)* **65** 649
[4] Kanaya K and Okayama S 1972 *J. Phys. D: Appl. Phys.* **5** 43
[5] Oelgart Q and Werner U 1984 *Phys. Stat. Sol. (a)* **85** 205
[6] Mikheev N N, Petrov V I and Stepovich M A 1991 *Bulletin of the Academy of Sciences of the USSR. Physical Series* **55** 1
[7] Mikheev N N and Stepovich M A 1996 *Industrial Laboratory* **62** 221
[8] Amrastanov A N, Ginzgeymer S A, Stepovich M A and Filippov M N 2016 *Bulletin of the Russian Academy of Sciences: Physics* **80** 1290